Student Companion with
Complete Solutions for

An Introduction
to Genetic Analysis

Sixth Edition

by
Anthony J. F. Griffiths, Jeffrey H. Miller,
David T. Suzuki, Richard C. Lewontin,
and William M. Gelbart

Diane K. Lavett

Emory University

W. H. Freeman and Company
New York

ISBN: 0-7167-2801-X

Printed in the United States of America

Second printing 1996, VB

Contents

Preface to the First Edition

Unlike many solutions manuals, the *Student Companion with Complete Solutions for An Introduction to Genetic Analysis*, Fourth Edition, attempts to provide a logical approach to solving genetics problems. While explaining the reasoning behind each answer given, the book also recognizes that what is obvious to geneticists is seldom apparent to the beginning student.

The *Companion* was written from the perspective of a teacher who is standing in front of a blackboard, trying to explain a problem to a class of beginning genetics students. Because I have been in that position many times, my students have taught me exactly where it is that difficulties will occur. I have tried to anticipate each of these potential obstacles in the explanations that follow.

The *Companion* has been tested in the classrooms at the State University of New York at Cortland and at Emory University. In addition, Dr. Anthony J. Pelletier of the Department of Molecular, Cellular, and Developmental Biology of the University of Colorado at Boulder has independently worked all the problems presented in the text. His careful, thorough work resulted in the detection of many errors, and he has my deepest respect and gratitude for his magnificent effort. Although Dr. Pelletier provided this invaluable service, he should not, however, be held responsible for any errors that may still remain. Those errors are mine, and I hope that all users of the *Companion* will feel free to communicate directly with me about any mistakes that they detect.

Diane K. Lavett
Department of Biological Science
The State University of New York
at Cortland

Preface to the Second Edition

As with the first edition, all new problems in this second edition have been checked by another geneticist, in this case, Michael T. Lewis and Mignon C. Fogarty of the Department of Biology at the University of California, Santa Cruz. Their sincere desire to see an answer book with no wrong answers has been much appreciated, and I thank them for their dedication to our common goal. They, however should not be held responsible for any errors that still remain; the errors are mine.

To the users of this text: Do not assume that just because an answer appears in a printed text that it is correct. If you do not follow an answer in this book and you have an answer of your own that you think is correct, please write me directly and give me the benefit of your thinking. You could be right, you know.

This edition has been written while on leave from SUNY Cortland and while a Visiting Scholar in the Department of Biology at Emory University. I am indebted to the former for time and the latter for space. I especially thank Ms. Joyce Woodward and Ms. Anne Kirk for their assistance so freely given and so unexpected because they had no obligation to do anything at all.

The people at Freeman are a wonderful group, dedicated to turning out the best books possible. As they turn to their next projects, I wish them all authors who pay attention to deadlines and respond to requests. Janet Tannenbaum and Erica Seifert bore the brunt of the task, and they have my deep appreciation. I owe Patrick Shriner special thanks for his patience with me.

Now, to correct an omission from the last edition: for reasons I cannot comprehend, I failed to acknowledge Patrick Fitzgerald's guidance throughout the entire project. He was the one who talked me into doing the writing, the one who gave me support and encouragement at each point that my frustrations reached a new high, and the one who somehow became a friend in the process.

Diane K. Lavett
Department of Biological Science
The State University of New York
at Cortland
and
Department of Biology
Emory University

Preface to the Third Edition

I am very excited about this edition because of the concept maps that were drawn for each chapter of the text. They are such a powerful way of learning material and integrating concepts! If I could learn from drawing them, then surely students can learn from contemplating them or, even better, by doing their own maps of the same concepts. Just as "new math" taught everyone the "secrets" that the very best of students discovered on their own before the development of new math, the concept maps visualize what the very best of geneticists have done mentally as they have struggled to master this material. Now, the only secrets that remain are the ones that need to be discovered by experimentation. My hope is that all students will take advantage of this method for learning.

This edition was written while I was on the faculty of the Biology Department of Emory University. The encouragement of my colleagues with it, especially in the development of the concept maps, has been very important to me. Likewise, the assistance provided by the office staff has been invaluable. I specifically thank Ms. Anne Kirk, Ms. Joyce Woodward, and Ms. Barbara Shannon for their cheerful help at every step.

The same people are at Freeman as with the last edition; they remain unchanged in their dedication and they have all become even better at their tasks than with the last edition. Erica Seifert, especially, has put much of herself into this edition, and I am very grateful to her.

As with the previous editions, if you discover an error, it is all mine, and I would appreciate it if you wrote to me at the address below so that any errors can be corrected.

Diane K. Lavett
Department of Biology
Emory University
Atlanta, Georgia 30322

To my teacher,
Dr. Charles Ray, Jr.
Professor Emeritus of Genetics
Biology Department, Emory University

Introduction

► HOW TO THINK LIKE A GENETICIST

You will sharpen a number of skills as you work through your textbook and this *Companion*. Because genetics requires very careful reading, you should become a more careful reader by the end of your course of study. For example, a great deal of information is conveyed by the sentence, "Two mutants were crossed and a wild-type phenotype was observed in the male offspring." If you doubt the amount of information meant to be conveyed by that sentence, I will list it, even though the information may have little meaning for you until later:

1. Two separate genes are involved.

2. Both mutants are recessive.

3. The mutant gene in the female is located on an autosome.

There are seldom superfluous words in a genetics problem. Consequently, you must learn to think about each word that is provided.

A second skill that will be sharpened as you progress through this course is systematic thought. To approach a problem in genetics in a haphazard manner is to enter quickly into the realm of chaos and confusion. I hope that this book will be of real assistance in sharpening your ability to think systematically.

A third skill that you should learn is what I call being gentle with your so-called mistakes. Our educational system has labeled as a mistake the failure to arrive at the correct answer on the first try. Yet, in genetics, as in all of science, progress is made by learning through trial and error what the explanation is *not*. No hypothesis can ever be proved in science, it can only be disproved or tentatively accepted. When you think in this way, I hope that you will begin to view your attempts to solve a problem as hypotheses that are being rejected rather than as errors. If you are not able to answer a particular problem, you need only go back to your initial assumptions and revise them. Perhaps you have misread the problem. Perhaps you have not understood what the question is. Perhaps you have thought correctly but have made a simple mathematical error. An important point to keep in mind is that in any trial and error learning, there often must be a number of "trials." The only true mistake that you can make is to stop generating hypotheses and to give up.

► HOW TO READ LIKE A GENETICIST

There is a big difference between studying and learning. Studying is the review of material already known; learning is an increase in the level of understanding. Most students confuse these two activities and equate all time spent reading the textbook with learning, when, in some instances, not even studying is occurring.

When I first pick up a book, I have to have a reason for opening it. When I was a student, very frequently my reason was simply that I was taking a course for which the book was required and I wanted to pass the course. Now, the first level of my reason is that I want to know what the authors are saying about the topic. I may, of course, have many other levels for wanting to read a specific book. If I do not want to know what the authors are saying, I will put down the book unopened. If your response to the textbook is that you do not want to know what is contained in it at any level, you should seriously think about why you are beginning a course of learning that holds no interest for you.

Once I open the book to a specific chapter, I skim through that chapter. I read the titles of the different sections. I look at the pictures and read the captions. I then read the chapter introduction and summary. I read some of the problems at the end of the chapter. With this cursory examination, I now know the general information that is presented in the chapter and the structure in which it will be presented. More importantly, I know whether the chapter contains the information that I am seeking. As a student reading a required text, I knew in a very general way the information that I was expected to master. Then and now, I ask myself what I already know about the material and ask myself questions that, from my little knowledge of the material, I would like to have answered. Only at this point am I ready to begin learning the material covered in the chapter.

Learning requires active reading, and most students are not accustomed to reading as carefully as is required for the learning of genetics and solving genetics problems. Ideally, as you read each sentence in the textbook, you should ask yourself the following questions:

1. What did the author say?

2. What did the author mean?

3. What is implied by what the author said and meant?

4. Do I agree with both what the author meant and what is implied by what she or he meant?

5. How does it connect with what I already know?

Until you can answer each of these questions with regard to a sentence, you should not read any further.

This type of reading is an exhausting process that at first will seem very artificial to you. However, there is no substitute for reading in this fashion if you wish to learn at anything but the superficial level. With practice, these questions will become automatic, will be asked and answered very rapidly, and from that point on you will always be learning instead of simply reading.

To get to the level of automatic questioning, I suggest you work with one or more classmates, reading out loud a sentence at a time and discussing it thoroughly before proceeding to the next. You might find that active reading is fun after a while, as you begin to anticipate a point that the author is trying to make or as you discover implications that the author does not realize. As you become increasingly skilled in active reading, you may find that you have begun to generalize these skills to other parts of your life, such as listening to a friend or the news on television, your own writing, and nonverbal events in your life. If this generalization occurs, you are well on your way to becoming a person who learns from all aspects of your existence.

▶ HOW TO SOLVE
PROBLEMS LIKE A
GENETICIST

Genetics is not a spectator sport; you cannot learn genetics without solving problems. Each problem should first be read in the same way that you read the textbook. Once you have read the problem, *write* the answers to the following questions:

1. What question is being asked?

2. What information is known?

3. What information is missing?

4. What information is extraneous?

5. How will I symbolize the genes?

6. What assumptions am I making?

7. What are the possible hypotheses that will answer the question being asked?

Using this approach, the problems will literally solve themselves.

A serious threat to your ability to solve the problems in your text is misuse of this *Companion*. You can convince yourself far too easily that you understand a problem as you read the solution to it, when, in fact, you do not understand it at all. Let me suggest the proper way to use this book:

1. Read the section entitled "Important Terms and Concepts" first. If any term or concept does not cause you to recall exactly what the text said about it, reread that section of the text.

2. Work with the concept map for the chapter until you have gained all that you can from it. "Using Concept Maps" is discussed later in this introduction.

3. Work on a problem without reading the *Companion* until you are truly stuck.

4. Read the explanation of the solution.

5. Without consulting the *Companion*, immediately rework the problem.

6. Two or three days later, work the problem without consulting the *Companion*. If you cannot do it at this point, you probably did not understand the problem earlier.

7. If you cannot work the problem without consulting the *Companion*, repeat steps 3 to 5 once. If you cannot work the problem at that point, consult your teacher or a friend who can explain the problem to you.

8. Throughout the problem-solving process consult the section entitled "Tips on Problem Solving" as needed. There are a limited number of types of problems. For each type of problem, there is a pattern to the method of solution. Each problem solved in this book has followed the pattern best-suited for the problem. Learn to duplicate the patterns.

If you can eventually solve all the problems at the back of a chapter correctly without referring to the answers supplied, you should be in a good position to handle whatever your teacher may ask you in a test.

► HOW TO LEARN
GENETICS

In addition to all that has been outlined above, as a student you have the task of integrating what you learn from the textbook with the lectures that you attend. That is your task as a student.

In order to achieve that task, you need to work at it. Actively read the assigned chapter and try at least some of the problems before going to the class that deals with that chapter. Define for yourself what you understand and do not understand about the chapter material. Ask as many questions as is necessary in class to clarify any difficulties. Interrupt your teacher as often as is necessary when he or she says something that you do not understand. If your questions are framed in such a way and asked in such a manner that they demonstrate that you are struggling with this material and not simply harassing your teacher, all your questions will be welcomed. There is no such thing as a stupid question, except the question that is not asked.

A student frequently does not ask questions in class, thinking that she or he is the only one who could be so dumb as not to understand a specific point. The reality is that if one student does not understand a point, most students in the class do not understand the point, and will be grateful that you have asked the question. Your teacher needs to know what you do not understand in order to do the best job of teaching possible. I urge you to engage with your teacher in active learning. Do yourself, all your classmates, and your teacher a favor by asking those questions that you are convinced are dumb. The first time, asking will be difficult; it becomes easier with practice.

Once class is over, do not put away your notebook until the next class. Go home and review every note that you made in class. You might try having a second notebook for each class into which you write out your class notes more fully and add to them while the class is still fresh in your mind. After doing this, each time read all the previous notes that you have taken since the beginning of the course. Identify the material that you know thoroughly and the material that you need to assimilate. Right then, as you are reminded of what you do not yet know, learn that material. If you go through this process after every class, you will not need to study at exam time.

The process of learning outlined above obviously cannot be completed in a "cram" session the night before a test or a final exam. The attempt to learn genetics in that fashion is doomed to failure. The best approach is to study genetics almost every day, weekends included. Keep your sessions short, not more than two or three hours at a time. Pick a quiet place where you will not be interrupted or distracted. If you find your concentration wavering, take a short break. If you find that anxiety is interfering, do some physical exercise. Avoid caffeine, both while studying and, most importantly, before taking a test. If the night before a test you are forced to make a choice between a good night's sleep and trying to learn far too much material for the time available, choose sleep.

Many students will have some difficulty with the material and problems in Chapters 2 through 6. Thereafter, the material will be easier to conceptualize, and the problems will be easier to do. The reason for the difficulty with the earlier chapters is that they require a level of abstract thought not usually demanded in undergraduate courses. Beginning with Chapter 7, however, the material becomes more descriptive and, simultaneously, more consistent with the skills required for success in other biology courses. Be aware that you will have to work quite hard in dealing with this early material; also be aware that you are not alone in your difficulty.

Generations of students have struggled with genetics, and the vast majority have been successful in their struggle. Their reward has been that they have learned a new way to view the universe. It is my sincere hope that this will be your reward, too.

► USING CONCEPT MAPS

You are introduced to concept maps in the first chapter of the text, and at the end of each chapter there is a list of terms that you are to arrange in a concept map. In this *Companion*, there is at least one concept map for each chapter. These differ from the terms at the back of each text chapter in that this *Companion* includes many more terms, the terms used are a more complete presentation of the ideas in the chapter than the suggested terms, and the terms are presented as drawings rather than as a list.

There is not just one right concept map; 10 people could take the same terms and produce at least 10 different maps from them. Each map would be a valid visual representation of the linking of concepts. Ideally, each student will make his or her own concept map because the active learning that occurs with making a concept map cannot be duplicated in any other fashion. It was decided, however, to produce concept maps for each chapter as an alternative way of presentation of the material for the sake of those students who are unwilling to invest the time required for making them.

The first step in using each concept map should be simply attempting to understand it. Look at it. Recall the meaning of each term. Ask yourself why two or more terms are linked in the concept map. If you cannot work at this level with a concept map, you need to return to the text and reread the chapter.

► CONCEPT MAP I-1:
SICK ENVIRONMENTS
PRODUCE SICK
PEOPLE

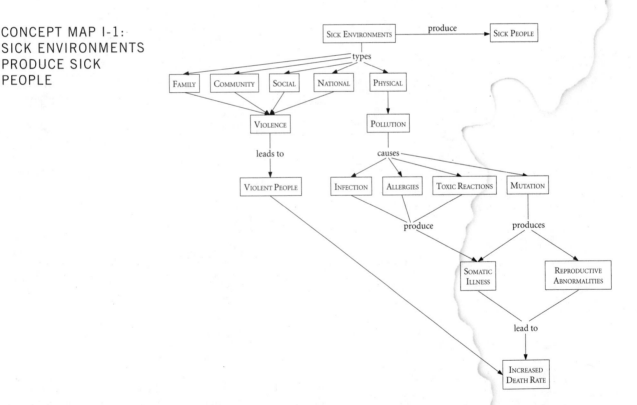

Each concept map in this *Companion* is accompanied by a series of simple questions that check whether the basics that were presented in the chapter have been learned. If you cannot answer one or more of these questions, you need to reread the chapter before proceeding to attempt the problems.

Once you can answer all the questions without resorting to looking at the answers, return to the concept map. Now begin to fill in details for which there were no questions.

As an illustration of this process, look at Concept Map I-1: Sick Environments Produce Sick People. Probably no one would argue with the title, but it is very unspecific. However, the concept map lists a number of different types of environments. The list is incomplete, for one could also add "International Environment" and perhaps others. The list is also mostly irrelevant to genetics because most of the terms are the province of the fields of psychology, sociology, and anthropology. You might ask yourself what you recall from these other fields before focusing on the physical environment, however.

When focusing on the physical environment, first ask yourself what "sick" means with regard to the environment. Now, ask yourself if the environment can be sick in ways other than by pollution. If so, explore those ways. Next, you might focus on any of the subtopics such as infection: how does pollution cause infection? To illustrate this process further, focus on the concept that pollution causes mutation. Look at Concept Map I-2: Gene Mutation (taken from Chapter 19).

► CONCEPT MAP I-2:
  GENE MUTATION

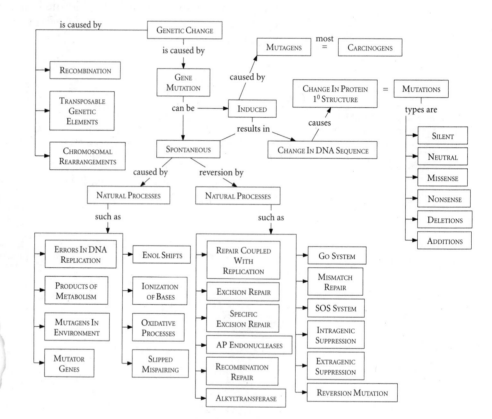

In this map, you see that one form of mutation, gene mutation, is one of four different mechanisms for genetic change. What are the other mechanisms? What types of gene mutations occur? What happens to the DNA sequence with a gene mutation? Again, ask yourself these questions and many more besides. Each question you ask yourself will lead to others and help you integrate the facts presented in the text into a coherent body of knowledge.

Remember that each term in this concept map, as with all terms in all concept maps, stands for a large body of knowledge that you are expected to learn. For example, the concept map dealing with mutation states that genetic change is also caused by chromosomal rearrangements. Concept Map I-3: Chromosome Mutations (taken from Chapter 8) is a detailed look at the chromosomal rearrangements. What types of chromosomal rearrangements exist? What are their consequences? Can you draw them? Can you draw synapsis and crossing-over for heterozygotes with a specific type of chromosomal rearrangement?

The list of questions is limited only by your engagement with this material. To help facilitate engagement and the learning that can occur with it, work on a concept map with a friend, take turns asking each other questions. Discuss each concept, and its ramifications, as deeply as you can. Return to each concept map periodically, as you learn new material.

► CONCEPT MAP I-3:
CHROMOSOME
MUTATIONS

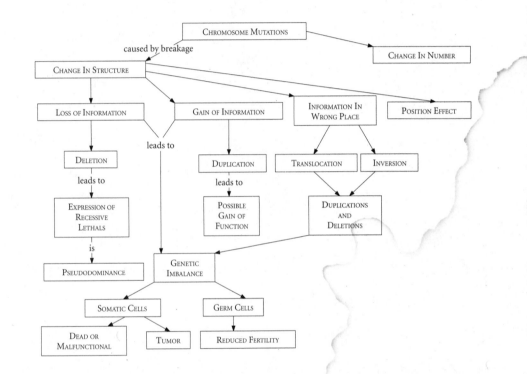

I would like you to point out to your teacher that a projected concept map is an excellent way in which to teach this material. Ask your teacher to try teaching a class once simply by asking questions based on a concept map. If your teacher does try this, be tolerant *and* read the assignment before going to class. Teaching in this manner is an exhausting yet exhilarating way to teach. It is also somewhat scary because the teacher loses control of the classroom as discussion takes your teacher in unknown, unpredicted directions. It is also a lot of fun.

In summary, if you follow the earlier directions on how to think like a geneticist, read like a geneticist, solve problems like a geneticist, and how to learn genetics, and if you work with the concept maps for each chapter, you will discover that, although it takes a lot of work, you will have learned genetics, and, perhaps more important, you will have learned how to learn.

1

Genetics and the Organism

► IMPORTANT TERMS
AND CONCEPTS

Genetics is the study of the inheritance of traits by means of the examination of their variation.

Genes are the basic functional units of heredity. They are composed of DNA and contain information that determines specific traits.

Chromosomes consist of long DNA molecules complexed with protein. They contain many genes.

All life-forms can be divided into **eukaryotes** and **prokaryotes.** Eukaryotes are organisms in which the genetic material is contained within a membrane-bound nucleus. Prokaryotes are organisms that do not have their genetic material contained within a membrane-bound nucleus. In both eukaryotes and prokaryotes, **the flow of information** is from DNA to RNA to protein.

Protein can be either structural or enzymatic. **Structural proteins** result in the physical forms of life. **Enzymatic proteins** result in the biochemical processes of life.

The life of any particular organism results from the interaction of its inherited material with the historical sequence of environments that it encounters. The **genotype** refers to the inherited genes in an organism. The **phenotype** refers to the physical appearance of an organism. The **norm of reaction** refers to the environment-phenotype relationship for a specific genotype. **Developmental noise** is the random variation that occurs in phenotype when both genotype and environment are held constant.

Genetic dissection is the process of identifying the specific hereditary components of a biological system. This process is aided by the use of **markers,** which are specific phenotypes produced by specific genotypes that allow the researcher to keep track of chromosomes, cells, or individuals.

Questions for Concept Map 1-1

1. What does the abbreviation *DNA* mean?

2. How many different DNA molecules are there in an organism that has 10 pairs of chromosomes?

3. Is the DNA in your father identical with the DNA in your mother?

4. Is your DNA totally different from the DNA found in a dog? An oak tree?

▶ CONCEPT MAP 1-1:
FROM GENOTYPE TO
PHENOTYPE

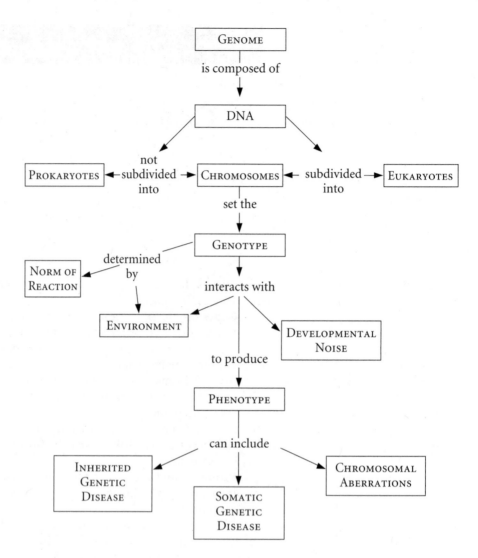

5. Distinguish between genotype and phenotype.

6. Is there always a one-to-one relationship between genotype and phenotype?

7. Which is more important to an organism, its genetic information or the
environment in which it develops?

8. If two genetically identical organisms develop in identical environments, will they necessarily be identical for a specific trait?

9. If two genetically different organisms develop in identical environments, will they necessarily be nonidentical for a specific trait?

10. If two genetically different organisms develop in different environments, will they necessarily be nonidentical for a specific trait?

11. What is the definition of a genome?

12. How many genomes does each human have?

13. What is the definition of a chromosome?

14. Distinguish between eukaryotes and prokaryotes.

15. Define *genotype*, *norm of reaction*, *developmental noise*, and *phenotype*.

16. What is another term for *developmental noise*?

17. Distinguish between inherited and somatic genetic disease.

Answers to Concept Map 1-1 Questions

1. Deoxyribonucleic acid.

2. Because each chromosome contains one long DNA molecule, there would be $10 \times 2 = 20$ DNA molecules in the organism.

3. No. Although your father and mother have genes that control the same functions, not all the genes are identical. Because these different genes are composed of DNA, the DNA must also be different (i. e., not identical).

4. No. All organisms must solve similar problems in order to survive. As an example, the Krebs cycle occurs in humans, dogs, and trees. Because these metabolic changes take place in all three organisms, identical or very similar enzymes must exist in all three organisms. Because the information for the enzymes resides in the DNA, the DNA responsible for these enzymes must be identical or very similar.

5. Genotype is the inherited information, while phenotype is the end result of an interaction of that information with the environment and developmental noise.

6. No. For most traits there is a many-to-one relationship between genotype and phenotype.

7. Both are very important. The inherited material initially sets the limits for the organism, while the development of an organism in a specific environment determines where along the spectrum of possibilities the organism will be. However, some genotypes are lethal in all environments, and some environments are lethal for all genotypes.

8. No. Developmental noise will produce variation.

9. No. The response to the environment may produce the same phenotype from two different genotypes.

10. No. The response to differing environments may produce the same phenotype from two different genotypes.

11. A genome is the contribution of genetic material from either a father or a mother.

12. Each human has two genomes, one from each parent.

13. A chromosome is composed of DNA and protein. It consists of a linear array of genes.

14. Eukaryotes have true linear chromosomes, nuclear membranes, and membrane-bound vesicles. Prokaryotes have naked DNA, and sometimes the DNA is circular. They also have no nuclear membrane or membrane-bound vesicles.

15. The genotype is the inherited genetic material. The norm of reaction is

the variation seen in phenotype from varying the environment with a specific genotype. Developmental noise is the variation produced in phenotype due to random events at the molecular level. The phenotype is the physical appearance of an organism.

16. Developmental noise is basically chance.

17. An inherited genetic disease is the product of the genotype, predominately. A somatic genetic disease is the product of a sudden change (mutation) in the genetic material in those cells that compose the body.

▶ SOLUTIONS TO PROBLEMS

Be sure that you have thoroughly read the entire chapter before you attempt any of the problems.

1. Genetics is the scientific study of heredity. The ancient Egyptian racehorse breeders can be only loosely classified as geneticists because their interests were highly focused on producing fast horses, rather than on attempting to understand the mechanisms of heredity; their understanding of the processes involved in producing fast horses was very incorrect; and their methods were not analytic in the modern sense of the word. Nevertheless, they did produce very fast horses through a combination of observation, trial and error matings, and artificial selection.

   If the Egyptian racehorse breeders can be classified as geneticists, so too can the much earlier people who, in the process of switching from hunting and gathering to agriculture, selected crop seeds based on the phenotype of their products. In fact, these people were the first geneticists, preceding horse breeders by millennia. Their current-day equivalents are still conducting genetic experiments at the same level in remote areas of the world where modern technology has made almost no inroads.

2. Genetics has affected modern society in law (copyrights, paternity suits, criminal cases), medicine in areas too numerous to list, agriculture, and virtually any other general area that exists. For example, agricultural pest control makes use of mutations to produce sterile males who then mate with females, producing no new generation. Flea control has become biological through the use of nematodes that search out and eat flea eggs outside, and currently genetic engineering is being used to develop nematodes that can live in your rugs for as long as a supply of flea eggs exists. The limit seems to be the imagination of those geneticists who have a focus on the practical uses of the ever-expanding base of genetic knowledge.

3. DNA determines all the specific attributes of a species (shape, size, form, behavioral characteristics, biochemical processes, etc.) and sets the limits for possible variation that is environmentally induced.

4. Two properties of DNA that are vital to its being the hereditary molecule are its ability to replicate and its stability that nevertheless contains within it the ability to change (mutate). Alien life-forms might utilize RNA, just as some viruses do, as a hereditary molecule. However, of the types of molecules that can exist on earth, only the nucleic acids possess the characteristics necessary for a hereditary molecule.

5. The norm of reaction is the phenotypic variation that exists for a species of a fixed genotype within a varying environment. The variability itself can become vital to a species with a change in environment, for it allows for the possibility that some individuals may be able to survive under the new environmental conditions long enough to reproduce.

6. Phenotypic variation within a species can be due to genotype, environmental effects, and pure chance (random noise).

7. The formula *genotype + environment = phenotype* is both accurate and inaccurate simultaneously. While phenotype is a product of the genotype and the environment, there is also an inherent variation due to random noise. Given complete information regarding both genotype and the environment, it is still impossible to specify the phenotype completely, although a close approximation can be achieved.

8. Following are five ethical dilemmas in genetics: (1) The Green Revolution has produced an increased population and pollution of the world owing in part to better understanding of genetics, and it has caused a cultural disruption as farming techniques have been forced to adapt to the requirements of the seeds used. Was it all "worth" it? (2) What does an individual do with the knowledge (discovered through genetic testing) that she or he may die in 10 or 20 years from a genetic disorder that has yet to produce any symptoms? Does the individual even want such knowledge? (3) What does society do with the knowledge that an individual is carrying a genetic disorder and is reproducing, and that the society will ultimately have to take care of the offspring of that person? (4) How can society protect the rights of an individual to privacy as more and more is known about that individual genetically? Does society have the right to prohibit reproduction? Do insurance companies have the right to the information about an individual whom they may insure? Does the individual have the right not to know of a genetic problem? (5) Genetic knowledge has been utilized in our legal system for decades. However, there have been many cases of framing (that is, the making up of test results or changing of test results to make the defendant match with crime scene evidence left by the perpetrator) produced with genetic knowledge over that period. There have also been numerous laboratory errors, resulting in indictment of the innocent and exculpation of the guilty. In addition, juries are usually confused by the details of genetic evidence so that they very frequently either ignore it completely or accept it without any questioning. Finally, there have been recent cases in which the finding of a laboratory technician has resulted in the overturning of the conclusion of a jury that heard all the evidence. This is, in effect, an event that completely undermines our legal system, making a technician the sole determiner of guilt or innocence and making a jury irrelevant to the legal process. Is this acceptable?

9. The goal of genetic dissection is to understand the functioning of all the genes that affect a specific trait, to know the DNA sequence of each gene, and to understand the regulation during development of each of the genes.

10. Genetics has unified biology because genetics provides the mechanisms with which each area of biology deals. For example, evolution makes no sense without genetic knowledge. Embryonic development is, in

essence, genes in action. Ecology is the population genetics of each species interacting within a changing environment. Each area of biology ultimately explains the details of its discipline at least in part with genetic knowledge.

2

Mendelian Analysis

▶ IMPORTANT TERMS AND CONCEPTS

A **gene** controls one characteristic, or trait. **Alleles** are alternative forms of a gene. A good analogy is a coin. You can have pennies, nickels, and dimes, all "genes." You can have 1956, 1971, and 1992 pennies, all "alleles" of the penny gene. Alleles determine alternative types (tall, short) for the gene (height) being studied.

A **pure line** is a strain that breeds true for the trait being studied. It is also called a **true-breeding line.**

The **parental generation**, P, is the generation from which the first cross in a series is taken. Usually, but not always, the parents are true-breeding.

The **first filial generation**, F_1, consists of the progeny from the parental generation.

The **second filial generation**, F_2, consists of progeny from the first filial generation.

Dominant refers to an allele that is expressed regardless of the other alleles present for the trait being studied. **Recessive** refers to an allele that is expressed only when it is the only type of allele present in an organism for the trait being studied.

A **heterozygous** line, or **hybrid** line, contains two different alleles for the trait being studied. The phenotype of the heterozygote indicates which allele is dominant.

A **homozygous** line contains identical alleles for the trait being studied. The line may be either homozygous recessive or homozygous dominant, depending on which allele is present. It is a true-breeding line.

Monohybrid refers to a single gene pair. F_1 monohybrid crosses lead to 1:2:1 genotypic ratios.

Dihybrid refers to two gene pairs being studied simultaneously. F_1 dihybrid crosses result in a 9:3:3:1 ratio.

Equal segregation of gene pairs refers to the separation of the two alleles of a gene into gametes, which contain only one allele each. This is the **first law of Mendel.**

Independent assortment of gene pairs refers to equal segregation of one allelic pair independently of another allelic pair. This is the **second law of Mendel.**

A **testcross** is a cross of a homozygous recessive organism with an organism of dominant appearance. It results in a 1:1 ratio if the organism is heterozygous and a 1:0 ratio if it is homozygous.

A **reciprocal cross** involves a pair of crosses that switch the trait being studied with the sex of the parent carrying that trait.

Continuous variation is seen in traits that exist on a spectrum, such as height and weight. **Discontinuous variation** is seen in traits that have discrete phenotypes, such as yellow versus brown and red versus blue.

Polymorphism, literally meaning "many forms," is the existence in a population of two or more common phenotypic variations for a trait. The different variations are called **morphs**.

Pedigree analysis is a tool for determining the genetic status of related individuals over several generations.

The **propositus** is the individual that first called attention to the family being studied.

▶ WORKING WITH
PROBABILITY

Probability is the number of times an event is expected to happen divided by the total number of times it could have happened.

"And" statement: indicates the need to multiply.

Example: the probability of a black female horse

$$= p(\text{black horse}) \text{ "and" } p(\text{female horse})$$

$$= p(\text{black horse}) \times p(\text{female horse})$$

Example: the probability of *AA Bb*

$$= p(A \text{ "and" } A) \text{ "and" } p(B \text{ "and" } b)$$

$$= p(A \times A) \times p(B \times b)$$

"Or" statement: indicates the need to add.

Example: the probability of a black or brown horse

$$= p(\text{black horse}) \text{ "or" } p(\text{brown horse})$$

$$= p(\text{black horse}) + p(\text{brown horse})$$

Example: the probability of *AA* or *Aa*

$$= p(A \text{ "and" } A) \text{ "or" } p(A \text{ "and" } a)$$

$$= p(A \times A) + p(A \times a)$$

"At least" statement: indicates the need to add.

Example: the probability of at least three out of five

$$= p(3 \text{ out of } 5) \text{ "or" } p(4 \text{ out of } 5) \text{ "or" } p(5 \text{ out of } 5)$$

$$= p(3 \text{ out of } 5) + p(4 \text{ out of } 5) + p(5 \text{ out of } 5)$$

Remember that the probability of at least three out of five = 1 − the probability of 0 or 1 or 2 out of 5.

Combinations statement: indicates the need to use the formula for combinations. This formula is

$$\frac{n!}{p! \, q!} (r)^P (s)^q$$

where *n* is the total number, *p* is the number of one kind, *q* is the number of the alternative, *r* is the probability of *p* occurring, and *s* is the probability of *q* occurring. The exclamation point indicates *factorial.* 5! = (5)(4)(3)(2)(1).

Example: If two heterozygous individuals have eight children, the probability of exactly two being homozygous recessive is

$$\frac{8!}{2! \; 6!} \, (1/4)^2 (3/4)^6 = \frac{8 \times 7 \times 6 \times 5 \times 4 \times 3 \times 2 \times 1}{2 \times 1 \times 6 \times 5 \times 4 \times 3 \times 2 \times 1} \, (1/4)^2 (3/4)^6$$

▶ CONCEPT MAP 2-1:
   BASIC RELATIONSHIPS

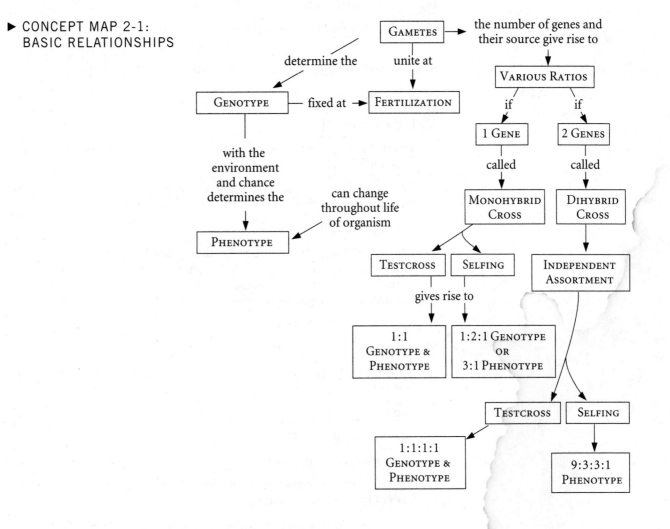

Questions for Concept Map 2-1

1. How many gametes unite?

2. What are the names of the gametes in humans?

3. Can one individual ever give rise to all gametes in a cross?

4. What is the definition of *genotype*? *Phenotype*?

5. What is another term for *chance*?

6. How can the phenotype change with time? What are some examples?

7. Are all genotypes and phenotypes viable?

8. What is the F_2 genotypic ratio of a selfed dihybrid?

9. What is the meaning of *hybrid*?

10. What is being hybridized?

11. What segregate?

12. What independently assort?

13. Distinguish between *gene* and *allele*.

14. What is a testcross?

15. Under what circumstances can a 1:2:1 ratio appear to be a 3:1 ratio?

Answers to Concept Map 2-1 Questions

1. 2.

2. Egg and sperm.

3. Yes, in those species that contain male and female gametes within the same individual.

4. Genotype is the inherited material received from both parents. Phenotype is the physical appearance of the individual.

5. Developmental noise.

6. The physical appearance in most animal species changes with aging. Hair changes color. The back becomes bent and usually shorter. Wrinkles appear in the face, and cartilaginous structures such as the ears and nose elongate somewhat.

7. No.

8. 9 (1:2:2:4):3 (1:2):3 (1:2):1

9. A hybrid is the product of a cross between two individuals that differ only with respect to one gene pair.

10. Two pure-breeding lines are being hybridized.

11. Alleles segregate.

12. Genes independently assort.

13. A gene controls a function (wing length), while an allele is a specific form of the gene (long wing).

14. A testcross utilizes a homozygous recessive individual to reveal if the individual being tested is homozygous or heterozygous.

15. With classical dominance, a 1:2:1 ratio appears to be a 3:1 ratio.

Questions for Concept Map 2-2

1. Do genes or alleles segregate?

2. Do genes or alleles independently assort?

3. What is the first law of Mendel? The second law?

4. Give three examples of continuous variation. Do the same for discontinuous variation.

► CONCEPT MAP 2-2:
  GENES

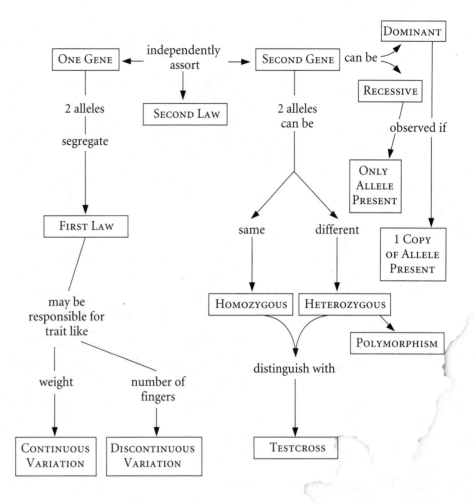

5. Is the level of enzyme activity continuous or discontinuous?

6. What is the definition of *dominant*? *Recessive*?

7. Are alleles or genes dominant and recessive?

8. What is the definition of *homozygous*? *Heterozygous*?

9. Give three examples of polymorphisms involving different genes.

10. What ratio do you expect in a testcross of a heterozygous organism? Of a homozygous organism?

11. What is the relationship between the first law and gamete formation?

12. What is the relationship between the second law and gamete formation?

13. How do heterozygotes give rise to polymorphisms?

Answers to Concept Map 2-2 Questions

1. Alleles segregate.

2. Genes independently assort.

3. The first law of Mendel is that alleles segregate. The second law of Mendel is that genes independently assort.

4. Height, weight, and stomach volume are examples of continuous variation. Number of legs, number of eyes, and number of stomachs (in cows) are examples of discontinuous variation.

5. Enzyme activity can be both continuous and discontinuous, depending on the enzyme.

6. *Dominant* means that one copy of the allele is sufficient for expression. *Recessive* means that the particular allele is not expressed unless it is the only allele present.

7. Alleles are dominant and recessive.

8. *Homozygous* means that both alleles are identical. *Heterozygous* means that two different alleles are present.

9. The enzyme phosphoglucomutase has three common forms. Eye color in *Drosophila* can be red, white, plum, scarlet, and other colors. Some people develop baldness while others do not.

10. A heterozygous organism would yield a 1:1 ratio in a testcross. A homozygous organism would yield a 1:0 ratio in a testcross.

11. The first law is enacted during the onset of anaphase I.

12. The second law is enacted during anaphase I.

13. Heterozygotes contain two differing alleles which, if homozygous, would yield two alternative phenotypes for the gene in question.

▶ SOLUTIONS TO PROBLEMS

Be sure that you have thoroughly read the entire chapter before you attempt any of the problems.

1. Mendel's first law states that alleles segregate during meiosis. Mendel's second law states that genes independently assort during meiosis.

2. To determine whether the *Drosophila* is *AA* or *Aa*, a testcross should be done. By definition, this means using a fly that is *aa*. If the original fly is *AA*, all progeny will have the *A* phenotype. If the original fly is *Aa*, half the progeny will have the *A* phenotype and half will have the *a* phenotype.

3. The progeny ratio is approximately 3:1, indicating a classic heterozygous-by-heterozygous mating. The parents must be *Bb* × *Bb*. Their black progeny must be *BB* and *Bb* in a 1:2 ratio, and their white progeny must be *bb*.

4. **a.** This is simply a matter of counting genotypes; there are 9 genotypes in the Punnett square.

   b. Again, simply count. The ratio is

1 *RRYY*	2 *RrYY*	1 *rrYY*
2 *RRYy*	4 *RrYy*	2 *rrYy*
1 *RRyy*	2 *Rryy*	1 *rryy*

   c. To find a formula for the number of genotypes, first make a chart.

NUMBER OF GENES	NUMBER OF GENOTYPES	NUMBER OF PHENOTYPES
1	$3 = 3^1$	$2 = 2^1$
2	$9 = 3^2$	$4 = 2^2$
3	$27 = 3^3$	$8 = 2^3$

Now note that the number of genotypes is 3 raised to some power in each case. In other words, a general formula for the number of genotypes is 3^n, where "n" equals the number of genes.

The number of phenotypes is 2 raised to some power in each case. The general formula for the number of phenotypes observed is 2^n, where "n" equals the number of genes.

d. The round, yellow phenotype is $R-Y-$. Two ways to determine the exact genotype of a specific plant are through selfing and conducting a testcross.

With selfing, complete heterozygosity will yield a 9:3:3:1 phenotypic ratio. Homozygosity at one locus will yield a 3:1 phenotypic ratio, while homozygosity at both loci will yield only one phenotypic class.

With a testcross, complete heterozygosity will yield a 1:1:1:1 phenotypic ratio. Homozygosity at one locus will yield a 1:1 phenotypic ratio, while homozygosity at both loci will yield only one phenotypic class.

5. Each die has six sides, so the probability of any one side (number) is 1/6. To get specific red, green, and blue numbers involves "and" statements.

a. $(1/6)(1/6)(1/6) = (1/6)^3$

b. $(1/6)(1/6)(1/6) = (1/6)^3$

c. $(1/6)(1/6)(1/6) = (1/6)^3$

d. To get no sixes is the same as getting anything but sixes:

$(1 - 1/6)(1 - 1/6)(1 - 1/6) = (5/6)^3.$

e. There are three ways to get two sixes and one five:

6R, 6G, 5B	$(1/6)(1/6)(1/6)$
or	+
6R, 5G, 6B	$(1/6)(1/6)(1/6)$
or	+
5R, 6G, 6B	$(1/6)(1/6)(1/6)$
	$= \quad 3(1/6)^3$

f. Here there are "and" and "or" statements:

p(three sixes "or" three fives)

$= p$(6R "and" 6G "and" 6B "or" 5R "and" 5G "and" 5B)

$= (1/6)^3 + (1/6)^3 = 2(1/6)^3$

g. There are six ways to fulfill this:

$6(1/6)^3 = (1/6)^2$

h. The easiest way to approach this problem is to consider each die separately:

The first die thrown can be any number. Therefore, the probability for it is 1.0.

The second die can be any number except the number obtained on the first die. Therefore, the probability of not duplicating the first die is $1.0 - p(\text{first die duplicated}) = 1.0 - 1/6 = 5/6$.

The third die can be any number except the numbers obtained on the first two dice. Therefore, the probability is $1.0 - p(\text{first two dice duplicated}) = 1.0 - 2/6 = 4/6$.

Therefore, the probability of all different dice is $(1.0)(5/6)(4/6) = 20/36 = 5/9$.

6. a. Before beginning the specific problems, write the probabilities associated with each jar.

jar 1 $p(R) = 600/(600 + 400) = 0.6$

$p(W) = 400/(600 + 400) = 0.4$

jar 2 $p(B) = 900/(900 + 100) = 0.9$

$p(W) = 100/(900 + 100) = 0.1$

jar 3 $p(G) = 10/(10 + 990) = 0.01$

$p(W) = 990/(10 + 990) = 0.99$

(1) $p(R, B, G) = (0.6)(0.9)(0.01) = 0.0054$

(2) $p(W, W, W) = (0.4)(0.1)(0.99) = 0.0396$

(3) Before plugging into the formula, you should realize that, while white can come from any jar, red and green must come from specific jars (jar 1 and jar 3). Therefore, white must come from jar 2:

$p(R, W, G) = (0.6)(0.1)(0.01) = 0.0006$

(4) $p(R, W, W) = (0.6)(0.1)(0.99) = 0.0594$

(5) There are three ways to satisfy this:

R, W, W   *or*   W, B, W   *or*   W, W, G

$= (0.6)(0.1)(0.99) + (0.4)(0.9)(0.99) + (0.4)(0.1)(0.01)$

$= 0.0594 + 0.3564 + 0.0004 = 0.4162$

(6) At least one white is the same as 1 minus no whites:

$p(\text{at least 1 W}) = 1 - p(\text{no W}) = 1 - p(R, B, G)$

$= 1 - (0.6)(0.9)(0.01) = 1 - 0.0054 = 0.9946$

b. The cross is $Rr \times Rr$. The probability of red ($R-$) is 3/4, and the probability of white (rr) is 1/4. Because only one white is needed, the only unacceptable result is all red.

In n trials, the probability of all red is $(3/4)^n$. Because the probability of failure must be 5 percent

$(3/4)^n = 0.05$

$n = 10.41$, or 11 seeds.

c. Because the eggs are implanted simultaneously the $p(\text{failure}) = 0.8$ for each egg.

$$p(5 \text{ failures}) = (0.8)^5$$

$$p(\text{at least one success}) = 1 - (0.8)^5 = 1 - 0.328 = 0.672$$

7. By drawing the pedigree, you will see that the cross in question is $Tt \times Tt$, which has a 3:1 phenotypic ratio.

   The probability of being a taster is 3/4, and the probability of being a nontaster is 1/4.

   The probability of being a boy equals the probability of being a girl equals 1/2.

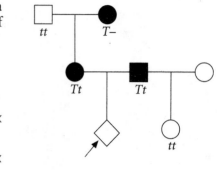

   a. (1) $p(\text{nontaster girl})$
       $= p(\text{nontaster}) \times p(\text{girl})$
       $= 1/4 \times 1/2 = 1/8$

   (2) $p(\text{taster girl}) = p(\text{taster}) \times$
       $p(\text{girl}) = 3/4 \times 1/2 = 3/8$

   (3) $p(\text{taster boy}) = p(\text{taster}) \times$
       $p(\text{boy}) = 3/4 \times 1/2 = 3/8$

   b. $p(\text{taster for first two children})$

   $= p(\text{taster for first child}) \times p(\text{taster for second child})$

   $=$ 3/4 $\times$ 3/4

   $= 9/16$

8. *Unpacking the Problem*

   a. Yes. The pedigree is given below.

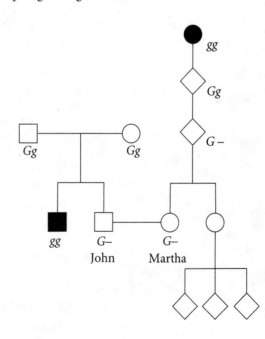

   b. In order to state this problem as a Punnett square, you must first know the genotypes of John and Martha. The genotypes can be determined only through drawing the pedigree. Even with the pedigree, however, the genotypes can be stated only as *G*–for both John and Martha, as is shown by the following discussion.

The probability that John is carrying the allele for galactosemia is 2/3, rather than the 1/2 that you might guess. To understand this, recall that John's parents must be heterozygous in order to have a child with the recessive disorder while still being normal themselves (the assumption of normalcy is based on the information given in the problem). John's parents, were both *Gg*. A Punnett square for their mating would be:

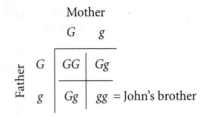

The cross is

P   *Gg* × *Gg*

F$_1$ *gg*   John's brother

     G–   John

The expected ratio of the F$_1$ is 1 *GG*:2 *Gg*:1 *gg*. Because John does not have galactosemia (an assumption based on the information given in the problem), he can be either *GG or Gg*, which occurs at a ratio of 1:2. Therefore, his probability of carrying the *g* allele is 2/3.

The probability that Martha is carrying the *g* allele is based on the following chain of logic. Her great-grandmother had galactosemia, which means that she had to pass the allele to Martha's grandparent. Because the problem states nothing with regard to the grandparent's phenotype, it must be assumed that the grandparent was normal, or *Gg*. The probability, if the grandparent received the *g* allele, that the grandparent passed it to Martha's parent is 1/2. Therefore, the probability that the grandparent received the allele and passed it to Martha's parent is 1 × 1/2, or 1/2. Next, the probability that Martha's parent passed the allele to Martha is 1/2, assuming that the parent actually has it. Therefore, the probability that Martha's parent has the allele and passed it to Martha is 1/2 × 1/2, or 1/4.

In summary:

John   *p*(*GG*) = 1/3

         p(*Gg*) = 2/3

Martha   *p*(*GG*) = 3/4

         p(*Gg*) = 1/4

This information does not fit easily into a Punnett square.

c. While the above information could be put into a branch diagram, it does not easily fit into one and overcomplicates the problem, just as a Punnett square would.

d. The mating between John's parents illustrates Mendel's first law.

e. The scientific words in this problem are *galactosemia*, *autosomal*, and *recessive*.

Galactosemia is a metabolic disorder characterized by the absence of the enzyme galactose-1-phosphate uridyl transferase, which results in an accumulation of galactose. In the vast majority of cases, galactosemia results in an enlarged liver, jaundice, vomiting, anorexia, lethargy, and very early death if galactose is not omitted from the diet (initially, the child obtains galactose from milk).

Autosomal refers to genes that are on the autosomes.

Recessive means that in order for an allele to be expressed, it must be the only form of the gene present in an organism.

f. The major assumption is that if nothing is stated about a person's phenotype, the person is of normal phenotype. Another assumption that may be of value, but is not actually needed, is that all people marrying into these two families are normal and do not carry the allele for galactosemia.

g. The people unmentioned in the problem, but who must be considered, are John's parents and Martha's grandparent and parent descended from her affected great-grandmother.

h. The major statistical rule needed to solve the problem is the product rule ("and" rule).

i. Autosomal recessive disorders are assumed to be rare and assumed to occur equally frequently in males and females. They are also assumed to be expressed if the person has an autosomal recessive genotype.

j. Rareness leads to the assumption that people who marry into a family that is being studied do not carry the allele, which was assumed in entry f above.

k. The only certain genotypes in the pedigree are John's parents, John's brother, and Martha's great-grandmother. All other individuals have uncertain genotypes.

l. John's family can be treated simply as a heterozygous-by-heterozygous cross, with John having a 2/3 probability of being a carrier, while it is unknown if either of Martha's parents carry the allele.

m. The information regarding Martha's sister and her children is irrelevant to the problem.

n. The problem contains a number of assumptions that have not been necessary in problem solving until now.

o. I can think of a number. Can you?

Solution to the Problem

p(child has galactosemia) = p(John is *Gg*) $\times$ p(Martha is *Gg*) $\times$ p(both parents passed *g* to the child) = (2/3)(1/4)(1/4) = 2/48 = 1/24

9. Charlie, his mate, or both, obviously were not pure-breeding, because his F_2 progeny were of two phenotypes. Let A = black and white, and a = red and white. If both parents were heterozygous, then red and white would have been expected in the F_1 generation. Red and white were not observed in the F_1 generation, so only one of the parents was heterozygous. The cross is

P   *Aa* $\times$ *AA*

F_1 1 *Aa*:1 *AA*

Two F_1 heterozygotes (*Aa*) when crossed would give 1 *AA* (black and white):2 *Aa* (black and white):1 *aa* (red and white).

If the red and white F_2 progeny were from more than one mate of Charlie's, then the farmer acted correctly. However, if the F_2 progeny came only from one mate, the farmer may have acted too quickly.

10. Because the parents are heterozygous, both are *Aa*. Both twins could be albino or both twins could be normal ("and" "or" "and" = multiply add multiply). The probability of being normal (*A*–) is 3/4, and the probability of being albino (*aa*) is 1/4.

$$p(\text{both normal}) \qquad + \qquad p(\text{both albino})$$
$$= p(\text{first normal}) \times p(\text{second normal}) + p(\text{first albino}) \times p(\text{second albino})$$
$$= \qquad (3/4)(3/4) \qquad + \qquad (1/4)(1/4)$$
$$= \qquad 9/16 \qquad + \qquad 1/16 = 5/8$$

11. The plants are approximately 3 blotched:1 unblotched. This suggests that blotched is dominant to unblotched and that the original plant which was selfed was a heterozygote.

   a. Let *A* = blotched, *a* = unblotched.

   P   *Aa* (blotched) × *Aa* (blotched)

   F_1 1 *AA*:2 *Aa*:1 *aa*

   3 *A*– (blotched):1 *aa* (unblotched)

   b. All unblotched plants should be pure-breeding in a testcross with an unblotched plant (*aa*), and one-third of the blotched plants should be pure-breeding.

12. In theory, it cannot be proved that an animal is not a carrier for a recessive allele. However, in an *A*– × *aa* cross, the more dominant phenotype progeny produced, the less likely it is that one parent is *Aa*. In such a cross half the progeny would be *aa* and half would be *Aa* if the parent were *Aa*. With *n* dominant phenotype progeny, the probability that the parent is *Aa* is $(1/2)^n$.

13. The results suggest that winged (*A*–) is dominant to wingless (*aa*) (cross 2 gives a 3:1 ratio). If that is correct, the crosses become

		NUMBER OF PROGENY PLANTS	
POLLINATION	GENOTYPES	WINGED	WINGLESS
Winged (selfed)	$AA \times AA$	91	1[*]
Winged (selfed)	$Aa \times Aa$	90	30
Wingless (selfed)	$aa \times aa$	4[*]	80
Winged × wingless	$AA \times aa$	161	0
Winged × wingless	$Aa \times aa$	29	31
Winged × wingless	$AA \times aa$	46	0
Winged × winged	$AA \times A-$	44	0
Winged × winged	$AA \times A-$	24	0

The five unusual plants are most likely due either to human error in classification or to contamination. Alternatively, they could result from environmental effects on development. For example, too little water may have prevented the seed pods from becoming winged even though they are genetically winged.

14. **a.** The disorder appears to be dominant because all affected individuals have an affected parent.

b. Assuming dominance, the genotypes are

   I: *dd, Dd*

   II: *Dd, dd, Dd, dd*

   III: *dd, Dd, dd, Dd, dd, dd, Dd, dd*

   IV: *Dd, dd, Dd, dd, dd, dd, dd, Dd, dd*

c. The mating is $Dd \times dd$. The probability of Dd equals 1/2, and the probability of dd equals 1/2. The formula to use is the one for combinations:

$$\frac{4!}{4!\,0!}(1/2)^4(1/2)^0 = (1/2)^4 = 1/16 = 0.063$$

15. **a.** *Pedigree 1:* The best answer is recessive because the disorder skips generations and appears in a mating between two related individuals.

   Pedigree 2: The best answer is dominant because it appears in each generation, roughly half the progeny are affected, and affected individuals have an affected parent.

   Pedigree 3: The best answer is dominant for the reasons stated for pedigree 2. Inbreeding, while present in the pedigree, does not allow an explanation of recessive because it cannot account for individuals in the second generation.

   Pedigree 4: The best answer is recessive even though the disorder appears in each generation. Two unaffected individuals had one-fourth affected progeny, and the affected individuals in the third generation had an affected father and a mother who could be carriers.

b. *Genotypes of pedigree 1:*

   Generation I: *AA, aa*

   Generation II: *Aa, Aa, Aa, A–, A–, Aa*

   Generation III: *Aa, Aa*

   Generation IV: *aa*

   Genotypes of pedigree 2:

   Generation I: *Aa, aa, Aa, aa*

   Generation II: *aa, aa, Aa, Aa, aa, aa, Aa, Aa, aa*

   Generation III: *aa, aa, aa, aa, aa, A–, A–, A–, Aa, aa*

   Generation IV: *aa, aa, aa*

Genotypes of pedigree 3:

  Generation I: *A–, aa*

  Generation II: *Aa, aa, aa, Aa*

  Generation III: *aa, Aa, aa, aa, Aa, aa*

  Generation IV: *aa, Aa, Aa, Aa, aa, aa*

Genotypes of pedigree 4:

  Generation I: *aa, A–, Aa, Aa*

  Generation II: *Aa, Aa, Aa, aa, A–, aa, A–, A–, A–, A–, A–*

  Generation III: *Aa, aa, Aa, Aa, aa, Aa*

16. a. The pedigree is

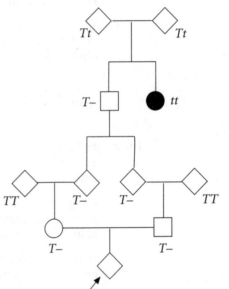

b. The probability that the child of the two first cousins will have Tay-Sachs disease is a function of three probabilities:

p(the woman is *Tt*) × p(the man is *Tt*) × p(both donate *t*)

= (2/3)(1/2)(1/2) × (2/3)(1/2)(1/2) × 1/4 = 1/144 = 0.0069

17. a. The most likely possibility is that the disorder is recessive because skipping of generations is present and affected individuals do not have affected parents.

b. The probability that the first child from individuals 1 and 2 will have the kidney disease is the probability that each has the gene times the probability that both pass it to their child. The father of individual 1 had to have received the gene from his mother. The chance that the father passed it to individual 1 is 1/2. Individual 2 had to have received the gene from her father. The final probability that the first child of individuals 1 and 2 will have the kidney disease is

p(1 received gene) × p(2 received gene) × p(both passed gene to child)

=   1/2   ×   1   ×   1/4

= 1/8

18. a. Yes.

b. Susan is highly unlikely to have Huntington's disease, because her great-grandmother is 75 years old and has yet to develop it, when nearly 100% of people carrying the allele will have developed the disease by that age. If her great-grandmother does not have it, Susan cannot inherit it.

Alan is somewhat more likely than Susan to develop Huntington's disease. His grandfather is approximately 50 years old, and approximately 20% of the people with the allele have yet to develop the disease by that age. If Alan's grandfather eventually develops Huntington's disease, there is a probability of 50% that Alan's father inherited it from him, and there is a probability of 50% that Alan received that allele from his father. Therefore, Alan has a $1/2 \times 1/2 = 1/4$ probability of developing Huntington's disease *if* his grandfather eventually develops it.

19. a.   I: *A–, aa, aa, A–*

II: *AA, Aa, Aa, Aa, Aa*

III: *AA, A–, A–, AA*

IV: *A–, A–*

b. The probability that individual A has the PKU allele is derived from individual II-2. II-2 must be *Aa*. Therefore, the probability that II-2 passed the PKU allele to individual III-2 is 1/2. If III-2 received the allele, the probability that he passed it to individual IV-1 (A) is 1/2. Therefore, the probability that A is a carrier is $1/2 \times 1/2 = 1/4$.

The probability that individual B has the allele goes back to the mating of II-3 and II-4, both of whom are heterozygous. Their child, III-3, has a 2/3 probability of having received the PKU allele and a probability of 1/2 of passing it to IV-2 (B). Therefore, the probability that B has the PKU allele is $2/3 \times 1/2 = 1/3$.

If both parents are heterozygous, they have a 1/4 chance of both passing the *a* allele to their child.

$$p(\text{child has PKU}) = p(\text{A is } Aa) \times p(\text{B is } Aa) \times p(\text{both parents donate } a)$$

$$= \quad 1/4 \quad \times \quad 1/3 \quad \times \quad 1/4$$

$$= 1/48$$

c. If the first child is normal, no additional information has been gained and the probability that the second child will have PKU is the same as the probability that the first child will have PKU, or 1/48.

d. If the first child has PKU, both parents are heterozygous. The probability of having an affected child is 1/4, and the probability of having an unaffected child is 3/4.

20. a. In order to draw this pedigree, you should realize that if an individual's status is not mentioned, then there is no way to assign a genotype to that person. The parents of the boy in question had a genotype that differed from his. Therefore, both parents were heterozygous and the boy, who is a non-roller, is homozygous recessive. Let *R* stand for the ability to roll the tongue and *r* stand for the inability to roll the tongue. The pedigree becomes

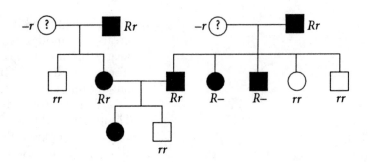

b. If the seven discordant pairs were not truly identical, that could account for the observation. If the seven discordant pairs were truly identical, that indicates that the ability to roll the tongue cannot be determined solely by one gene. Assuming the twins are identical, there must be either an environmental component to the expression of that gene or developmental noise (see Chapter 1) may play a role. There is also a learning aspect to tongue rolling, which means there must be the ability to learn.

21. Let *C* stand for the normal allele and *c* stand for the allele that causes cystic fibrosis.

a.

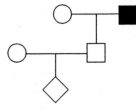

b. The man has a probability of 1.0 of having the *c* allele. His wife, who is from the general population, has a 1/50 chance of having the *c* allele. If both have the allele, then 1/4 of their children will have cystic fibrosis. The probability that their first child will have cystic fibrosis is

p(man has *c*) × p(woman has *c*) × p(both pass *c* to the child)

=   1.0   ×   1/50   ×   1/4 = 1/200 = 0.005

c. If the first child does have cystic fibrosis, then the woman is a carrier of the *c* allele. Both parents are *Cc*. The chance that the second child will be normal is the probability of a normal child in a heterozygous × heterozygous mating, or 3/4.

22. a. The inheritance pattern for red hair suggested by this pedigree is recessive.

b. The allele appears to be somewhat rare.

23. *Taster by taster cross:* Tasters can be either *PP* or *Pp*, and the genotypic status cannot be determined until a large number of progeny are observed. A failure to obtain a 3:1 ratio in the matings of two tasters would be expected because there are three types of matings:

| MATING | CHILDREN | |
	GENOTYPES	PHENOTYPES
$PP \times PP$	all PP	all tasters
$PP \times Pp$	1/2 PP:1/2 Pp	all tasters
$Pp \times Pp$	1/4 PP:1/2 Pp : 1/4 pp	3/4 tasters:1/4 nontasters

Taster by nontaster cross: There are two types of matings that resulted in the observed progeny:

| MATING | CHILDREN | |
	GENOTYPES	PHENOTYPES
$PP \times pp$	all Pp	all tasters
$Pp \times pp$	1/2 Pp:1/2 pp	1/2 tasters:1/2 nontasters

Again, the failure to obtain either a 1:0 ratio or a 1:1 ratio would be expected because of the two mating types.

Nontaster by nontaster cross: There is only one mating that is nontaster by nontaster ($pp \times pp$), and 100% of the progeny would be expected to be nontasters. Of 223 children 5 were classified as tasters. Some could be the result of mutation (unlikely), some could be the result of misclassification (likely), some could be the result of a second gene that affects the expression of the gene in question (possible), some could be the result of developmental noise (possible), and some could be due to illegitimacy (possible).

24. Use the following symbols:

GENE FUNCTION	DOMINANT ALLELE	RECESSIVE ALLELE
Color	R = red	r = yellow
Loculed	L = two	l = many
Height	H = tall	h = dwarf

The starting plants are pure-breeding, so their genotypes are

red, two-loculed, dwarf: *RR LL hh* and
yellow, many-loculed, tall: *rr ll HH*.

The farmer wants to produce a pure-breeding line that is yellow, two-loculed, and tall, which would have the genotype *rr LL HH*.

The two pure-breeding starting lines will produce an F_1 that will be *Rr Ll Hh*. By doing an F_1 cross and selecting yellow, two-loculed, and tall plants, the known genotype will be *rr L– H–*. The task then will be to do sequential testcrosses for both the *L–* and *H–* genes among these yellow, two-loculed, and tall plants. Because the two genes in question are homozygous recessive in different plants, each plant that is yellow, two-loculed, and tall will have to be testcrossed twice.

For each testcross, the plant will obviously be discarded if the test-cross reveals a heterozygous state for the gene in question. If no recessive allele is detected, then the minimum number of progeny that must be examined to be 95% confident that the plant is homozygous is based on the frequency of the dominant phenotype, which is 3/4. In n progeny, the probability of obtaining all dominant progeny given that the plant is heterozygous is $(3/4)^n$. To be 95% confident of homozygosity, the following formula is used, where 5% is the probability that it is not homozygous:

$$(3/4)^n = 0.05$$

$n = 10.41$, or 11 dominant phenotype progeny must be obtained from each testcross to be 95% confident that the plant is homozygous.

25. Let A represent achondroplasia and a represent normal height. Let N represent neurofibromatosis and n represent the normal allele. Because both conditions are extremely rare, the affected individuals are assumed to be heterozygous. The genes are also assumed to assort independently. The cross is

P $Aa\ nn \times aa\ Nn$

F_1 1 $Aa\ nn$:1 $aa\ nn$:1 $aa\ Nn$:1 $Aa\ Nn$

1 dwarf:1 normal:1 neurofibromatosis:1 dwarf, neurofibromatosis

26. a. The ratio of dark:albino and short:long is 3:1. Therefore, each gene is heterozygous in the parents. The cross is $Cc\ Ss \times Cc\ Ss$.

b. Because all progeny are dark, one of the parents is CC. The other is C–. The ratio of short:long is 1:1, a testcross. The cross is C– $Ss \times C$– ss, with one of the parents CC. Assuming homozygosity, the cross is $CC\ Ss \times CC\ ss$.

c. All progeny are short (S–), and the ratio of dark to albino is 1:1, indicating a testcross. Therefore, the cross is $Cc\ S$– $\times cc\ S$–, with one of the parents SS. Assuming homozygosity, the cross is $Cc\ SS \times cc\ SS$.

d. All progeny are albino (cc), and the ratio of short to long is 3:1, indicating a heterozygous $\times$ heterozygous cross. Therefore, the cross is $cc\ Ss \times cc\ Ss$.

e. The dark:albino ratio is 3:1, indicating a $Cc \times Cc$ cross. All animals are long (ss). The cross is $Cc\ ss \times Cc\ ss$.

f. All animals are dark, indicating at least one parent is homozygous for color. Short:long = 3:1, indicating $Ss \times Ss$. The cross is C– $Ss \times C$– Ss, with one parent CC. Assuming homozygosity, the cross is $CC\ Ss \times CC\ Ss$.

g. Dark:albino = 3:1. Short:long = 1:1, a testcross. The cross is $Cc\ Ss \times Cc\ ss$.

27. a. From cross 3, purple is dominant to green. From cross 4, cut is dominant to potato.

b. Let A stand for color and B stand for shape.

Cross 1: Immediately, the cross can be written A– B– $\times aa\ B$–. A 1:1 ratio for color indicates a testcross, and a 3:1 ratio for shape indicates a heterozygous cross. The cross is $Aa\ Bb \times aa\ Bb$.

Cross 2: Immediately, the cross can be written $A- B- \times A- bb$. A 3:1 ratio for color and a 1:1 ratio for shape exists. The cross is $Aa\ Bb \times Aa\ bb$.

Cross 3: Because no green plants exist, the purple parent is homozygous. There is a 3:1 ratio for shape. The cross is $AA\ Bb \times aa\ Bb$.

Cross 4: No potato is seen, therefore cut is homozygous. There is a 1:1 ratio for color, a testcross. The cross is $Aa\ BB \times aa\ bb$.

Cross 5: A 1:1:1:1 ratio indicates a testcross for each gene. The cross is $Aa\ bb \times aa\ Bb$.

28. a. Look at each gene separately.

(1) $A- B- C- D- E- = (1/2)(3/4)(1/2)(3/4)(1/2) = 9/128$

(2) $aa\ B- cc\ D- ee = (1/2)(3/4)(1/2)(3/4)(1/2) = 9/128$

(3) (question 1) + (question 2) = $9/128 + 9/128 = 9/64$

(4) $1 -$ (question 1) $-$ (question 2) = $1 - 9/128 - 9/128 = 55/64$

b. (1) $Aa\ Bb\ Cc\ Dd\ Ee = (1/2)(1/2)(1/2)(1/2)(1/2) = 1/32$

(2) $aa\ Bb\ cc\ Dd\ ee = (1/2)(1/2)(1/2)(1/2)(1/2) = 1/32$

(3) (question 1) + (question 2) = $1/32 + 1/32 = 1/16$

(4) $1 -$ (question 1) $-$ (question 2) = $1 - 1/32 - 1/32 = 15/16$

29. The single yellow fly may be either recessive *yy* or may be yellow due to a diet of silver salts. Cross that fly with a known recessive yellow fly and raise half the larvae on a diet with no silver salts and half the larvae on a diet with silver salts. A true recessive will result in flies with yellow bodies on both diets, while a phenocopy that is genetically *Y–* will produce flies with brown or yellow bodies, depending on the diet.

Phenocopies are caused by environmental factors. Many drugs used by pregnant women result in genetic defects that are phenocopies. One example is cleft lip or palate caused by Valium taken before the fetal face is completely formed. Retardation caused by the consumption of alcohol during pregnancy is another phenocopy effect.

30. a. Cataracts appear to be caused by a dominant allele because affected people have affected parents. Dwarfism appears to be caused by a recessive allele because affected people have unaffected parents.

b. Indicate genotypes on the pedigree, using *A* for cataracts and *B* for dwarfism. The genotypes are

I: *AaB–, aabb*

II: *aaBb, aaBb, aaBb, AaBb, AaBb, AaBb, aaB–*

III: *aaB–, aaBb, AaB–, aaB–, AaBb, AaBb, aaB–, aaB–, AaB–*

IV: *aabb, aaB–, aaB–, aaB–, A–B–, A–bb, aaB–, A–B–, aabb*

c. The mating is $aabb \times A-B-$. Recall that the probability of a child's being affected by any disease is a function of the probability of each parent's carrying the allele in question and the probability that one parent (for a dominant disorder) or both parents (for a recessive disorder) donate it to the child. Individual IV-1 must donate *ab* to the child in question. Therefore, the first task is to determine the probabilities associated with individual IV-5.

The probability that individual IV-5 is heterozygous for dwarfism is 2/3. The probability that she does not carry the allele is 1/3. In other words, the probability that she has the allele and will pass it to her child is $2/3 \times 1/2 = 1/3$.

The probability that individual IV-5 is homozygous for cataracts is 1/3, and the probability that she is heterozygous is 2/3. If she is homozygous for the allele that causes cataracts, she must pass it to her child. If she is heterozygous for cataracts, she has a probability of $2/3 \times 1/2 = 1/3$ of passing it to her child. There is no way to tell if she is homozygous or heterozygous for cataracts, however; this leads to two differing probabilities rather than a specific probability.

The probability that the first child is a dwarf with cataracts is the probability that the child is A–bb. Because the father must donate ab, the question reduces to the probability of the mother's donating Ab, which is $1/3 \times 1/3$ to $1 \times 1/3$, or 1/9 to 1/3.

The probability of having a phenotypically normal child is the probability that the mother donates aB, which is either 0, if the mother is homozygous, or $1/3 \times 1/3 = 1/9$.

31. a. and b. Begin with any two of the three lines and cross them. If, for example, you began with $aaBBCC \times AAbbCC$, the progeny would all be $AaBbCC$. Crossing two of these would yield

9	A–B–CC
3	aaB–CC
3	A–$bbCC$
1	$aabbCC$

The $aabbCC$ genotype has two of the genes in a homozygous recessive state and occurs in 1/16 of the offspring. If that were crossed with $AABBcc$, the progeny would all be $AaBbCc$. Crossing two of them would lead to a 27:9:9:9:3:3:3:1 ratio, and the plant occurring in 1/64 of the progeny would be the desired $aabbcc$..

c. There are several different routes to obtaining $aabbcc$, but the one outlined above requires only four crosses.

32. a. At year 1, the genotypic ratio is

AA	0.55
Aa	0.40
aa	0.05

Because all plants self-pollinate, the three crosses are

$AA \times AA \rightarrow 100\%\ AA$

$Aa \times Aa \rightarrow 25\%\ AA,\ 50\%\ Aa,\ 25\%\ aa$

$aa \times aa \rightarrow 100\%\ aa$

In other words, the two homozygous classes grow at the expense of the heterozygous class. At year 2, the genotypic ratio would be

$AA \quad 0.55 + 1/4(0.40) = 0.65$

Aa $1/2(0.40) = 0.20$

aa $0.5 + 1/4(0.40) = 0.15$

Again, when this population self-pollinates, the homozygous classes will grow, by the same percentages, at the expense of the heterozygous class. At year 3, the genotypic ratio would be

AA $0.65 + 1/4(0.20) = 0.70$

Aa $1/2(0.20) = 0.10$

aa $0.15 + 1/4(0.20) = 0.20$

A third year of selfing will produce the following genotypic ratio:

AA $0.70 + 1/4(0.10) = 0.725$

Aa $1/2(0.10) = 0.05$

aa $0.20 + 1/4(0.10) = 0.225$

b. Consider the cross *Aa* × *Aa*. Only 1/2 the progeny would be expected to be heterozygous. This will hold for each locus that began as heterozygous. Therefore, overall heterozygosity will be reduced by 50% with each generation of selfing. After the third generation of selfing, only $(1/2)^3 \times 40\% = 1/8(40\%) = 5\%$ of the loci would remain heterozygous.

c. The same mathematical concepts apply in each situation. The generalized formula for a reduction of heterozygosity with selfing is $(1/2)^n X$, where n = the number of generations and X = the initial percentage of heterozygosity.

► TIPS ON PROBLEM SOLVING

Ratios: A 1:2:1 (or 3:1) ratio indicates that one gene is involved (see Problem 3). A 9:3:3:1 ratio, or some modification of it, indicates that two genes are involved (see Problem 27a). A testcross results in a 1:1 ratio if the organism being tested is heterozygous and a 1:0 ratio if it is homozygous (see Problem 2).

Pedigrees: Normal parents have affected offspring in recessive disorders (see Problem 16). Normal parents have normal offspring and affected parents have affected offspring in dominant disorders (see Problems 16, 26). If phenotypically identical parents produce progeny with two phenotypes, the parents were both heterozygous (see Problems 16, 18, 21).

Probability: When dealing with two or more independently assorting genes, consider each gene separately (see Problems 26, 27, 29).

► A SYSTEMATIC APPROACH TO PROBLEM SOLVING

Now that you have struggled with a number of genetics problems, it may be worthwhile to make some generalizations about problem solving beyond what has been presented for each chapter so far.

The first task always is to determine exactly what information has been presented and what is being asked. Frequently, it is necessary to rewrite the problem or to symbolize the presented information in some way.

The second task is to formulate and test hypotheses. If the results generated by a hypothesis contradict some aspect of the problem, then the hypothesis is rejected. If the hypothesis generates data compatible with the problem, then it is retained.

A systematic approach is the only safe approach in working genetics problems. Shortcuts in thought processes usually lead to an incorrect answer.

Consider the following two types of problems.

1. When analyzing pedigrees, there are usually only four possibilities (hypotheses) to be considered: autosomal dominant, autosomal recessive, X-linked dominant, and X-linked recessive. The criteria for each should be checked against the data. Additional factors that should be kept in mind are epistasis, penetrance, expressivity, age of onset, incorrect diagnosis in earlier generations, adultery, adoptions that are not mentioned, and inaccurate information in general. All these factors can be expected in real life, although few will be encountered in the problems presented here (see Chapters 3 and 4 for X linkage and other factors).

2. When studying matings, frequently the first task is to decide whether you are dealing with one gene, two genes, or more than two genes (hypotheses). The location of the gene or genes may or may not be important. If location is important, then there are two hypotheses: autosomal and X-linked. If there are two or more genes, then you may have to decide on linkage relationships between them. There are two hypotheses: unlinked and linked.

If ratios are presented, then 1:2:1 (or some modification signaling dominance) indicates one gene, 9:3:3:1 (or some modification reflecting epistasis) indicates two genes, and 27:9:9:9:3:3:3:1 (or some modification signaling epistasis) indicates three genes. If ratios are presented that bear no relationship to the above, such as 35:35:15:15, then you are dealing with two linked genes (see Chapter 5 for a discussion of linkage).

If phenotypes rather than ratios are emphasized in the problem, then a cross of two mutants that results in wild type indicates the involvement of two genes rather than alleles of the same gene. Both mutants are recessive to wild type. A correlation of sex with phenotype indicates X linkage for the gene mutant in the female parent, while a lack of correlation indicates autosomal location.

If the problem involves X linkage, frequently the only way to solve it is to focus on the male progeny.

Once you determine the number of genes being followed and their location, the problem essentially solves itself if you make a systematic listing of genotype and phenotype.

Sometimes, the final portion of a problem will give additional information that requires you to adjust all the work that you have done up to that point. As an example, in Problem 52 of Chapter 4, crosses 1-3 lead you to assume that you are working with one gene. In cross 4, data incompatible with this assumption are presented. Your initial assumption of one gene is correct for the information given in the first three crosses; it is not a mistake.

Other than a lack of systematic thought, the greatest mistake that a student can make is to label a rejected hypothesis an error. This decreases self-confidence and increases anxiety, with the result that real mistakes will likely follow. The beginner needs to keep in mind that science progresses by the rejection of hypotheses. When a hypothesis is rejected, something concrete is known: the proposed hypothesis does not explain the results. An unrejected hypothesis may be right or it may be wrong, and there is no way to know without further experimentation.

A very generalized flowchart for problem solving would look like this:

1. Determine what information is being presented and what is being asked.

2. Formulate all possible hypotheses.

3. Check the consequences of each hypothesis against the data (the given information).

4. Reject all hypotheses that are incompatible with the data. Retain all hypotheses that are compatible with the data.

5. If no hypothesis is compatible with the data, return to step 1.

3

Chromosome Theory
of Inheritance

► IMPORTANT TERMS
AND CONCEPTS

The **chromosome theory of heredity** means that genes are located on chromosomes and that the behavior of genes during mitosis and meiosis parallels the behavior of chromosomes during mitosis and meiosis.

Mitosis is the orderly distribution of one chromatid from each chromosome to each of two daughter cells. The chromosome number remains constant between cell generations. The stages of the mitotic cycle are interphase, prophase, metaphase, anaphase, and telophase. Interphase is divided into three stages: G1, S, and G2. G1 is the "gap" before chromosome replication. "S" is the stage of chromosome replication and DNA synthesis. G2 is the stage after chromosome replication.

Meiosis consists of two sequential cell divisions. Usually meiosis I (reductional division) is the orderly distribution of one chromosome from each chromosome pair to each of two daughter cells. The chromosome number is reduced by half. The stage between meiosis I and II is called *interkinesis*. Usually meiosis II (equational division) is the orderly distribution of one chromatid from each chromosome to each of two daughter cells. In form, meiosis II is identical with mitosis. The chromosome number remains constant between cell generations. Both meiotic divisions are subdivided into a number of stages.

The **nuclear spindle** is a protein structure that aids in proper chromosome movement during mitosis and meiosis.

Homologs (also spelled **homologues**) are two morphologically identical chromosomes that carry genes for the same functions. One comes from each parent.

Synapsis is the pairing of homologous chromosomes before the first meiotic division. The process occurs with the aid of the **synaptonemal complex**, a protein structure.

A **chiasma** is the chromosomal structure that is assumed to be the visible proof of the molecular event of crossing-over.

A **genome** (usually $1n$) consists of one set of unique chromosomes. The number of chromosomes per genome varies with the species.

A **diploid** ($2n$) cell or organism contains two homologous genomes, one from each parent.

A **haploid** ($1n$) cell or organism has one genome.

The number of sets of genomes present in an organism varies with the species, as does which sex is the heterogametic sex (if sexual differentiation exists). Some organisms are mostly diploid, some mostly haploid, and some alternate between haploid and diploid. Other organisms may routinely have more than two genomes (see Chapter 9). In all species that have meiosis, the laws of Mendel apply.

A **heteromorphic pair** is a pair of sex chromosomes that are nonidentical in shape or size and presumably have only partial homology. Genes present only on the X or Y chromosome in the heterogametic sex are called **hemizygous.**

The **heterogametic** sex has two nonidentical sex chromosomes (XY males in humans). The **homogametic** sex has two identical sex chromosomes (XX females in humans).

X linkage refers to genes located on the X chromosome.

Y linkage refers to genes located on the Y chromosome.

Autosomes are the paired (in diploids) nonsex chromosomes of both sexes.

Pseudoautosomal refers to the behavior of genes located on both the X and Y chromosomes.

X-inactivation occurs during early development of all mammals. It involves most of one of the two X chromosomes in females and is detectable by the presence of the **Barr body** during interphase. Because the highly condensed Barr body is genetically inactive, and because inactivation is random with respect to which chromosome is inactivated in each cell, females have two functional cell lines for each heterozygous gene on the X chromosome. Females are therefore **mosaics.**

Wild type refers to the most frequent allelic form of a gene found in natural (or laboratory) populations. It is indicated by a superscript + in the *Drosophila* system of symbolism. Deviations from the wild type are used to name the gene. Dominant deviations from wild type use an uppercase letter while recessive deviations use a lowercase letter. Thus curly, a dominant deviation, is *Cy*, and the wild-type form of the gene is Cy^+. The white-eye allele, a recessive deviation from wild type, is *w*, and the wild-type allele of that gene is w^+.

Questions for Concept Map 3-1

1. Name three organisms that spend most of their life cycle in the diploid state.

2. How many chromosomes, total, are in humans?

3. How many pairs of chromosomes are in humans?

4. What is the diploid number in humans?

5. What is the haploid number in humans?

6. If the two alleles in humans for a gene are different, what is the term used to describe that locus genetically? If the alleles are the same?

7. Are the two alleles of a gene on sister chromatids or homologous chromosomes?

8. What types of cells are considered somatic cells? Germ cells?

9. If the allelic content in a cell undergoing mitosis is symbolized by *Aa*, how do you symbolize each of the daughter cells? Draw the chromosomes at metaphase and anaphase.

► CONCEPT MAP 3-1:
   DIPLOID ORGANISMS

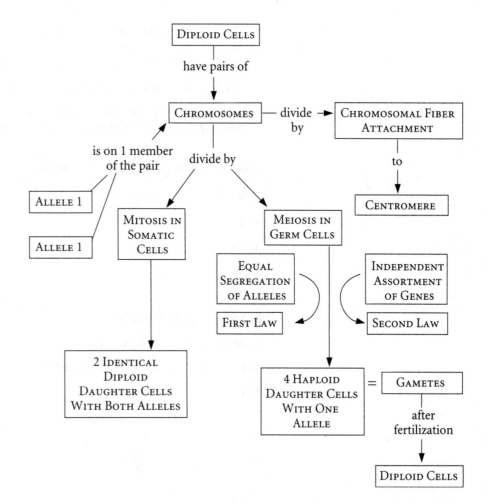

10. If the allelic content in a cell undergoing meiosis is symbolized by *Aa*, how do you symbolize each of the daughter cells after one cell division? Draw the chromosomes at metaphase and anaphase.

11. If the allelic content in a cell undergoing meiosis is symbolized by *Aa*, how do you symbolize each of the daughter cells after two cell divisions? Draw the chromosomes at the second metaphase and anaphase for each cell.

12. When does equal segregation of alleles occur? Be very specific.

13. When does independent assortment occur? Be very specific.

14. What is the full structure of which the chromosomal fiber is one part?

15. How many DNA molecules would there be per cell in a mature sperm from an organism in which $2n = 18$?

16. How many chromatids per cell would there be in the first polar body of a mammal with $2n = 24$?

17. Outline a decision-making flowchart for deciding on the mode of inheritance from data in a human pedigree.

18. In some organisms, the male is the homogametic sex. For a recessive sex chromosome–linked disorder, which sex would be affected at a higher rate?

19. Two heterozygotes for independently assorting genes *A* and *B* mate. Among their offspring, there are only two phenotypic classes, *A– B–* and *aa bb*, in a 9:7 ratio. How can you account for this?

20. If a person is *Dd E/Y*, what gametes will he produce?

21. Genes *D*, *E*, and *F* are independent. What will be the progeny and their ratios in a mating of the following individuals: *Dd EE Ff × Dd ee FY*?

Answers to Concept Map 3-1 Questions

1. Tree, dogs, humans.
2. 46.
3. 23.
4. 46.
5. 23.
6. Heterozygous, homozygous.
7. Homologous chromosomes.
8. Lung, heart, brain. Eggs and sperm.
9. Each is *Aa*.

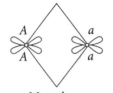
Metaphase

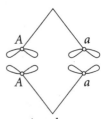
Anaphase

10. *A* and *a*.

Metaphase I

Anaphase I

11. Two are *A* and two are *a*.

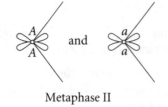
Metaphase II

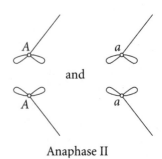
Anaphase II

12. Equal segregation of alleles occurs at the transition from metaphase I to anaphase I.

13. Independent assortment of genes occurs at the transition from metaphase I to anaphase I.

14. The full structure of which the chromosomal fiber is one part is the chromosomal division apparatus, also known as the *mitotic apparatus.*

15. If $2n = 18$, a mature sperm would have 9 chromosomes, each with one chromatid. There would be one DNA molecule per chromosome, or nine DNA molecules.

16. The first polar body in mammals contains a $1n$ complement of chromosomes, each with two chromatids. If $2n = 24$, $1n = 12$. Therefore, there would be 24 chromatids in the polar body.

17. There are two decisions that need to be made in most situations: (1) whether the disorder is dominant or recessive and (2) whether it is X-linked or autosomal. The decisions can be made in either order, but it is often easier to decide between X-linked and autosomal first.

    a. Is the disorder X-linked or autosomal?

        (1) What is the sex ratio among affected individuals? If equal, then may be autosomal. If clearly skewed, X-linked.

        (2) Is there male-to-male transmission? If yes, then autosomal. If no, may be X-linked.

    b. If the disorder is X-linked:

        (1) If females are affected at a greater rate than males, the disorder is dominant.

        (2) If males are affected at a greater rate than females, the disorder is recessive.

    c. If the disorder is autosomal:

        (1) Do affected individuals have affected parents? If yes, possibly dominant. If no, recessive.

        (2) Do normal parents have affected offspring? If yes, recessive. If no, dominant.
             There are three other possibilities: Y linkage, autosomal with expression limited to one sex, and pseudoautosomal inheritance. However, they are so rare that they normally will not be encountered.

18. Females.

19. The two genes somehow affect the expression of each other. *A– B–* is one phenotypic class; all other gene combinations constitute the other phenotypic class.

20. *DE, dE, DY, d*Y.

21.

1/16 *DD Ee FF*	1/8 *Dd Ee FF*	1/16 *dd Ee FF*
1/16 *DD Ee FY*	1/8 *Dd Ee FY*	1/16 *dd Ee FY*
1/16 *DD Ee Ff*	1/8 *Dd Ee Ff*	1/16 *dd Ee Ff*
1/16 *DD Ee fY*	1/8 *Dd Ee fY*	1/16 *dd Ee fY*

▶ CONCEPT MAP 3-2:
  HAPLOID ORGANISMS

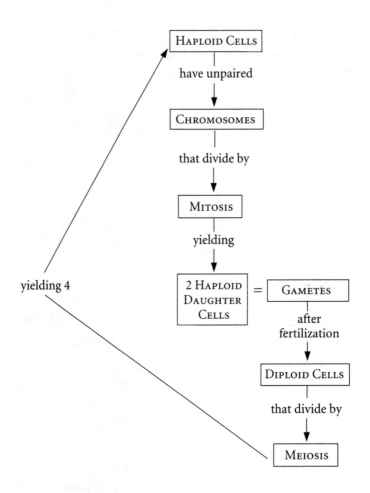

Questions for Concept Map 3-2

1. Name three organisms that spend most of their life cycle in the haploid state.

2. How does haploid mitosis differ from diploid mitosis?

3. Do haploid cells have a mitotic apparatus?

4. What are the components of the mitotic apparatus? What is the function of each?

5. Do cells undergoing meiosis have a mitotic apparatus? If yes, is there some problem with that name for the structure?

6. Assume two alleles, *A* and *a*. Take these alleles through mitosis, fertilization, and meiosis in a haploid organism. Do the same for a diploid organism.

7. In a haploid organism in which $1n = 15$, how many chromosomes would there be in a cell beginning meiosis I?

Answers to Concept Map 3-2 Questions

1. *Neurospora*, bacteria, *Aspergillus*.

2. It is identical even though there are no pairs of chromosomes.

3. Yes.

4. The chromosomal fibers connect the chromosomes by their centromeres to the mitotic apparatus. Under the guidance of the fibers, the chromosomes move toward the poles. The spindle fibers form the center of the mitotic apparatus, passing from pole to pole. They elongate during anaphase and telophase. The poles are the points toward which the chromosomes move. Their role in that process is unknown.

5. Yes. A better name is the division apparatus.

6.

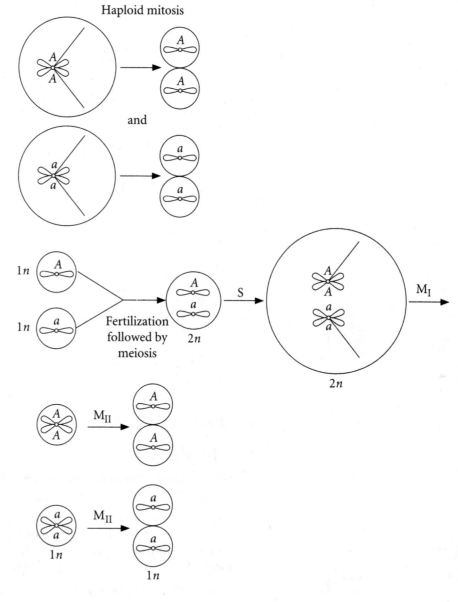

Haploid mitosis

Diploid mitosis

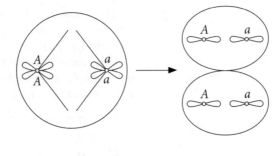

Fertilization

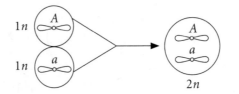

Diploid meiosis

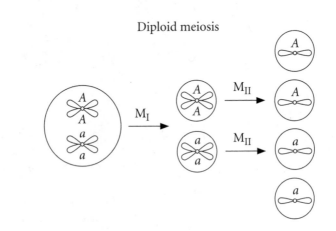

7. 30.

Questions for Concept Map 3-3

1. Define *sex determination.*

2. Why is dosage compensation needed when sex is determined through sex chromosomes?

3. There are consequences for sex determination when mediated by sex chromosomes; what are some possible consequences for sex determination when autosomal genes are the determining factors? When environmental factors are at play? Hormonal factors?

4. What is the potential genetic imbalance associated with chromosomal sex determination?

5. What is being compensated in dosage compensation?

6. If either changing the rate of synthesis of enzymes or changing the rate of function of enzymes is used as a means of dosage compensation, would you expect that very few genes would be involved in the regulation of X chromosome gene expression between the sexes?

► CONCEPT MAP 3-3:
  SEX DETERMINATION

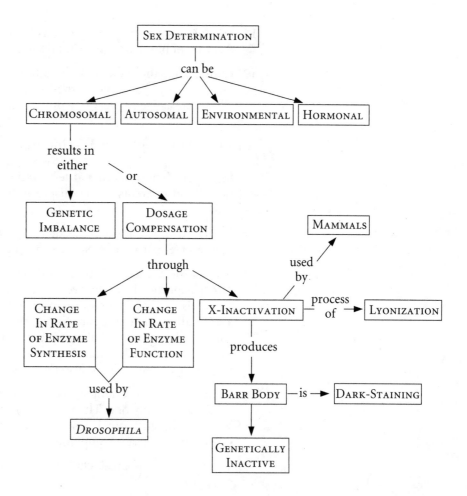

7. What would be the most likely location for the genes that regulate the X chromosome genes?

8. What would be a likely trigger for the regulation of these genes that in turn regulate the X chromosome genes?

9. What are the changes that occur during X-inactivation?

10. What is the source of the term *lyonization*?

11. The Barr body is dark-staining. What does that imply about its physical structure?

12. Is the entire X chromosome inactivated?

13. If the normal female has one active X, why do you think that individuals who are XO, which means that they have only one sex chromosome, die at a very high rate before birth?

14. Could two differently colored eyes, which occurs rarely, be a result of X-inactivation?

15. If the answer is yes to question 14, what would be the sex of that person?

16. Some males are XXY. Would you expect them to have a Barr body?

17. Some females are XXX. How many Barr bodies would you expect to see in their cells?

18. Suppose a woman is heterozygous for hemophilia, a recessive X-linked disorder of blood clotting. Could she exhibit symptoms of hemophilia?

19. Considering the woman in question 18, what would be the consequences if, by chance alone, the X chromosome carrying the hemophilia allele were active in every one of her lung cells? Her bone marrow cells?

Answers to Concept Map 3-3 Questions

1. Sex determination occurs at fertilization through the genetic contribution of both parents to the offspring. It is the setting of the sex genetically.

2. Dosage compensation is needed to restore genetic balance between male and female in those species in which there are sex chromosomes.

3. An autosomal mechanism of sex determination could lead to a low rate of hermaphrodism when two loci are involved. Chromosome breakage could also unlink the two or more autosomal genes involved, thereby also leading to hermaphrodism. Environmental determinism of sex could lead to a very unbalanced sex ratio. Hormonal sex determination, assuming that the hormones come from outside the developing organism (pheromones), would also be prone to an unbalanced sex ratio.

4. In humans, as an example, the female has two X chromosomes while the male has only one. The ratio of gene products from the X chromosome to the autosomal gene products would be different in the two sexes without dosage compensation.

5. The ratio of gene products from the X chromosome to the autosomal gene products.

6. Both mechanisms would require a large percentage of the genome for the complex regulation that would be involved.

7. Autosomal.

8. Sex hormones released or not released, as the case may be.

9. One X chromosome in each somatic female cell becomes heavily methylated and tightly condensed. All genes in this region are turned off.

10. Mary Lyon first recognized that the Barr body represented an inactive X chromosome.

11. The DNA and protein remain constant per unit length, whether or not the region is active. However, the condensation packs this DNA-protein complex more tightly so that the stain appears darker than in the surrounding active regions, even though the complex is, in fact, picking up the same number of molecules of stain per unit length of DNA and protein.

12. No.

13. The normal female has two active X chromosomes prior to day 16 of development; she also has two copies of the small portion that is not inactivated during her entire life span. Females who are XO do not have two active X's at any point in their life. While pseudodominance can account for some of the lethality of the XO genotype, it does not account for all of it. Very likely, there is some process during early development,

perhaps involving the equivalent of the bobbed locus, that requires two copies for normal development.

14. Yes.

15. The person would have to have two X chromosomes. The sex could be either female or male (Klinefelter syndrome).

16. Yes.

17. Two.

18. Yes.

19. If they were active in her lung cells, there would be no consequences, because the lungs do not produce clotting factors. However, her bone marrow cells do produce clotting factor, so she would have symptoms of hemophilia.

▶ SOLUTIONS TO PROBLEMS

Be sure that you have thoroughly read the entire chapter before you attempt any of the problems.

1. a. In mitosis the chromosome number remains unchanged, but in meiosis the chromosome number is halved.

    b. In mitosis sister chromatids separate from each other, but in meiosis both homologous chromosomes and sister chromatids separate from each other.

    c. Mitosis leads to two cells; meiosis leads to four cells.

    d. Homologous pairing occurs only in meiosis.

    e. Recombination is much more frequent in meiosis than in mitosis.

2. Because mitosis does not involve the separation of allelic alternatives, the daughter cells will both be $Aa\ Bb\ Cc$.

3. P $ad^-\ a \times ad^+\ \propto$

    Transient diploid $ad^+/ad^-\ a/\propto$

    F_1 1 $ad^+\ a$, white

         1 $ad^-\ a$, purple

         1 $ad^+\ \propto$, white

         1 $ad^-\ \propto$, purple

4. P $s^+/s^+ \times s/Y$

    F_1 1 s^+/s normal female

         1 s^+/Y normal male

    F_2 1 s^+/s^+ normal female

         1 s^+/s normal female

         1 s^+/Y normal male

         1 s/Y small male

    In the cross $s^+/s \times s/Y$, the progeny will be

$1\ s^+/s$ normal female

$1\ s/s$ small female

$1\ s^+/Y$ normal male

$1\ s/Y$ small male

5.

	MITOSIS	MEIOSIS
fern	sporophyte gametophyte	prothallus
moss	sporophyte gametophyte	archegonia antheridia
flowering plant	sporophyte gametophyte	flowers
pine tree	sporophyte gametophyte	pinecones
mushroom	sporophyte gametophyte	hyphae
frog	somatic cells	gonads
butterfly	somatic cells	gonads
snail	somatic cells	gonads

6. This problem is tricky because the answers depend on how a cell is defined. In general, geneticists consider the transition from one cell to two cells to occur with the onset of anaphase in both mitosis and meiosis even though cytoplasmic division occurs at a later stage.

  a. 46 physically separate chromosomes, each with 2 chromatids = 92 chromatids

  b. 46 physically separate chromosomes, each with 2 chromatids = 92 chromatids

  c. 46 physically separate chromosomes in each of 2 about-to-be-formed cells, each with 1 chromatid = 92 chromatids

  d. 23 physically separate chromosomes in each of 2 about-to-be-formed cells, each with 2 chromatids = 92 chromatids

  e. 23 physically separate chromosomes in each of 2 about-to-be-formed cells, each with 1 chromatid = 46 chromatids

7. e. chromosome pairing

8. His children will have to inherit the satellite 4 (probability = 1/2), the abnormally staining 7 (probability = 1/2), and the Y chromosome (probability = 1/2). To get all three, the probability is (1/2)(1/2)(1/2) = 1/8.

9. The parental set of centromeres can match either parent, which means there are two ways to satisfy the problem. For any one pair, the probability of a centromere from one parent going into a specific gamete is 1/2. For n pairs, the probability of all the centromeres being from one parent is $(1/2)^n$. Therefore, the total probability of having a haploid complement of centromeres from either parent is $2(1/2)^n = (1/2)^{n-1}$.

10. a. Your mother gives you 23 chromosomes, half of all that she has and half of all that you have. Therefore, you have half of all your genes in common with your mother.

   b. For any heterozygous gene (*Aa*) in your mother,

YOU	BROTHER
A	*A*
A	*a*
a	*A*
a	*a*

Thus, there are two of four combinations between you and your brother that will be a match. The same holds for any heterozygous gene in your father. Therefore, you and your brother will have half your genes in common for each gene that is heterozygous in your parents. You and your brother will have identical alleles from your parents for each gene that is homozygous in your parents. The degree of homozygosity in a parent is unknown. For this reason, the final answer can be stated only for heterozygosity in the parents, and homozygosity must be ignored.

11. First assume that the cross is autosomal and that the male is the heterogametic sex:

P   *G*– (graceful female) × *gg* (gruesome male)

F$_1$ 1 *Gg* graceful female

      1 *g*– ? female

      1 *Gg* graceful male

      1 *g*– ? male

This cross does not meet the observation, so it must be wrong. The cross also cannot be *G*– × *g*Y (X-linked), because graceful females will result.

Next assume that the female is the heterogametic sex:

P   *GO* (graceful female) × *gg* (gruesome male)

F$_1$ 1 *gO* gruesome female

      1 *Gg* graceful male

Notice that this outcome matches the observed results. Therefore, in the schmoo the female is the heterogametic sex. The *O* can be a female-determining chromosome or no chromosome.

12. The disease cannot be autosomal because there is a sexual split in the progeny. It cannot be X-linked recessive for two reasons: (1) males get their Y from their fathers and couldn't have the disease, and (2) females get a normal X from their mother, who is unlikely to be a carrier. If the mother *is* a carrier, it is highly unlikely she would pass the gene to all her daughters and none of her sons. The daughters will show any dominant X-linked gene that their father has because he must pass it to them. He cannot pass an X chromosome to his sons. The answer is d, X-linked dominant.

13. Let H = hypophosphatemia and h = normal. The cross is $HY \times hh$, yielding Hh (females) and hY (males). The answer is e, 0.

14. *If* the historical record is accurate, the data suggest Y linkage. Another explanation is an autosomal gene that is dominant in males and recessive in females. This has been observed for other genes in both humans and other species.

15. You should draw pedigrees for this question.

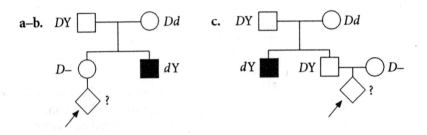

   a. The probability that the woman inherited the d allele from her mother is 1/2. The probability that she passes it to her child is 1/2. The probability that the child is male is 1/2. The total probability of the woman's having an affected child is $(1/2)(1/2)(1/2) = 1/8$.

   b. Your maternal grandmother had to be a carrier, Dd. The probability that your mother received the allele is 1/2. The probability that your mother passed it to you is 1/2. The total probability is $(1/2)(1/2) = 1/4$.

   c. Because your father does not have the disease, you cannot inherit the allele from him. The probability is 0.

16. a. Because neither parent shows it, the disease must be recessive. Because of the sexual split in the progeny, it is most likely X-linked. If it were autosomal, all three parents would have to carry it, which is rather unlikely.

   b. P AY, Aa, AY

   F$_1$ $AY, A-, aY, A-, AY, aY, aY, A-, aY, A-$

17. You should draw the pedigree before beginning.

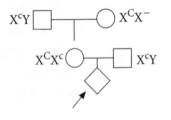

   a. $X^C X^c, X^c X^c$

   b. p(colorblind) p(male) $= (1/2)(1/2) = 1/4$

   c. The girls will be 1 normal ($X^C X^c$):1 colorblind ($X^c X^c$)

   d. The cross is $X^C X^c \times X^c Y$, yielding 1 normal:1 colorblind for both sexes

18. a. This problem involves X-inactivation. Let B = black and b = orange.

FEMALES	MALES
$X^B X^B = BB$ = black	$X^B Y = BY$ = black
$X^b X^b = bb$ = orange	$X^b Y = bY$ = orange
$X^B X^b = Bb$ = calico	

b. P $X^b X^b$ (orange) $\times$ $X^B Y$ (black) or $bb \times BY$

F_1 $X^B X^b$ calico female   or Bb

     $X^b Y$ orange male   or bY

c. P $X^B X^B$ (black) $\times$ $X^b Y$ (orange) or $BB \times bY$

F_1 $X^B X^b$ calico female   or Bb

     $X^B Y$ black male   or BY

d. Because the males are black or orange, the mother had to have been calico. Half the daughters are black, indicating homozygosity, which means that their father was black.

e. Males were orange or black, indicating that the mothers were calico. Orange females, indicating homozygosity, mean that the father was orange.

19. e. 1/4

20. a. X-linked recessive

b. *Generation I:* X^+/Y, X^+/X^x

Generation II: X^+-, X^x/Y, X^+/Y, X^+-, X^+/X^x, X^+/Y

Generation III: X^+-, X^+/Y, X^+/X^x, X^+/X^x, X^+/Y, X^+-, X^x/Y, X^+/Y, X^+-

c. The first couple has no chance of an affected child because the son received his Y chromosome from his father. The second couple has a 50 percent chance of having affected sons and no chance of having affected daughters. The third couple has no chance of having an affected child.

21. a. Autosomal recessive: excluded by unaffected female in third generation.

b. Autosomal dominant: consistent.

c. X-linked recessive: excluded by affected female with unaffected father.

d. X-linked dominant: excluded by unaffected female in third generation.

e. Y-linked: excluded by affected females.

22. a. Cross 6, bent $\times$ bent, leads to some normal progeny, indicating that bent is dominant.

b. The sexual differences in phenotype in cross 6 indicate that it is X-linked.

c. Let B = bent and b = normal.

CROSS	PARENTS		PROGENY	
	FEMALE	MALE	FEMALE	MALE
1	bb	BY	Bb	bY
2	Bb	bY	Bb, bb	BY, bY
3	BB	bY	Bb	BY
4	bb	bY	bb	bY
5	BB	BY	BB	BY
6	Bb	BY	BB, Bb	BY, bY

23. *Unpacking the Problem*

   a. *Normal* is wild type, or red eye color and long wings.

   b. Both "line" and "strain" are used to denote pure-breeding fly stocks, and the words are interchangeable.

   c. Your choice.

   d. Three characters are being followed: eye color, wing length, and sex.

   e. For eye color, there are two phenotypes: red and brown. For wing length there are two phenotypes: long and short. For sex there are two phenotypes: male and female.

   f. The F_1 females designated *normal* have red eyes and long wings.

   g. The F_1 males that are called *short-winged* have red eyes and short wings.

   h. The F_2 ratio is

   3/16 red eyes, long wings, female
   3/16 red eyes, long wings, male

   3/16 red eyes, short wings, female
   3/16 red eyes, short wings, male

   1/16 brown eyes, long wings, female
   1/16 brown eyes, long wings, male

   1/16 brown eyes, short wings, female
   1/16 brown eyes, short wings, male

   i. Ignoring sex, because there is not a 9:3:3:1 ratio, one of the factors that can distort the expected dihybrid ratio is present. Such factors can be sex linkage, autosomal linkage, epistasis, environmental effect, reduced penetrance, or a lack of complete dominance in one or both genes.

   j. With sex linkage, males and females can have different phenotypes, depending on the initial cross. With autosomal inheritance, males and females have the same phenotypes.

   k. The F_2 suggests no sex linkage.

l. The F_1 data do suggest sex linkage.

m. The F_1 suggests that long is dominant to short and red is dominant to brown. The F_2 data simply confirm the F_1 suggestion.

n. If Mendelian notation is used, then red and long need to be designated with uppercase letters, while brown and short need to be designated with lowercase letters. If *Drosophila* notation is used, then the eye-color gene needs to be designated with a lowercase "*b*," with a superscript "+" symbol for *red*, and the wing-length gene needs to be designated with a lower case "*s*," with a superscript "+" symbol for *long*.

o. To deduce the inheritance of these phenotypes means to provide all genotypes for all animals in the three generations discussed and account for the ratios observed.

Solution to the Problem

Start this problem by writing the crosses and results so that all the details are clear.

P   brown, short female × red, long male

F_1 red, long females

　　　 red, short males

These results tell you that long is dominant to short and that the chromosome carrying the gene is X-linked, because males differ from females in their genotype with regard to wing length. The results also tell you that eye color is autosomal, because males do not differ in phenotype from females with regard to eye color, and that red is dominant to brown. Let B = red, b = brown, S = long, and s = short. The cross can be rewritten as follows:

P $b/b\ s/s \times B/B\ S/Y$

F_1 1/2 $B/b\ S/s$ females

　　　 1/2 $B/b\ s/Y$ males

F_1 gametes

　　　 female:　　　　 1/4 *BS*:1/4 *Bs* : 1/4 *bS* :1/4 *bs*

　　　 male:　　　　　 1/4 *Bs*:1/4 *bs* :1/4 *BY*:1/4 *bY*

F_2 females

1/16 $B/B\ S/s$	red, long
1/16 $B/B\ s/s$	red, short
2/16 $B/b\ S/s$	red, long
2/16 $B/b\ s/s$	red, short
1/16 $b/b\ S/s$	brown, long
1/16 $b/b\ s/s$	brown, short

F_2 males

1/16 $B/B\ S/Y$	red, long
1/16 $B/B\ s/Y$	red, short
2/16 $B/b\ S/Y$	red, long

2/16 *B/b s*/Y   red, short

1/16 *b/b S*/Y   brown, long

1/16 *b/b s*/Y   brown, short

The final phenotypic ratio is

3/8 red, long   1/8 brown, short

3/8 red, short   1/8 brown, long

24. Notice that F$_2$ males differ in phenotype from the females in the first cross. The sexual difference in the F$_2$ suggests that the gene is sex-linked. The first cross also indicates that the wild-type large spots are dominant over the lacticolor small spots. Let *A* = wild type and *a* = lacti.

   Cross 1: If the male is assumed to be the hemizygous sex, then it soon becomes clear that the assumption is incorrect because the predictions do not match what was observed:

P   *aa* female × *A*Y male

F$_1$ *Aa* wild-type females

   *a*Y lacti males

Therefore, assume that the female is the hemizygous sex. Let Z stand for the sex-determining chromosome in females. The cross becomes:

P   *a*Z female × *AA* male

F$_1$ *Aa*   wild-type male

   *A*Z   wild-type female

F$_2$ 1/4 *A*Z   wild-type females

   1/2 *A*–   wild-type males

   1/4 *a*Z   lacti females

Cross 2:

P   *A*Z female × *aa* male

F$_1$ *a*Z   lacti females

   Aa   wild-type males

F$_2$ 1/4 *A*Z   wild-type females

   1/4 *Aa*   wild-type males

   1/4 *a*Z   lacti females

   1/4 *aa*   lacti males

25. On the basis of phenotype, the woman appears to have two different cell lines for G6PD activity in her red blood cells. If *G* = normal enzyme activity and *g* = reduced enzyme activity and malaria resistance, then the woman appears to have *G*– and *gg* cells. This can most easily be explained by X-inactivation in the woman's cells. Assume that she is *Gg*. In approximately half her cells the *G* allele will be inactivated, leaving only a functional *g* allele. Those cells will be resistant to the malaria parasite. In the other half of her cells the *g* allele will be inactivated. Those cells will have a functional *G* allele and will be susceptible to the parasite.

26. The suggestion is that the woman was a carrier for testicular feminization. Testicular feminization is an X-linked recessive disorder that renders individuals unresponsive to androgens. Chromosomal males with the disorder exhibit an almost idealized feminine appearance, with large breasts, little to no body hair, extremely smooth skin; they are sterile. Females usually show no effect in the heterozygous state.

The gene for testicular feminization is on the X chromosome, and therefore would be subjected to inactivation an expected 50 percent of the time. If testicular feminization is responsible for the woman's unusual phenotype, then the allele that results in this disorder in males would be functioning in approximately 50 percent of her cells. Those cells would be unresponsive to androgens, which would account for the increased breast size and the lack of pubic hair. That these symptoms are confined to one side of her body suggests that she had an unusual midline split as to which X chromosome was inactivated. Her right side expressed the testicular feminization allele, while her left side expressed the normal allele. The menstrual irregularities are completely compatible with her being a carrier for testicular feminization.

If she were a carrier, then 50 percent of her sons would be expected to suffer from testicular feminization. Her brothers would also have a 50 percent risk of having the disorder. Also, her daughters would be expected to have a 50 percent chance of being carriers, which could lead to this disorder in 50 percent of her grandsons. In other words, the pedigree is completely compatible with the suggestion that she was a carrier for testicular feminization:

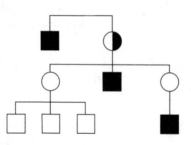

27. a. Note that only males are affected. For rare traits, this is what is expected for an X-linked recessive disorder.

   b. The mothers of all affected sons must be heterozygous for the disorder. In addition, the daughters of all affected men must be heterozygous. Finally, barring mutation, individual II 1 must have been heterozygous.

28. Note that only males are affected but that affected males have affected sons. This suggests that the disorder is caused by an autosomal dominant with expression limited to males.

29. Dear Monk Mendel:

I have recently read your most engrossing manuscript detailing the results of your most wise experiments with garden peas. I salute both your curiosity and your ingenuity in conducting said experiments, thereby opening up for scientific exploration an entire new area of our Maker's universe. Dear Sir, your findings are extraordinary!

While I do not pretend to compare myself to you in any fashion, I beg to bring to your attention certain findings I have made with the aid of that most fascinating and revealing instrument, the microscope. I have been turning my attention to the smallest of worlds with an instrument that I myself have built, and I have noticed some structures that may parallel in behavior the factors which you have postulated in the pea.

I have worked with grasshoppers, however, not your garden peas. Although you are a man of the cloth, you are also a man of science, and I pray that you will not be offended when I state that I have specifically studied the reproductive organs of male grasshoppers. Indeed, I did not limit myself to studying the organs themselves; instead, I also studied the smaller units that make up the male organs and have beheld structures most amazing within them.

These structures are contained within numerous small bags within the male organs. Each bag has a number of these structures, which are long and threadlike at some times and short and compact at other times. They come together in the middle of a bag, and then they appear to divide equally. Shortly thereafter, the bag itself divides, and what looks like half of the threadlike structures goes into each new bag. Could it be, Sir, that these threadlike structures are the very same as your factors? I know, of course, that garden peas do not have male organs in the same way that grasshoppers do, but it seems to me that you found it necessary to emasculate the garden peas in order to do some crosses, so I do not think it too far-fetched to postulate a similarity between grasshoppers and garden peas in this respect.

Pray, Sir, do not laugh at me and dismiss my thoughts on this subject even though I have neither your excellent training nor your astounding wisdom in the Sciences. I remain your humble servant to eternity!

30. First, draw the pedigree.

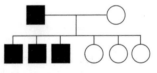

Let the genes be designated by the pigment produced by the normal allele: red pigment, *R*; green pigment, *G*; and blue pigment, *B*.

Recall that the sole X in males comes from the mother, while females obtain an X from each parent. Also recall that a difference in phenotype between sons and daughters is usually due to an X-linked gene. Because all the sons are colorblind and neither the mother nor the daughters are, the mother must carry one colorblind allele on each X chromosome. In other words, she is heterozygous for both X-linked genes, and they are in repulsion: *Rg/rG*. With regard to the autosomal gene, she must be *B–*.

Because all the daughters are normal, the father, who is colorblind, must be able to complement the defects in the mother with regard to his X chromosome. Because he has only one X with which to do so, his genotype must be *RG/Y bb*. Likewise, the mother must be able to complement the father's defect, so she must be *BB*.

The original cross is therefore

P $Rg/rG\ BB \times RG/Y\ bb$

and the F_1 are

F_1	FEMALES	MALES
	$Rg/RG\ Bb$	$Rg/Y\ Bb$
	$rG/RG\ Bb$	$rG/Y\ Bb$

31. (1) Impossible.

(2) Meiosis II.

(3) Meiosis II.

(4) Meiosis II.

(5) Mitosis.

(6) Impossible.

(7) Impossible.

(8) Impossible.

(9) Impossible.

(10) Meiosis I.

(11) Impossible.

(12) Impossible.

32. Because the disorder is X-linked recessive, the affected male had to have received the allele, a, from the female common ancestor in the first generation. The probability that the affected man's wife also carries the a allele is the probability that she also received it from the female common ancestor. That probability is 1/8.

The probability that the couple will have an affected boy

$= p(\text{father donates Y}) \times p(\text{the mother has } a) \times p(\text{mother donates } a)$

$=$ 1/2   1/8   1/2

$= 1/32$

The probability that the couple will have an affected girl is

$= p(\text{father donates } a) \times p(\text{the mother has } a) \times p(\text{mother donates } a)$

$=$ 1/2   1/8   1/2

$= 1/32$

The probability of normal children

$= 1 - p(\text{affected children})$

$= 1 - p(\text{affected male}) - p(\text{affected female})$

$= 1 -$ 1/32 $-$ 1/32

$= 30/32$

Half the normal children will be boys, with a probability of 15/32, and half will be girls, with a probability of 15/32.

▶ TIPS ON PROBLEM
  SOLVING

Problem 11 clearly illustrates the process of successive generation of hypotheses, which are then tested against the data.

X-linked or autosomal: if the male phenotype is different from the female phenotype, X linkage is involved for the allele carried by the female (see Problems 15, 27).

Inheritance patterns: There are only seven possible inheritance patterns for a gene (see Problem 21). Usually only numbers 1–4 will be encountered:

1. Autosomal dominant.

2. Autosomal recessive.

3. X-linked dominant.

4. X-linked recessive.

5. Autosomal with expression limited to one sex.

6. Y-linked.

7. X- and Y-linked (pseudoautosomal).

4

Extensions of
Mendelian Analysis

Incomplete dominance produces in the heterozygote a phenotype intermediate between those of the two homozygotes. A 1:2:1 ratio is observed. Examples are flower color (red and white → pink) and enzyme activity (high and low → medium).

Codominance is revealed when the heterozygote possesses the phenotype of both homozygotes. A 1:2:1 ratio is observed. Examples are hemoglobin variants (Hb^A and Hb^S → Hb^A + Hb^S) and phosphoglucomutase variants (PGM-1 and PGM-2 → PGM-1 + PGM-2), as determined by electrophoresis.

For allelic interactions, what you determine to be the mode of interaction depends quite frequently on the manner in which the trait is observed. Consider the heterozygote $Hb^A Hb^S$. If you look at the red blood cells under normal conditions, you conclude that Hb^A is dominant to Hb^S. Under conditions of low oxygen tension the cells sickle, leading to the opposite conclusion. At the level of electrophoresis of the globin proteins, however, the conclusion is codominance.

Multiple alleles lead to single-gene-pair ratios (1:1, 3:1, 1:2:1).

Multiple genes lead to ratios indicating two genes (9:3:3:1) or three genes (27:9:9:9:3:3:3:1), or some modification of the ratios, in heterozygote × heterozygote crosses.

Recessive lethal genes cause a distortion in the expected ratio. An example is a 2:1 ratio in a heterozygote × heterozygote cross.

Epistasis is the alteration of expression of one gene by the expression of another gene. Distortions of expected ratios are observed (12:4, 9:7, 9:6:1, etc.).

Modifier genes modify the expression of other genes, for example genes for light and dark color. They are detected in heterozygote × heterozygote crosses by a 9:3:3:1 ratio.

Complementary genes work together to produce a phenotype. Heterozygote × heterozygote crosses usually yield a 9:7 ratio.

Suppression occurs when one gene blocks the expression of another. In F_1 crosses, the ratios can be 13:3, 12:4, or some other variant of the expected ratio.

Penetrance is a populational term and is defined as the percentage of individuals of a given genotype that express the phenotype associated with the genotype. A lack of penetrance in an individual is due to epistatic relationships between genes. The environment may also be a factor.

Expressivity is the extent to which a phenotype is expressed in an individual. The range is from minimal to full, depending on the effects of epistatic genes and environmental factors.

Pleiotropy is the phenomenon of multiple effects from one gene. This results in a syndrome (a collection of symptoms associated with a particular disorder).

Dominance occurs within genes, between alleles, whereas epistasis occurs between genes.

▶ CONCEPT MAP 4-1:
GENETIC
INTERACTIONS

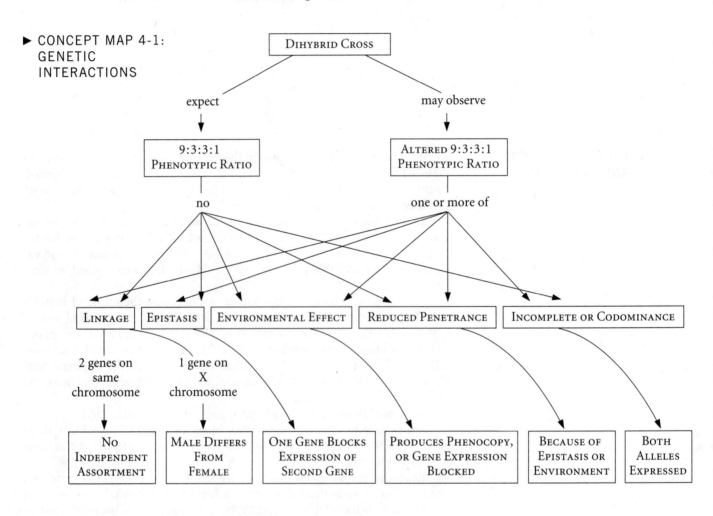

Questions for Concept Map 4-1

1. What is the formal definition of *penetrance*?

2. What is epistasis?

3. How can the environment produce reduced penetrance?

4. How can epistasis result in reduced penetrance?

5. What besides other genes and the environment could result in reduced penetrance?

6. If reduced penetrance occurs, how would that alter the expected 9:3:3:1 ratio in a dihybrid cross? What ratios are possible?

7. If there is an alteration of the expected 9:3:3:1 ratio in a dihybrid cross due to an environmental effect, what ratios could be observed?

8. Is a phenocopy inherited?

9. How can the environment produce a phenocopy?

10. What altered ratio would you observe from the expected 9:3:3:1 ratio in a dihybrid cross if phenocopies are being produced?

11. The expected 9:3:3:1 ratio in a dihybrid cross assumes that the two genes are on separate chromosomes (and therefore assort independently). Would you see the expected 9:3:3:1 ratio if the two genes are on the same chromosome (linked)?

12. What happens to the expected 9:3:3:1 ratio in a dihybrid cross if one gene is located on the X chromosome?

13. The expected 9:3:3:1 ratio in a dihybrid cross assumes independence physically and metabolically. If there is not metabolic independence, what is the term used for the effect of that on the F_2 ratio?

14. What ratio would you observe if there is not metabolic independence?

15. What ratio would you observe if there is not physical independence?

Answers to Concept Map 4-1 Questions

1. *Penetrance* is a populational concept. It is the percentage of people who have the genotype for a specific phenotype and express it.

2. In a very general way, epistasis is the phenomenon of one gene affecting the expression of a second gene.

3. One example would be if the environment provides a nutrient that the organism would otherwise need to produce itself but is unable to do because of a defective allele. Therefore, the associated phenotype of the defective allele is not expressed.

4. One gene could block the expression of a second gene that otherwise produces a variant phenotype.

5. Chance, or developmental noise, can result in reduced penetrance.

6. The altered ratio would be the function of the specific genotype that is not being expressed. The possibilities are set by the many ways the four numbers can be combined. Some examples are 15:1, 12:4, 12:3:1, 9:7.

7. Again, some of the ratios that could be observed are 15:1, 12:4, 12:3:1, 9:7.

8. Phenocopies are phenotypes produced by the environment and are therefore not inherited.

9. The environment could produce a phenocopy by containing some mineral that results in a phenotype indistinguishable from a genetically caused phenotype. One example is silver salts that produce yellow-bodied *Drosophila*. Another example is the drug thalidomide that produces phocomelia, or shortened limbs, that can be caused by a gene abnormality. A third example is environmental radiation that produces cancer of the same type for which there are specific alleles.

10. True phenocopies are not inherited. Therefore, there should be no alteration of the 9:3:3:1 ratio (even though phenotypic ratios might be changed).

11. Two genes that are on the same chromosome will produce a 9:3:3:1 ratio only if those genes are far apart (more than 50 map units).

12. An X-linked gene does not produce a 9:3:3:1 ratio. The specific ratio observed depends on the father's genotype.

13. A lack of metabolic independence produces epistasis.

14. The ratio would depend on the specific form of epistasis resulting from a lack of metabolic independence.

15. A lack of physical independence is linkage. The specific ratio observed depends on the tightness of the linkage.

▶ CONCEPT MAP 4-2:
   GENETIC RATIOS

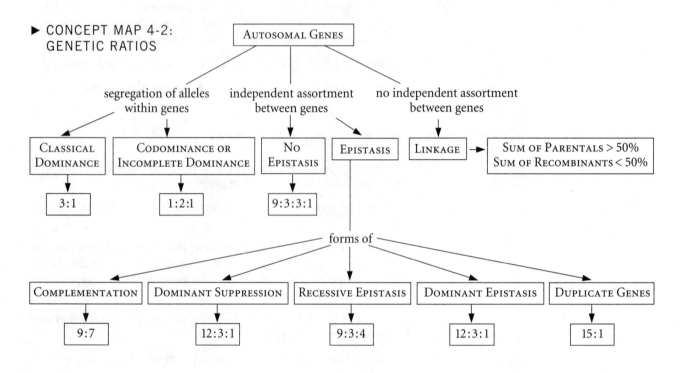

In the concept map, linkage and its expected phenotypic ratios have been included because of a desire to be thorough. Return to this map after reading Chapter 5.

Questions for Concept Map 4-2

1. What is the first law of Mendel?

2. What is the second law of Mendel?

3. What is the basis for classical dominance?

4. What is the basis for codominance and incomplete dominance?

5. Are codominance and incomplete dominance always different phenomena or are they a product of how the organism is studied?

6. Can a specific gene be classified as manifesting classical dominance, codominance, *and* incomplete dominance? If so, what is changing?

7. What determines the presence or absence of epistasis?

8. Are the ratios given for various forms of epistasis the only ones that could be observed?

9. How can duplicate genes arise?

10. Can a specific ratio be used as an indicator of a specific type of genetic interaction?

Answers to Concept Map 4-2 Questions

1. The first law of Mendel refers to the segregation of alleles of a gene.

2. The second law of Mendel refers to the independent assortment of two genes that have both physical and biochemical independence.

3. Classical dominance is usually seen when one allele produces the wild-type product and the other allele produces either nothing or a defective product.

4. In both codominance and incomplete dominance both alleles produce a product.

5. Both codominance and incomplete dominance can frequently be named as the mechanism of allele interaction, depending on how the organism is studied.

6. Yes, a specific gene can be classified as showing classical dominance, codominance, and incomplete dominance. This occurs as the method of assessment of phenotype varies.

7. Epistasis is seen when there is a lack of metabolic independence.

8. No, other ratios could be detected.

9. Duplicate genes can arise through faulty replication of DNA.

10. In general a specific ratio can be used as an indicator of a specific type of genetic interaction; however, one should not rely solely on ratios but should also examine the specific situation.

In Concept Map 4-3, the gene is designated by *A*. An arrow has been used to shorten the sequence of gene → enzyme → metabolic reaction. An arrow with an X in front of it signifies a gene that produces either no enzyme or a non-functional enzyme. Metabolic substrates and products are designated by D–F.

Questions for Concept Map 4-3

1. What is one type of classical dominance at the metabolic level? What other type could exist?

2. The metabolic sequences for codominance and incomplete dominance appear to be identical. What distinguishes the two?

▶ CONCEPT MAP 4-3:
MONOHYBRID
METABOLIC
SEQUENCES THAT
PRODUCE F_2 RATIOS

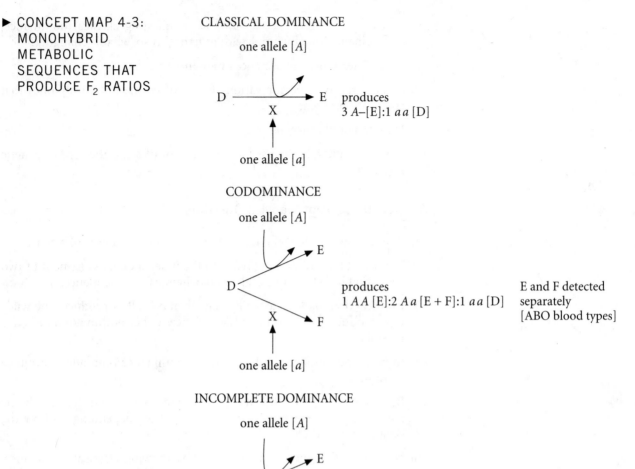

CLASSICAL DOMINANCE

one allele [A]

D ———————▶ E   produces
         X   3 A–[E]:1 aa [D]

one allele [a]

CODOMINANCE

one allele [A]

D ——————————▶ E
                X   ▶ F

produces   E and F detected
1 AA [E]:2 Aa [E + F]:1 aa [D]   separately
                                  [ABO blood types]

one allele [a]

INCOMPLETE DOMINANCE

one allele [A]

D ——————————▶ E
                X   ▶ F

produces   E and F detected as
1 AA [E]:2 Aa [EF]:1 aa [F]   a mixture
                                  [pink flowers]

one allele [a]

3. If blood types in the ABO system are codominant, how are they detected?

4. What, if any, would be the experimental design that would result in the ABO system's being classified as incomplete dominance?

5. Could the color of a specific flower that is detected as pink and classified as incomplete dominance be detected as red and white and classified as codominance under different experimental conditions? If so, what are those conditions?

Answers to Concept Map 4-3 Questions

1. One type of classical dominance at the metabolic level is that one allele makes no *functional* product. The other type is that no product is made, functional or otherwise.

2. Codominance and incomplete dominance are frequently distinguished by the method of study or assessment.

3. Blood types in the ABO system are detected through an agglutination test in which samples of the blood are exposed to different antibodies.

4. If the ABO system being classified were rated as incomplete dominance rather than codominance, the blood being tested would be exposed to antibodies against A and B simultaneously.

5. A pink flower could be detected as red and white and classified as codominance if the individual pigments were isolated.

▶ CONCEPT MAP 4-4: DIHYBRID METABOLIC SEQUENCES THAT PRODUCE F$_2$ RATIOS

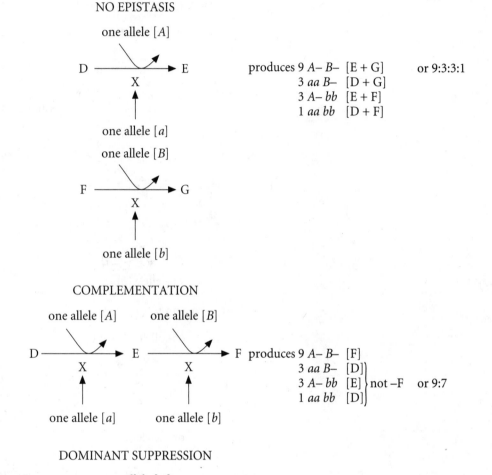

NO EPISTASIS

one allele [A]

D ————————→ E   produces 9 *A– B–* [E + G]   or 9:3:3:1
        X   3 *aa B–* [D + G]
        ↑   3 *A– bb* [E + F]
   one allele [a]   1 *aa bb* [D + F]

one allele [B]

F ————————→ G
        X
        ↑
   one allele [b]

COMPLEMENTATION

one allele [A]   one allele [B]

D ——————→ E ——————→ F produces 9 *A– B–* [F]
     X   X   3 *aa B–* [D]
     ↑   ↑   3 *A– bb* [E] } not –F   or 9:7
one allele [a] one allele [b]   1 *aa bb* [D]

DOMINANT SUPPRESSION

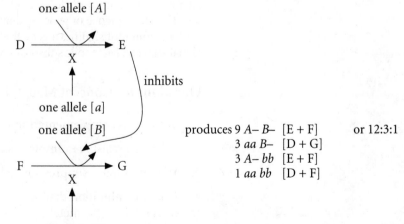

one allele [A]

D ————————→ E
        X   inhibits
        ↑
   one allele [a]

one allele [B]   produces 9 *A– B–* [E + F]   or 12:3:1
                                          3 *aa B–* [D + G]
F ————————→ G   3 *A– bb* [E + F]
        X   1 *aa bb* [D + F]
        ↑
   one allele [b]

► CONCEPT MAP 4-4:
DIHYBRID METABOLIC
SEQUENCES THAT
PRODUCE F$_2$ RATIOS
CONTINUED

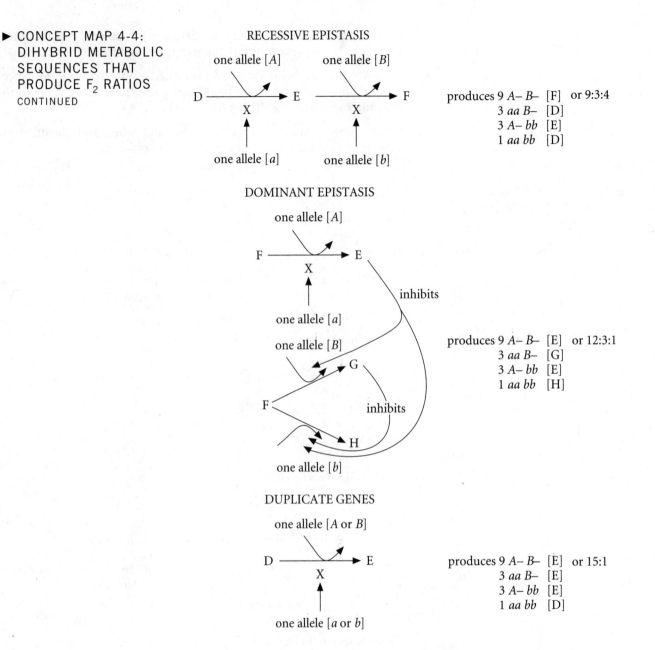

RECESSIVE EPISTASIS

produces 9 *A– B–* [F]   or 9:3:4
         3 *aa B–* [D]
         3 *A– bb* [E]
         1 *aa bb* [D]

DOMINANT EPISTASIS

produces 9 *A– B–* [E]   or 12:3:1
         3 *aa B–* [G]
         3 *A– bb* [E]
         1 *aa bb* [H]

DUPLICATE GENES

produces 9 *A– B–* [E]   or 15:1
         3 *aa B–* [E]
         3 *A– bb* [E]
         1 *aa bb* [D]

In this map, genes are designated by *A* and *B*. An arrow has been used to shorten the sequence of gene → enzyme → metabolic reaction. An arrow with an X in front of it signifies a gene that produces either no enzyme or a nonfunctional enzyme. Metabolic substrates and products are designated by D–H.

Questions for Concept Map 4-4

1. What does complementation mean?

2. What is a real-life example of complementation?

3. With dominant suppression, what specifically inhibits what?

4. Does the inhibition that occurs with dominant suppression happen more than rarely in the metabolic changes that occur within an organism?

5. What type of inhibition occurs with dominant suppression? NOTE: to answer this you must recall some biochemistry.

6. Would you expect that the type of inhibition which occurs in dominant suppression involves a gene active during development, with a time-limited expression, or that it involves a gene which is part of the metabolism sequence of an organism? Why?

7. In the diagram for recessive epistasis, which gene, and which allele, affects the expression of a second gene?

8. Does your answer change for Question 7 if the organisms are classified as F and not-F? E and not-E? D and not-D?

9. Two inhibitions are noted for the diagram for dominant epistasis: (1) E inhibiting the functioning of the enzymes produced by *B* and *b* and (2) G inhibiting the functioning of the enzyme produced by *b*. Which inhibition, if any, would you expect to be most likely to occur during development? Which inhibition is most likely a function of enzyme kinetics? Is there a biochemical term used to describe the second type of inhibition? The first type of inhibition?

10. Would a differing classification scheme for the progeny of a dominant epistatic relationship lead to a differing conclusion as to which gene is epistatic to which?

11. What distinguishes genes *A* and *B* other than physical location on a chromosome?

Answers to Concept Map 4-4 Questions

1. Complementation refers to two (or more) genes, each with a variant phenotype, together producing the wild type.

2. One example of complementation is a hearing child from the mating of two people with deaf-mutism.

3. With dominant suppression, the dominant allele of one gene blocks the expression of the dominant allele of a second gene.

4. Dominant suppression can be detected quite easily by studying the metabolic table. From that table, it appears to happen quite often.

5. One type of dominant suppression is feedback inhibition.

6. The type of inhibition that occurs in dominant suppression could involve a gene active during development, with a time-limited expression, and it could involve a gene that is part of the metabolic sequence of an organism.

7. In the diagram for recessive epistasis, each recessive allele of both genes can affect the expression of a second gene.

8. No, because the answer for Question 7 considered those possibilities.

9. The two inhibitions that are noted for the diagram for dominant epistasis [(1) E inhibiting the functioning of the enzymes produced by *B* and *b* and (2) G inhibiting the functioning of the enzyme produced by *b*] could both occur during development, but the second would be more likely to occur. The first is more likely a function of enzyme kinetics, and it is known as *feedback inhibition*. The second type of inhibition is known as *classical dominance*.

10. As a genetic interaction is assessed, the method of assessment determines the observed functioning.

11. Genes *A* and *B* can also be distinguished by the function that each controls.

▶ CONCEPT MAP 4-5: OTHER METABOLIC SEQUENCES PRODUCING F₂ RATIOS

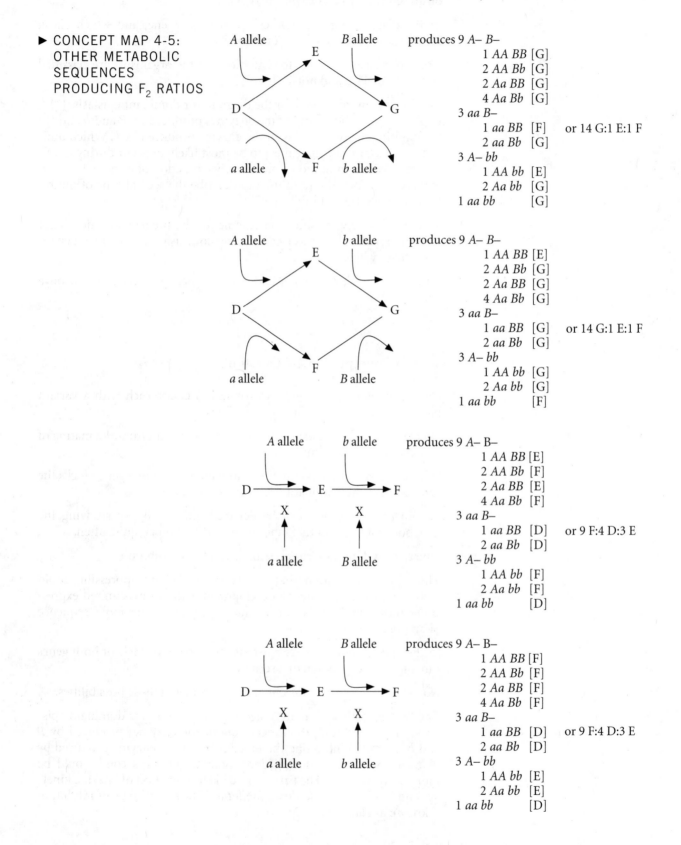

produces 9 *A– B–*
  1 *AA BB* [G]
  2 *AA Bb* [G]
  2 *Aa BB* [G]
  4 *Aa Bb* [G]
3 *aa B–*
  1 *aa BB* [F]   or 14 G:1 E:1 F
  2 *aa Bb* [G]
3 *A– bb*
  1 *AA bb* [E]
  2 *Aa bb* [G]
1 *aa bb*   [G]

produces 9 *A– B–*
  1 *AA BB* [E]
  2 *AA Bb* [G]
  2 *Aa BB* [G]
  4 *Aa Bb* [G]
3 *aa B–*
  1 *aa BB* [G]   or 14 G:1 E:1 F
  2 *aa Bb* [G]
3 *A– bb*
  1 *AA bb* [G]
  2 *Aa bb* [G]
1 *aa bb*   [F]

produces 9 *A– B–*
  1 *AA BB* [E]
  2 *AA Bb* [F]
  2 *Aa BB* [E]
  4 *Aa Bb* [F]
3 *aa B–*
  1 *aa BB* [D]   or 9 F:4 D:3 E
  2 *aa Bb* [D]
3 *A– bb*
  1 *AA bb* [F]
  2 *Aa bb* [F]
1 *aa bb*   [D]

produces 9 *A– B–*
  1 *AA BB* [F]
  2 *AA Bb* [F]
  2 *Aa BB* [F]
  4 *Aa Bb* [F]
3 *aa B–*
  1 *aa BB* [D]   or 9 F:4 D:3 E
  2 *aa Bb* [D]
3 *A– bb*
  1 *AA bb* [E]
  2 *Aa bb* [E]
1 *aa bb*   [D]

In this map, genes are designated by *A* and *B*. An arrow has been used to shorten the sequence of gene → enzyme → metabolic reaction. An arrow with an X in front of it signifies a gene that produces either no enzyme or a nonfunctional enzyme. Metabolic substrates and products are designated by D–G.

Questions for Concept Map 4-5

1. Do metabolic sequences exist that are like example 1?

2. What differs between examples 1 and 2?

3. Could examples 1 and 2 also be classified as producing a 14:2 ratio?

4. Do metabolic sequences exist that are like example 3?

5. What differs between examples 3 and 4?

6. Could examples 3 and 4 also be classified as producing a 14:2 ratio? A 12:4 ratio? A 13:3 ratio?

Answers to Concept Map 4-5 Questions

1. Yes.

2. Examples 1 and 2 differ by the specific enzymatic function of the alleles for gene *B*.

3. One could classify examples 1 and 2 as "G" and "not-G."

4. Yes.

5. Examples 3 and 4 differ as to the specific functioning of the *B* and *b* alleles.

6. Examples 3 and 4 could be classified as producing a 12:4 ratio or a 13:3 ratio.

Questions for Concept Map 4-6

1. What is pleiotropy?

2. How are pleiotropy and syndrome related?

3. If a gene has invariant expression, is it penetrant?

4. If a gene is nonpenetrant, can it be detected in an individual?

5. If other genes result in a gene's being nonpenetrant, what is another term for the phenomenon?

6. If a gene has variable expressivity, what is the term for the role of other genes in producing that variable expressivity?

7. What, besides other genes and the environment, can affect the penetrance and expressivity of a gene?

Answers Concept Map 4-6 Questions

1. Pleiotropy is the production of more than one trait by a specific allele.

► CONCEPT MAP 4-6:
  PRODUCING THE
  PHENOTYPE

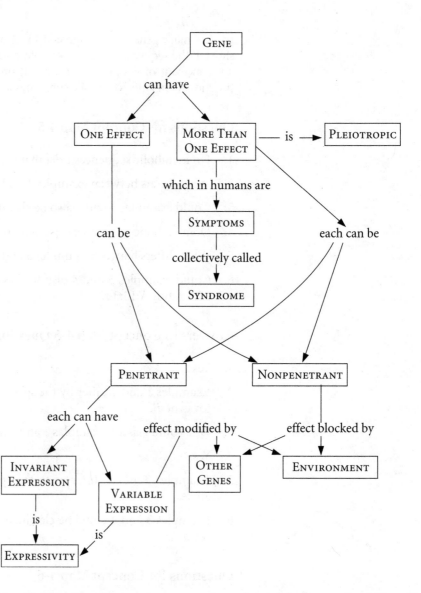

2. Pleiotropic genes produce a number of traits that are collectively called a *syndrome*.

3. Yes. A gene must be penetrant before it can have any expression, including invariant expression.

4. If a gene is nonpenetrant, it can frequently be detected in an individual. Whether the answer is yes or no depends on the specific gene.

5. Epistasis means that other genes result in a gene's being nonpenetrant.

6. Epistasis can produce variable expressivity.

7. Chance, or developmental noise, can affect the penetrance and expressivity of a gene.

► SOLUTIONS TO
  PROBLEMS

Be sure that you have thoroughly read the entire chapter before you attempt any of the problems.

1. The woman must be *AO*, so the mating is *AO* × *AB*. Their children will be

GENOTYPE	PHENOTYPE
1 *AA*	A
1 *AB*	AB
1 *AO*	A
1 *BO*	B

2. You are told that the cross of two erminette fowls results in 22 erminette, 14 black, and 12 pure white. Two facts are important: (1) the parents consist of only one phenotype, yet the offspring have three phenotypes, and (2) the progeny appear in an approximate ratio of 1:2:1. These facts should tell you immediately that you are dealing with a heterozygous × heterozygous cross involving one gene and that the erminette phenotype must be the heterozygous phenotype.

   When the heterozygote shows a different phenotype from either of the two homozygotes, the heterozygous phenotype results from incomplete dominance or codominance. Because two of the three phenotypes contain black, either fully or in an occasional feather, you might classify the erminette as an instance of incomplete dominance because it is intermediate between fully black and fully white. Alternatively, because the erminette has both black and white feathers, you might classify the phenotype as codominant. Your decision will rest on whether you look at the whole animal (incomplete dominance) or at individual feathers (codominance). This is yet another instance where what you conclude is determined by how you observe.

   To test the hypothesis that the erminette phenotype is a heterozygous phenotype, you could cross an erminette with either, or both, of the homozygotes. You should observe a 1:1 ratio in the progeny of both crosses.

3. **a.** The original cross and results were

   P   long, white × round, red

   F_1 oval, purple

   F_2

9 long, red	19 oval, red	8 round, white
15 long, purple	32 oval, purple	16 round, purple
8 long, white	16 oval, white	9 round, red
32 long	67 oval	33 round

   The data show that, when the results are rearranged by shape, a 1:2:1 ratio is observed for color within each shape category. Likewise, when the data are rearranged by color, a 1:2:1 ratio is observed for shape within each color category:

9 long, red	15 long, purple	8 round, white
19 oval, red	32 oval, purple	16 oval, white
9 round, red	16 round, purple	8 long, white
37 red	63 purple	32 white

   A 1:2:1 ratio is observed when there is a heterozygous × heterozygous cross. Therefore, the original cross was a dihybrid cross. Both oval and purple must represent an incomplete dominant phenotype.

Let L = long, L' = round, R = red and R' = white. The cross becomes

P $LL\ R'R' \times L'L'\ RR$

F_1 $LL'\ RR' \times LL'\ RR'$

F_2 1/4 LL
— 1/4 RR = 1/16 long, red
— 1/2 RR' = 1/8 long, purple
— 1/4 $R'R'$ = 1/16 long, white

1/2 LL'
— 1/4 RR = 1/8 oval, red
— 1/2 RR' = 1/4 oval, purple
— 1/4 $R'R'$ = 1/8 oval, white

1/4 $L'L'$
— 1/4 RR = 1/16 round, red
— 1/4 RR' = 1/8 round, purple
— 1/4 $R'R'$ = 1/16 round, white

b. A long, purple × oval, purple cross is as follows:

P $LL\ RR' \times LL'\ RR'$

F_1 1/2 LL
— 1/4 RR = 1/8 long, red
— 1/2 RR' = 1/4 long, purple
— 1/4 $R'R'$ = 1/8 long, white

1/2 LL'
— 1/4 RR = 1/8 oval, red
— 1/2 RR' = 1/4 oval, purple
— 1/4 $R'R'$ = 1/8 oval, white

4. The easy way to solve this problem is to note that the Himalayan phenotype is observed only in homozygotes. Because one parent does not have the Himalayan allele, it will be impossible to obtain progeny that are homozygous for it. The answer is e, 0 percent.

The more scholarly approach to this problem is to look at the progeny and their ratio before concluding that Himalayan will not be observed. The progeny are

1 C^+C^{ch} full color
1 C^+C^h full color
1 $C^{ch}C^{ch}$ chinchilla
1 $C^{ch}C^h$ chinchilla

5. **a.** You could begin this problem using one of several assumptions. Only one will be right, but there is no way to know in advance which that will be. The correct assumption is that there is one gene, with multiple alleles.

CROSS	PARENTS	PROGENY	CONCLUSION
1	$ba \times ba \longrightarrow$ 3 $b{-}$:1 aa		black is dominant to albino
2	$bs \times aa \longrightarrow$ 1 ba:1 sa		black is dominant to sepia

CROSS	PARENTS	PROGENY	CONCLUSION
3	$ca \times ca \longrightarrow$	$3\ c\text{--}{:}1\ aa$	cream is dominant to albino
4	$sa \times ca \longrightarrow$	$1\ ca{:}2\ s\text{--}{:}1\ aa$	sepia is dominant to albino
5	$bc \times aa \longrightarrow$	$1\ ba{:}1\ ca$	black is dominant to cream
6	$bs \times c\text{--} \longrightarrow$	$1\ b\text{--}{:}1\ s\text{--}$	black is dominant to sepia
7	$bs \times s\text{--} \longrightarrow$	$1\ b\text{--}{:}1\ s\text{--}$	black is dominant to sepia
8	$bc \times sc \longrightarrow$	$2\ b\text{--}{:}1\ sc{:}1\ cc$	sepia is dominant to cream
9	$sc \times sc \longrightarrow$	$3\ s\text{--}{:}1cc$	sepia is dominant to cream
10	$ca \times aa \longrightarrow$	$1\ ca : 1\ aa$	cream is dominant to albino

The order of dominance is $b > s > c > a$.

b. The cross between parents is $bs \times bc$. The progeny are

1 *bb* black   1 *bc* black

1 *bs* black   1 *sc* sepia

6. Both codominance (=) and classical dominance (>) are present in the multiple allelic series for blood type: $A = B$, $A > O$, $B > O$.

PARENTS' PHENOTYPE	PARENTS' POSSIBLE GENOTYPES	PARENTS' POSSIBLE CHILDREN
a. AB × O	$AB \times OO$	AO, BO
b. A × O	AA or $AO \times OO$	AO, OO
c. A × AB	AA or $AO \times AB$	AA, AB, AO, BO
d. O × O	$OO \times OO$	OO

The children are

PHENOTYPE	POSSIBLE GENOTYPES
1. O	OO
2. A	AA, AO
3. B	BB, BO
4. AB	AB

Using the assumption that each set of parents had to have had one child, the following combinations are the only ones that will work as a solution.

PARENTS	CHILD
a. AB × O	**3.** B
b. A × O	**2.** A
c. A × AB	**4.** AB
d. O × O	**1.** O

7. *M* and *N* are codominant alleles. The rhesus group is determined by classically dominant alleles. The *ABO* alleles are mixed codominance and classical dominance (see Problem 6).

PERSON	BLOOD GROUP			OBLIGATE	PATERNAL	DONATION
Husband	O	M	Rh$^+$	O	M	R or r
Wife's lover	AB	MN	Rh$^-$	A or B	M or N	r
Wife	A	N	Rh$^+$	–	–	–
Child 1	O	MN	Rh$^+$	O	M	R or r
Child 2	A	N	Rh$^+$	A or O	N	R or r
Child 3	A	MN	Rh$^-$	A or O	M	r

The wife is *AO NN Rr*. Only the husband could donate *O* to child 1. Only the lover could donate *A* and *N* to child 2. Both the husband and the lover could have donated the necessary alleles to child 3.

8. The key to solving this problem is in the statement that breeders cannot develop a pure-breeding stock and that a cross of two platinum foxes results in some normal progeny. Platinum must be dominant to normal color and heterozygous (*Aa*). An 82:38 ratio is very close to 2:1. Because a 1:2:1 ratio is expected in a heterozygous cross, one genotype is nonviable. It must be the *AA*, homozygous platinum, genotype that is nonviable, because the homozygous recessive genotype is normal color (*aa*). Therefore, the platinum allele is a pleiotropic allele that governs coat color in the heterozygous state and is lethal in the homozygous recessive state.

9. a. Because Pelger crossed with normal results in two phenotypes in a 1:1 ratio, either Pelger or normal is heterozygous (*Aa*) and the other is homozygous (*aa*) recessive. The problem states that normal is true-breeding, or *aa*. Pelger must be *Aa*.

   b. The cross of two Pelger rabbits results in three phenotypes. This means that the Pelger anomaly is dominant to normal. This cross is *Aa* × *Aa*, with an expected ratio of 1:2:1. Because the normal must be *aa*, the extremely abnormal progeny must be *AA*. There were only 39 extremely abnormal progeny because the others died before birth.

   c. The Pelger allele is pleiotropic. In the heterozygous state it is dominant for nuclear segmentation of white blood cells. In the homozygous state it is a recessive lethal.
      You could look for the nonsurviving fetuses in utero. Because the hypothesis of embryonic death of the homozygous dominant predicts a one-fourth reduction in litter size, you could also do an extensive statistical analysis of litter size, comparing normal × normal with Pelger × Pelger.

   d. By analogy with rabbits, the absence of a homozygous Pelger anomaly in humans can be explained as recessive lethality. Alternatively, because 1 in 1000 births results in a Pelger anomaly, a heterozygous × heterozygous mating would be expected in only 1 of 1 million

($1/1000 \times 1/1000$) matings, and only 1 in 4 of the progeny would be expected to be homozygous. Thus, the homozygous Pelger anomaly is expected in only 1 of 4 million births. This is extremely rare and might not be recognized.

e. By analogy with rabbits, among the children of a man and a woman with the Pelger anomaly two-thirds of the surviving progeny would be expected to show the Pelger anomaly and one-third would be expected to be normal. The child who is homozygous for the Pelger allele would be expected to have severe skeletal defects if he survived until birth.

10. a. The sex ratio is expected to be 1:1, but this was not observed. Approximately half the males are missing.

b. If the normal-looking female were heterozygous for an X-linked recessive allele that was lethal in either the homozygous or the hemizygous state, then all the female progeny and half the male progeny would survive.

c. In order to test this explanation, assume that the original cross was $Aa \times AY$. Half the females from this cross should be heterozygous and half should be homozygous. These F_1 females could be crossed individually with normal males and the sex ratio of the progeny could be determined for each cross.

11. Note that a cross of the short-bristled female with a normal male results in two phenotypes with regard to bristles and an abnormal sex ratio of 2 females:1 male. Furthermore, all the males are normal, while the females are normal and short in equal numbers. Whenever the sexes differ with respect to phenotype among the progeny, an X-linked gene is involved. Because only the normal phenotype is observed in males, the short-bristled phenotype must be heterozygous, and the allele must be a recessive lethal. Thus the first cross was $Aa \times aY$.

Long-bristled females (aa) were crossed with long-bristled males (aY). All their progeny would be expected to be long-bristled (aa or aY).

Short-bristled females (Aa) were crossed with long-bristled males (aY). The progeny expected are

1 Aa short-bristled females   1 aY long-bristled males

1 aa long-bristled females   1 AY nonviable

12. In order to do this problem, you need first to restate the information provided. The following two genes are independently assorting:

hh = hairy ss = no effect

Hh = hairless Ss suppresses Hh, giving hairy

HH = lethal SS = lethal

a. The cross is $Hh\ Ss \times Hh\ Ss$. Because this is a typical dihybrid cross, the expected ratio is 9:3:3:1. However, the problem cannot be worked in this simple fashion because of the epistatic relationship of these two genes. Therefore, the following approach should be used.

For the H gene, you expect 1/4 HH:1/2 Hh:1/4 hh. For the S gene, you expect 1/4 SS:1/2 Ss:1/4 ss. To get the final ratios, multiply the frequency of the first genotype by the frequency of the second genotype.

1/4 *HH*————————— all progeny die regardless of the *S* gene

2/4 *Hh* ⟨ 1/4 *SS* = 2/16 *Hh Ss* die
          1/2 *Ss* = 4/16 *Hh Ss* hairy
          1/4 *ss* = 2/16 *hh ss* hairless

1/4 *hh* ⟨ 1/4 *SS* = 1/16 *hh SS* die
          1/2 *Ss* = 2/16 *hh Ss* hairy
          1/4 *ss* = 1/16 *hh ss* hairy

Of the 9 living progeny, the ratio of hairy to hairless is 7:2.

b. This cross is *Hh ss* × *Hh Ss*. A 1:2:1 ratio is expected for the *H* gene and a 1:1 ratio is expected for the *S* gene.

1/4 *HH*————————— all progeny die regardless of the *S* gene

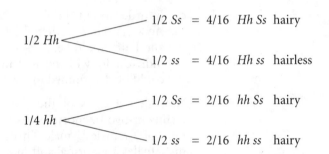

1/2 *Hh* ⟨ 1/2 *Ss* = 4/16 *Hh Ss* hairy
          1/2 *ss* = 4/16 *Hh ss* hairless

1/4 *hh* ⟨ 1/2 *Ss* = 2/16 *hh Ss* hairy
          1/2 *ss* = 2/16 *hh ss* hairy

Of the 12 living progeny, the ratio of hairy to hairless is 2:1.

13. Note that the F_2 are in a 9:6:1 ratio. This indicates a dihybrid cross in which *A– bb* has the same appearance as *aa B–*. Let the disc phenotype be the result of *A– B–* and the long phenotype be the result of *aa bb*. The crosses are

P   *AA BB* (disc) × *aa bb* (long)

F_1 *Aa Bb* (disc)

F_2 9 *A– B–*   disc

    3 *aa B–*   sphere

    3 *A– bb*   sphere

    1 *aa bb*   long

14. The suggestion from the data is that the two albino lines had defects in two different genes. When the extracts from the two lines were placed in the same test tube, they were capable of producing color because the gene product of one line was capable of compensating for the absence of a gene product from the second line.

a. The most obvious control is to cross the two pure-breeding lines. The cross would be *AA bb* × *aa BB*. The progeny will be *Aa Bb*, and all should be reddish purple. Alternatively, grind up leaves from each line separately and see if they become red.

b. The most likely explanation is that the red pigment is produced by the action of at least two different gene products, for example:

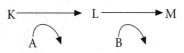

c. The genotypes of the two lines should be *AA bb* and *aa BB*.

d. The F_1 would all be *Aa Bb*, colored. The F_2 would be

9 *A– B–* colored

3 *A– bb* albino

3 *aa B–* albino

1 *aa bb* albino

15. a. This is yet another example where one phenotype in the parents gives rise to three phenotypes in the offspring. The frizzle fowl must be heterozygous with incomplete dominance. The cross is

P   *Aa × Aa*

F_1 1 *AA* : 2 *Aa* : 1 *aa*

The frizzle is *Aa*, with normal and woolly being homozygotes.

b. In order to produce only frizzle fowl, all crosses should involve a normal × woolly.

16. a. The second generation indicates that Marfan's syndrome is caused by a dominant autosomal allele.

b. The pedigree exhibits pleiotropy and variable expressivity. Reduced penetrance is not evident.

c. The pleiotropy indicates that a gene product that exists in a number of different organs is defective. The variable expressivity is due to modification by one or more other genes or to an environmental effect.

17. *Unpacking the Problem*

a. The character being studied is petal color.

b. The wild-type phenotype is blue.

c. A variant is a phenotypic difference from wild type that is observed.

d. There are two variants: pink and white.

e. "In nature" means that the variants did not appear in laboratory stock and, instead, were found growing wild.

f. Possibly the variants appeared as a small patch within a larger patch of wild type.

g. Seeds would be grown to check the outcome from each cross.

h. Given that no X linkage appears to exist (sex is not specified in parents or offspring), "blue × white" means the same as "white × blue." Similar results would be expected because the trait being studied appears to be autosomal.

i. The first two crosses show a 3:1 ratio in the F_2, suggesting the segregation of one gene. The third cross has a 9:4:3 ratio for the F_2, suggesting that two genes are segregating.

j. Blue is dominant to both white and pink.

k. Complementation is the production of wild type by two recessive variants.

l. The product of one gene interacts with the product of a second gene to produce wild type, or blue.

m. Blueness from a pink × white cross arises through complementation of two genes.

n. The following ratios are observed: 3:1, 9:4:3.

o. There are monohybrid ratios observed in the first two crosses.

p. There is a modified 9:3:3:1 ratio in the third cross.

q. A monohybrid ratio indicates that one gene is segregating, while a dihybrid ratio indicates that two genes are segregating.

r. 13:3, 12:3:1, 9:4:3, 9:7

s. There is a modified dihybrid ratio in the third cross.

t. A modified dihybrid ratio most frequently indicates the interaction of two or more genes.

u. Recessive epistasis is indicated by the modified dihybrid ratio.

v.

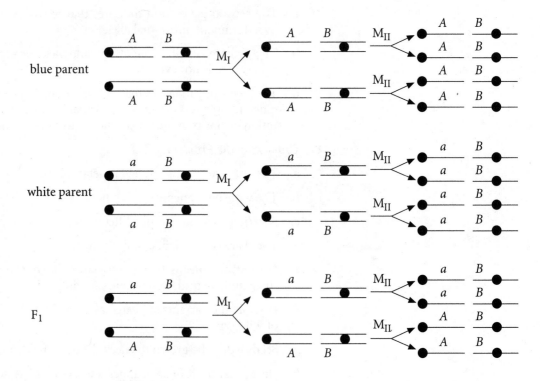

w.

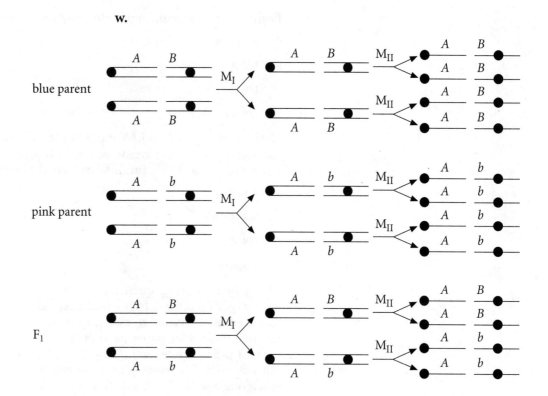

Solution to the Problem

a. Let: A = wild type, a = white, B = wild type, and b = pink.

Cross 1:	P	blue × white	$AA\ BB \times aa\ BB$
	F_1	all blue	all $Aa\ BB$
	F_2	3 blue:1 white	$3\ A\!-\ BB : 1\ aa\ BB$
Cross 2:	P	blue × pink	$AA\ BB \times AA\ bb$
	F_1	all blue	all $AA\ Bb$
	F_2	3 blue : 1 pink	$3\ AA\ B\!-: 1\ AA\ bb$
Cross 3:	P	pink × white	$AA\ bb \times aa\ BB$
	F_1	all blue	all $Aa\ Bb$
	F_2	9 blue	$9\ A\!-\ B\!-\ $ blue
		4 white	$3\ aa\ B\!-\ $ white
		3 pink	$3\ A\!-\ bb\ $ pink
			$1\ aa\ bb\ $ white

The allele a blocks the expression of alleles B and b, while the A allele allows the expression of both B and b.

b. The cross is

F_2	blue × white
F_3	3/8 blue
	1/8 pink
	4/8 white

Begin by writing as much of each genotype as can be assumed:

F$_2$ $A\!- B\!- \times aa\,-\,-$

F$_3$ 3/8 $A\!- B\!-$

   1/8 $A\!- bb$

   4/8 $aa\,-\,-$

Notice that both *aa* and *bb* appear in the F$_3$ progeny. In order for the homozygous recessives to occur, each parent must have at least one *a* and one *b* . Using this information, the cross becomes

F$_2$ $Aa\,Bb \times aa\,-b$

F$_3$ 3/8 $Aa\,Bb$

   1/8 $Aa\,bb$

   4/8 $aa\,-b$

The only remaining question is whether the white parent was homozygous recessive, *bb*, or heterozygous, *Bb*. If the white parent had been homozygous recessive, then the cross would have been a testcross of the blue parent, and the progeny ratio would have been 1 blue:1 pink:2 white, or 1 *Aa Bb*:1 *Aa bb*:1 *aa Bb*:1 *aa bb*. This was not observed. Therefore, the white parent had to have been heterozygous, and the F$_2$ cross was $Aa\,Bb \times aa\,Bb$.

18. **a.**   Note that the third cross has a 13:3 F$_2$ ratio. This is a modified dihybrid ratio, indicating two genes. Let the deviation from wild type in line 1 be symbolized by *a* and the deviation from wild type in line 2 be symbolized by *B*. Whether an uppercase or a lowercase symbol was chosen for the deviation was dictated by the F$_1$ result in the first two crosses. That is, if the F$_1$ is wild-type, then the deviation is recessive. However, if the F$_1$ is not wild-type, then the deviation is dominant.

   Obviously, epistasis is involved in the production of pisatin. In line 1, the wild-type allele makes pisatin, and the recessive deviation does not produce pisatin. In line 2, the wild-type allele allows the expression of the pisatin that is normally made by the first gene, while the dominant deviation blocks expression of that wild-type product.

   b.   *Cross 1:*   P $aa\,bb \times AA\,bb$

         F$_1$ $Aa\,bb$

         F$_2$ 3 $A\!- bb$:1 $aa\,bb$

      Cross 2:   P $AA\,BB \times AA\,bb$

         F$_1$ $AA\,Bb$

         F$_2$ 3 $AA\,B\!-$:1 $AA\,bb$

      Cross 3:   P $aa\,bb \times AA\,BB$

         F$_1$ $Aa\,Bb$

         F$_2$ 9 $A\!- B\!-$ no pisatin

            3 $A\!- bb$ pisatin

3 *aa B–*   no pisatin

1 *aa bb*   no pisatin

c. Line 1 does not make pisatin, while line 2 blocks the expression of pisatin.

19. It is possible to produce black offspring from two pure-breeding recessive albino parents if albinism results from alleles of two different genes. If the cross is designated

AA bb × *aa BB*

and epistasis is assumed, then all the offspring would be

Aa Bb

and they would have a black phenotype because of complementation.

20. The data indicate that white is dominant to solid purple. Note that the F_2 are in a 12:3:1 ratio. In order to achieve such a ratio, epistasis must be involved.

a. Because a modified 9:3:3:1 ratio was obtained in the F_2, the F_1 had to be a double heterozygote. Solid purple occurred at one-third the rate of white, which means that it will be in the form of either *D– ee* or *dd E–*. In order to achieve a double heterozygote in the F_1, the original white parent also has to be either *D– ee* or *dd E–*.

Arbitrarily assume that the original cross was *DD ee* (white) × *dd EE* (purple). The F_1 would all be *Dd Ee*. The F_2 would be

9 *D– E–*   white, by definition

3 *dd E–*   purple, by definition

3 *D– ee*   white, by logical deduction

1 *dd ee*   spotted purple, by logical deduction

Under these assumptions, *D* blocks the expression of both *E* and *e*. The *d* allele has no effect on the expression of *E* and *e*. *E* results in solid purple, while *e* results in spotted purple. It would also be correct, of course, to assume the opposite set of epistatic relationships (*E* blocks the expression of *D* or *d*, *D* results in solid purple, and *d* results in spotted purple).

b. The cross is white × solid purple. While the solid purple genotype must be *dd E–*, as defined in part a, the white genotype can be one of several possibilities. Note that the progeny phenotypes are in a 1:2:1 ratio and that one of the phenotypes, spotted, must be *dd ee*. In order to achieve such an outcome, the purple genotype must be *dd Ee*. The white genotype of the parent must contain both a *D* and a *d* allele in order to produce both white (*D–*) and spotted plants (*dd*). At this point, the cross has been deduced to be *Dd – –* (white) × *dd Ee* (purple).

If the white plant is *EE*, the progeny will be

1/2 *Dd E–*   white

1/2 *dd E–*   solid purple

This was not observed. If the white plant is *Ee*, the progeny will be

3/8 *Dd E–*   white

1/8 *Dd ee*   white

3/8 *dd E–*   solid purple

1/8 *dd ee*   spotted purple

The phenotypes were observed, but in a different ratio. If the white plant is *ee*, the progeny will be

1/4 *Dd Ee*   white

1/4 *Dd ee*   white

1/4 *dd Ee*   solid purple

1/4 *dd ee*   spotted purple

This was observed in the progeny. Therefore, the parents were *Dd ee* (white) × *dd Ee* (purple).

21.a. and b. Crosses 1–3 show a 3:1 ratio, indicating that brown, black, and yellow are all alleles of one gene. Crosses 4–6 show a modified 9:3:3:1 ratio, indicating that at least two genes are involved. Those crosses also indicate that the presence of color is dominant to its absence. Furthermore, epistasis must be involved for there to be a modified 9:3:3:1 ratio.

By looking at the F_1 of crosses 1–3, the following allelic dominance relationships can be seen easily: black > brown > yellow. Arbitrarily assign the following genotypes for homozygotes: B^lB^l = black, B^rB^r = brown, B^yB^y = yellow.

By looking at the F_2 of crosses 4–6, a white phenotype is composed of two categories: the double homozygote and one class of the mixed homozygote-heterozygote. Let lack of color be caused by *cc*. Color will therefore be *C–*.

CROSS	PARENTS	F_1	F_2
1	$B^rB^r CC \times B^yB^y CC$	$B^rB^y CC$	3 B^r– *CC*:1 $B^yB^y CC$
2	$B^lB^l CC \times B^rB^r CC$	$B^lB^r CC$	3 B^l– *CC*:1 $B^rB^r CC$
3	$B^lB^l CC \times B^yB^y CC$	$B^lB^y CC$	3 B^l– *CC*:1 $B^yB^y CC$
4	$B^lB^l cc \times B^yB^y CC$	$B^lB^y Cc$	9 B^l– *C–*:3 B^yB^y *C–*: 3 B^l– *cc*:1 $B^yB^y cc$
5	$B^lB^l cc \times B^rB^r CC$	$B^lB^r Cc$	9 B^l– *C–*:3 B^rB^r *C–*: 3 B^l– *cc*:1 $B^rB^r cc$
6	$B^lB^l CC \times B^yB^y cc$	$B^lB^y Cc$	9 B^l– *C–*:3 B^yB^y *C–*: 3 B^l– *cc*:1 $B^yB^y cc$

22. Whenever a cross involving two deviants from normal results in a normal phenotype, more than one gene is involved in producing the phenotype, the normal is dominant to the deviation, and the two parents are abnormal for different genes. Thus, one parent could be *aa BB,* and the other parent could be *AA bb.* All offspring would be *Aa Bb* (normal). The doubly heterozygous offspring have one copy of a functional allele for each gene, whereas each of the two parents is lacking a functional allele for one of the genes.

23. a. To solve this problem you will have to use a trial-and-error approach.

The first decision regards the number of genes involved. Cross

2 tells you that there are at least two genes because white × white yields a new phenotype. Cross 5, however, indicates that there are at least three genes involved because a 1:7 ratio is not observed with two genes.

Look at cross 1. The two lines cannot compensate for each other, suggesting a shared homozygous defective gene.

Compare crosses 2 and 3. Lines 1 and 3 can compensate for each other's defects, but lines 2 and 3 cannot. This suggests that line 2 shares a defective homozygous gene with line 3 and that line 1 is normal for the defective gene seen in line 3 but not in line 2.

At this point, you must arbitrarily make some assumptions and test them against the results. Let the three genes involved be *A*, *B*, and *D*. You could assume, for instance, that lines 1 and 2 share a defect in gene *A*, and that lines 2 and 3 share a defect in *B*. However, if line 3 can compensate for line 1 but not for line 2, there must be an additional defect in line 2. Thus far, we have determined the genotypes of the lines to be

line 1: *aa*   *BB*   *??*

line 2: *aa*   *bb*   *dd*

line 3: *AA*   *bb*   *DD*

Because lines 1 and 2 cannot produce color, line 1 must be *aa BB dd*. Now the crosses can be explained.

Cross 1: aa BB dd × *aa bb dd* ⟶ all *aa Bb dd*   white

Cross 2: aa BB dd × *AA bb DD* ⟶ all *Aa Bb Dd*   red

Cross 3: aa bb dd × *AA bb DD* ⟶ all *Aa bb Dd*   white

Cross 4: Aa Bb Dd × *aa BB dd* ⟶ 1/8 *Aa Bb Dd*   red

1/8 *Aa Bb dd*   white

1/8 *aa Bb Dd*   white

1/8 *aa Bb dd*   white

1/8 *Aa BB Dd*   red

1/8 *Aa BB dd*   white

1/8 *aa BB Dd*   white

1/8 *aa BB dd*   white

Cross 5: Aa Bb Dd × *aa bb dd* ⟶ 1/8 *Aa Bb Dd*   red

1/8 *Aa Bb dd*   white

1/8 *Aa bb Dd*   white

1/8 *Aa bb dd*   white

1/8 *aa Bb Dd*   white

1/8 *aa Bb dd*   white

1/8 *aa bb Dd*   white

1/8 *aa bb dd*   white

Cross 6: Aa Bb Dd × *AA bb DD* ⟶ 1/8 *AA Bb Dd*   red

1/8 *Aa Bb Dd*   red

1/8 *AA Bb DD*   red

1/8 *Aa Bb DD*   red

1/8 *AA bb DD*   white

1/8 *Aa bb DD*   white

1/8 *AA bb Dd*   white

1/8 *Aa bb Dd*   white

b. The cross is *Aa Bb Dd* × *Aa bb Dd*. The red progeny will have to be *A– B– D–*, which equals $(3/4)(1/2)(3/4) = 9/32$.

24. The first step in each cross is to write as much of the genotype as possible from the phenotype.

Cross 1:   *A– B–* × *aa bb* ⟶   1 *A– B–*:2 ?- ?-:1 *aa bb*

Because the double recessive appears, the blue parent must be *Aa Bb*. The two purple then must be *Aa bb* and *aa Bb*.

Cross 2:   ?? ?? × ?? ?? ⟶   1 *A– B–*:2 ?? ??:1 *aa bb*

The two parents must be, in either order, *Aa bb* and *aa Bb*. The two purple progeny must be the same. The blue progeny are *Aa Bb*.

Cross 3:   *A– B–* × *A– B–* ⟶ 3 *A– B–*:1 ?? ??

The only conclusions possible here are that one parent is either *AA* or *BB* and the other parent is *Bb* if the first is *AA* or *Aa* if the first is *BB*.

Cross 4:   *A– B–* × ?? ?? ⟶   3 *A– B–* : 4 ?? ??:1 *aa bb*

The purple parent can be either *Aa bb* or *aa Bb* for this answer. Assume the purple parent is *Aa bb*. The blue parent must be *Aa Bb*. The progeny are

3/4 *A–*
  — 1/2 *Bb* = 3/8 *A– Bb* blue
  — 1/2 *bb* = 3/8 *A– bb* purple

1/4 *aa*
  — 1/2 *Bb* = 1/8 *aa Bb* purple
  — 1/2 *bb* = 1/8 *aa bb* scarlet

Cross 5:   *A– bb* × *aa bb* ⟶   1 *A– bb*:1 *aa bb*

As written this is a testcross for gene *A*. The purple parent and progeny are *Aa bb*. Alternatively, the purple parent and progeny could be *aa Bb*.

25. The F_1 progeny of cross 1 indicate that sun-red is dominant to pink. The F_2 progeny, which are approximately in a 3:1 ratio, support this. The same pattern is seen in crosses 2 and 3, with sun-red dominant to orange and orange dominant to pink. Thus, we have a multiple allelic series with sun-red > orange > pink. In all three crosses, the parents must be homozygous.

If c^{sr} = sun-red, c^o = orange, and c^p = pink, then the crosses and the results are

CROSS	PARENTS	F_1	F_2
1	$c^{sr}c^{sr} \times c^p c^p$	$c^{sr}c^p$	$3\ c^{sr}\text{--}{:}1\ c^p c^p$
2	$c^o c^o \times c^{sr}c^{sr}$	$c^{sr}c^o$	$3\ c^{sr}\text{--}{:}1\ c^o c^o$
3	$c^o c^o \times c^p c^p$	$c^o c^p$	$3\ c^o\text{--}{:}1\ c^p c^p$

Cross 4 presents a new situation. The color of the F_1 differs from that of either parent, suggesting that two separate genes are involved. An alternative explanation is either codominance or incomplete dominance. If either codominance or incomplete dominance is involved, then the F_2 will appear in a 1:2:1 ratio. If two genes are involved, then a 9:3:3:1 ratio, or some variant of it, will be observed. Because the wild-type phenotype appears in the F_1 and F_2, complementation is occurring. This requires two genes. The progeny actually are in a 9:4:3 ratio. This means that two genes are involved and that there is epistasis. Furthermore, for three phenotypes to be present in the F_2, the two F_1 parents must have been heterozygous.

Let *a* stand for the scarlet gene and *A* for its colorless allele, and assume that there is a dominant allele, *C*, that blocks the expression of the alleles that we have been studying to this point.

Cross 4. P $c^o c^o\ AA \times CC\ aa$

F$_1$ Cc^o Aa

F$_2$ 9 *C– A–*   yellow

3 *C– aa*   scarlet

3 $c^o c^o$ *A–*   orange

1 $c^o c^o$ *aa*   orange (epistasis, with c^o blocking the expression of *aa*)

26. a. P   *AA* (agouti) × *aa* (nonagouti)

gametes   *A* and *a*

F$_1$ *Aa* (agouti)

gametes   *A* and *a*

F$_2$ 1 *AA* (agouti):2 *Aa* (agouti):1 *aa* (nonagouti)

b. P   *BB* (wild type) × *bb* (cinnamon)

gametes   *B* and *b*

F$_1$ *Bb* (wild type)

gametes   *B* and *b*

F$_2$ 1 *BB* (wild type):2 *Bb* (wild type):1 bb (cinnamon)

c. P   *AA bb* (cinnamon or brown agouti) × *aa BB* (black nonagouti)

gametes   *Ab* and *aB*

F$_1$ *Aa Bb* (wild type, or black agouti)

d. 9 *A– B–*   black agouti

3 *aa B–*   black nonagouti

3 *A– bb*   cinnamon

1 aa *bb*   chocolate

e. P   *AA bb* (cinnamon) × *aa BB* (black nonagouti)

gametes   *Ab* and *aB*

F$_1$ *Aa Bb* (wild type)

gametes   *AB, Ab, aB,* and *ab*

F$_2$ 9 *A– B–* wild type

    1 *AA BB*

    2 *Aa BB*

    2 *AA Bb*

    4 *Aa Bb*

3 *aa B–* black nonagouti

    1 *aa BB*

    2 *aa Bb*

3 *A– bb* cinnamon

    1 *AA bb*

    2 *Aa bb*

1 *aa bb* chocolate

f. P   *Aa Bb* × *AA bb*   *Aa Bb* × *aa BB*

(wild type) (cinnamon)   (wild type) (black nonagouti)

F$_1$ 1 *AA B* wild type   1 *Aa BB* wild type

    1 *Aa Bb* wild type   1 *Aa Bb* wild type

    1 *AA bb* cinnamon   1 *aa BB* black nonagouti

    1 *Aa bb* cinnamon   1 *aa Bb* black nonagouti

g. P   *Aa Bb* × *aa bb*

(wild type) (chocolate)

F$_1$ 1 *Aa Bb* wild type

    1 *Aa bb* cinnamon

    1 *aa Bb* black nonagouti

    1 *aa bb* chocolate

h. To be albino, the mice must be *cc*, but the genotype with regard to the *A* and *B* genes can be determined only by realizing that the wild type is *AA BB* and looking at the F$_2$ progeny.

Cross 1:   P   *cc ?? ??* × *CC AA BB*

F$_1$ *Cc A– B–*

F$_2$ 87 wild type   *C– A– B–*

    32 cinnamon   *C– A– bb*

    39 albino   *cc ?? ??*

For cinnamon to appear in the F_2, the F_1 parents must be *Bb*. Because the wild type is *BB*, the albino parent must have been *bb*. Now the F_1 parent can be written *Cc A– Bb*. With such a cross, one-fourth of the progeny would be expected to be albino (*cc*), which is what is observed. Three-fourths of the remaining progeny would be black, either agouti or nonagouti, and one-fourth would be either cinnamon, if agouti, or chocolate, if nonagouti. Because chocolate is not observed, the F_1 parent must not carry the allele for nonagouti. Therefore, the F_1 parent is *AA* and the original albino must have been *cc AA bb*.

Cross 2:   P   *cc ?? ?? × CC AA BB*

        F_1 *Cc A– B–*

        F_2 62 wild type   *C– A– B–*

               18 albino   *cc ?? ??*

This is a 3:1 ratio, indicating that only one gene is heterozygous in the F_1. That gene must be *Cc*. Therefore, the albino parent must be *cc AA BB*.

Cross 3:   P   *cc ?? ?? × CC AA BB*

        F_1 *Cc A– B–*

        F_2 96 wild type   *C– A– B–*

               30 black   *C– aa B–*

               41 albino   *cc ?? ??*

For a black nonagouti phenotype to appear in the F_2, the F_1 must have been heterozygous for the *A* gene. Therefore, its genotype can be written *Cc Aa B–* and the albino parent must be *cc aa ??*. Among the colored F_2 a 3:1 ratio is observed, indicating that only one of the two genes is heterozygous in the F_1. Therefore, the F_1 must be *Cc Aa BB* and the albino parent must be *cc aa BB*.

Cross 4:   P   *cc ?? ?? × CC AA BB*

        F_1 *Cc A– B–*

        F_2 287 wild type   *C– A– B–*

              86 black   *C– aa B–*

              92 cinnamon   *C– A– bb*

              29 chocolate   *C– aa bb*

           164 albino   *cc– ?? ??*

To get chocolate F_2 progeny the F_1 parent must be heterozygous for all genes and the albino parent must be *cc aa bb*.

27. To solve this problem, first restate the information:

    A– yellow   *A– R–* gray

    R– black   *aa rr* white

The cross is gray × yellow, or *A– R– × A– rr*. The F_1 progeny are

    3/8 yellow   1/8 black

    3/8 gray   1/8 white

To achieve white, both parents must carry an *r* and an *a* allele. Now the cross can be rewritten as *Aa Rr × Aa rr*.

28. a. P   single-combed × walnut-combed

$$(rr\,pp) \qquad\qquad (RR\,PP)$$

F$_1$ *Rr Pp*   walnut

F$_2$ 9 *R– P–*   walnut

3 *rr P–*   pea

3 *R– pp*   rose

1 *rr pp*   single

b. P   walnut-combed × rose-combed

$$(R–\,P–) \qquad\qquad (R–\,pp)$$

F$_1$ 3/8 *R– pp*   rose

3/8 *R– P–*   walnut

1/8 *rr P–*   pea

1/8 *rr pp*   single

The 3 *R–*:1 *rr* ratio indicates that the parents were heterozygous for the *R* gene. The 1 *P–*:1 *pp* ratio indicates a testcross for this gene. Therefore, the parents were *Rr Pp* and *Rr pp*.

c. P   walnut-combed × rose-combed

$$(R–\,P–) \qquad\qquad (R–\,pp)$$

F$_1$ walnut

$$(R–\,P–)$$

To get this result, one of the parents must be homozygous *R*, but both need not be, and the walnut parent must be homozygous *PP*.

d. *RR PP, Rr PP, RR Pp, Rr Pp*

29. Notice that the F$_1$ shows a difference in phenotype correlated with sex. At least one of the two genes is X-linked. The F$_2$ ratio suggests independent assortment between the two genes. Because purple is present in the F$_1$, the parental white-eyed male must have at least one *P* allele. The presence of white eyes in the F$_2$ suggests that the F$_1$ was heterozygous for pigment production, which means that the male also must carry the *a* allele. A start on the parental genotypes can now be made:

P   *AA pp × a–P–*, where "–" could be either a Y chromosome or a second allele.

The question now is, which gene is X-linked? If the *A* gene is X-linked, the cross is

P   *AA pp × aY PP*

F$_1$ *Aa Pp × AY Pp*

All F$_2$ females will inherit the *A* allele from their father. Under this circumstance, no white-eyed females would be observed. Therefore, the

A gene cannot be X-linked. The cross is

P   *AA pp* × *aa PY*

F$_1$ *Aa Pp*   purple-eyed females

      Aa pY   red-eyed males

F$_2$ females   males

   3/8 *A– Pp* purple   3/8 *A– PY* purple

   3/8 *A– pp* red   3/8 *A– pY* red

   1/8 *aa Pp* white   1/8 *aa PY* white

   1/8 *aa pp* white   1/8 *aa pY* white

30. The results indicate that two genes are involved (modified 9:3:3:1 ratio), with white blocking the expression of color by the other gene. The ratio of white:color is 3:1, indicating that the F$_1$ is heterozygous (*Ww*). Among colored dogs, the ratio is 3 black:1 brown, indicating that black is dominant to brown and the F$_1$ is heterozygous (*Bb*). The original brown dog is *ww bb* and the original white dog is *WW BB*. The F$_1$ progeny are *Ww Bb* and the F$_2$ progeny are

   9 *W– B–* white   3 *ww B–* black

   3 *W– bb* white   1 *ww bb* brown

31.

CROSS	RESULTS	CONCLUSION
A– C– R– × *aa cc RR*	50% colored	Colored or white will depend on the *A* and *C* genes. Because half the seeds are colored, one of the two genes is heterozygous.
A– C– R– × *aa CC rr*	25% colored	Color depends on *A* and *R* in this cross. If only one gene were heterozygous, 50% would be colored. Therefore, both *A* and *R* are heterozygous. The seed is *Aa CC Rr*.
A– C– R– × *AA cc rr*	50% colored	This supports the above conclusion.

32. **a.** The *AA CC RR prpr* parent produces pigment that is not converted to purple. The phenotype is red. The *aa cc rr PrPr* does not produce pigment. The phenotype is yellow.

  b. The F$_1$ will be *Aa Cc Rr Prpr*, which will produce pigment. The pigment will be converted to purple.

  c. The difficult way to determine the phenotypic ratios is to do a branch diagram, yielding the following results.

81/256 *A– C– R– Pr–*	purple		9/256 *aa C– R– Pr–*	yellow
27/256 *A– C– R– prpr*	red		9/256 *aa C– R– prpr*	yellow
27/256 *A– C– rr Pr–*	yellow		9/256 *aa C– rr Pr–*	yellow
9/256 *A– C– rr prp*	yellow		3/256 *aa C– rr prpr*	yellow

27/256	*A– cc R– Pr–*	yellow	27/256	*aa cc R– Pr–*	yellow
9/256	*A– cc R– prpr*	yellow	3/256	*aa cc R– prpr*	yellow
9/256	*A– cc rr Pr–*	yellow	3/256	*aa cc rr Pr–*	yellow
3/256	*A– cc rr prpr*	yellow	1/256	*aa cc rr prpr*	yellow

The final phenotypic ratio is 81 purple:27 red:148 yellow.

The easier method of determining phenotypic ratios is to recognize that four genes are involved in a heterozygous × heterozygous cross. Purple requires a dominant allele for each gene. The probability of all dominant alleles is $(3/4)^4 = 81/256$. Red results from all dominant alleles except for the *Pr* gene. The probability of that outcome is $(3/4)^3(1/4) = 27/256$. The remainder of the outcomes will produce no pigment, resulting in yellow. The probability is $256 - 81 - 27 = 148$.

d. The cross is *Aa Cc Rr Prpr* × *aa cc rr prpr*. Again, either a branch diagram or the easier method can be used. The final probabilities are

purple = $(1/2)^4 = 1/16$

red = $(1/2)^3(1/2) = 1/16$

yellow = $1 - 1/16 - 1/16 = 14/16$

33. a. The cross is

P   *td su* (wild type) × *td⁺ su⁺* (wild type)

F$_1$ 1 *td su*   wild type

1 *td su⁺*   requires tryptophan

1 *td⁺ su⁺*   wild type

1 *td⁺ su*   wild type

b. 1 tryptophan-dependent : 3 tryptophan-independent

34. a. This type of epistasis is called *suppression*.

b.   I.   *Bb Ee, Bb Ee*

II. *bb Ee, Bb Ee, – – ee , bb E–, Bb E–, bb Ee*

III. *Bb E–, b– ee, bb E–, Bb E–, bb E–, Bb E–, b– ee*

35. P   *AA BB CC DD SS* × *aa bb cc dd ss*

F$_1$ *Aa Bb Cc Dd Ss*

F_2

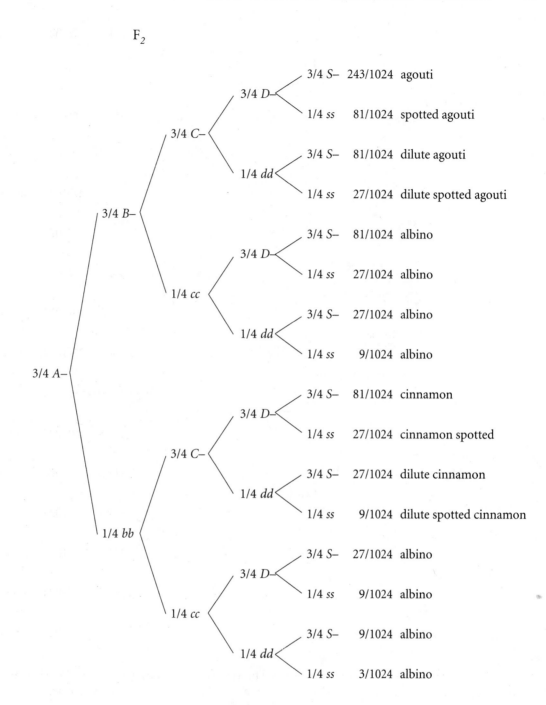

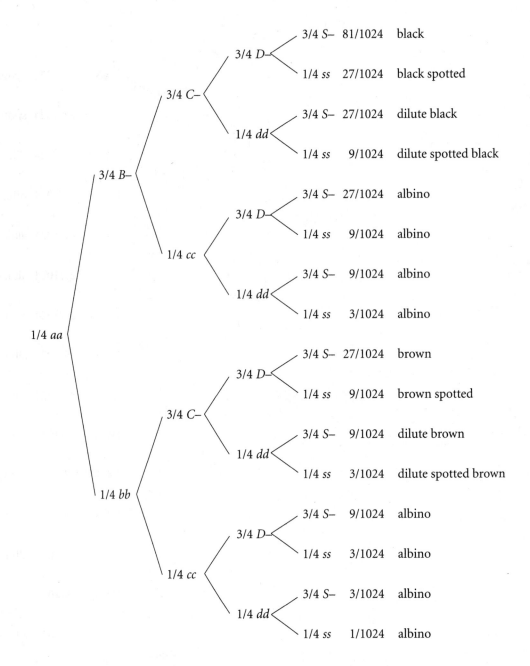

36. a. and b. The two starting lines are *ii DD MM WW* and *II dd mm ww*, and you are seeking *ii dd mm ww* . There are many ways to proceed, one of which follows below.

    I. *ii DD MM WW* × *II dd mm ww*

  II. *Ii Dd Mm Ww* × *Ii Dd Mm Ww*

 III. select *ii dd mm ww*, which has a probability of $(1/4)^4 = 1/256$.

c. and d. In the first cross, all progeny chickens will be of the desired genotype to proceed to the second cross. Therefore, the only problems are to be sure that progeny of both sexes are obtained for the second cross, which is relatively easy, and that enough females are obtained to make the time required for the desired genotype to appear feasible. Because chickens lay one to two eggs a day, the more females who are egg-laying, the faster the desired genotype will be obtained. In addition, it will be necessary to obtain a male and a female of the desired genotype in order to establish a pure-breeding line.

Assume that each female will lay two eggs a day, that money is no problem, and that excess males cause no problems. By hatching 200 eggs from the first cross, approximately 100 females will be available for the second cross. These 100 females will produce 200 eggs each day. Thus, in one week a total of 1,400 eggs will be produced. Of these 1,400 eggs, there will be approximately 700 of each sex. At a probability of 1/256, the desired genotype should be achieved 2.7 times for each sex within that first week.

37. Pedigrees like this are quite common. They indicate lack of penetrance due to epistasis or environmental effects. Individual A must have the dominant autosomal gene.

38. In cross 1, the following can be written immediately:

P   *M– D– ww* (dark reddish) × *mm ?? ??* (white with yellowish spots)

F_1 1/2 *M– D– ww*   dark reddish

       1/2 *M– dd ww*   light reddish

All progeny are colored, indicating that no *W* allele is present in the parents. Because the progeny are in a 1:1 ratio, only one of the genes in the parents is heterozygous. Also, the light reddish progeny, *dd*, indicates which gene that is. Therefore, the genotypes must be

P   *MM Dd ww × mm dd ww*

F_1 1 *Mm Dd ww*:1 *Mm dd ww*

In cross 2, the following can be written immediately:

P   *mm ?? ??* (white with yellowish spots) × *M– dd ww* (light reddish)

F_1 1/2 *M– ?? W–*   white with reddish spots

       1/4 *M– D– ww*   dark reddish

       1/4 *M– dd ww*   light reddish

For light and dark plants to appear in a 1:1 ratio among the colored plants, one of the parents must be heterozygous *Dd*. The ratio of white to colored is 1:1, a testcross, so one of the parents is heterozygous *Ww*. All plants are reddish, indicating that one parent is homozygous *MM*. Therefore, the genotypes are

P   *mm Dd Ww × MM dd ww*

F_1 1/2 *Mm Dd* (or *dd*) *Ww*

       1/4 *Mm Dd ww*

       1/4 *Mm dd ww*

Note that two genotypes were determined for white plants with yellowish spots in the two crosses: *mm dd ww* and *mm Dd ww*. Both are correct. Because the plants do not make pigment, the enhancer (*D–* or *dd*) gene is irrelevant to the phenotype.

39. a. This is a dihybrid cross with only one phenotype in the F_2 being colored. The ratio of white to red indicates that the double recessive is not the colored phenotype. Instead, the general formula for color is represented by $X- yy$.

Let line 1 be $AA\ BB$ and line 2 be $aa\ bb$. The F_1 is $Aa\ Bb$. Assume that A blocks color in line 1 and bb blocks color in line 2. The F_1 will be white because of the presence of A. The F_2 are

9 $A- B-$ white because of A

3 $A- bb$ white because of A

3 $aa\ B-$ red

1 $aa\ bb$ white because of bb

b. Cross 1: $AA\ BB \times Aa\ Bb \longrightarrow$ all $A- B-$ white

   Cross 2: $aa\ bb \times Aa\ Bb \longrightarrow$ 1/4 $Aa\ Bb$ white

                                       1/4 $Aa\ bb$ white

                                       1/4 $aa\ bb$ white

                                       1/4 $aa\ Bb$ red

40. a. Note that blue is always present, indicating EE (blue) in both parents. Because of the ratios that are observed neither C nor D is varying. In this case, the gene pairs that are involved are Aa and Bb. The F_1 is $Aa\ Bb$ and the F_2 is

9 $A- B-$ blue + red, or purple

3 $A- bb$ blue + yellow, or green

3 $aa\ B-$ blue + white$_2$, or blue

1 $aa\ bb$ blue + white$_2$, or blue

b. Blue is not always present, indicating Ee in the F_1. Because green never appears, the F_1 must be $BB\ CC\ DD$. The F_1 is $Aa\ Ee$, and the F_2 is

9 $A- E-$ red + blue, or purple

3 $A- ee$ red + white$_1$, or red

3 $aa\ E-$ white$_2$ + blue, or blue

1 $aa\ ee$ white$_2$ + white$_1$, or white

c. Blue is always present, indicating that the F_1 is EE. No green appears, indicating that the F_1 is BB. The two genes involved are A and D. The F_2 is

9 $A- D-$ blue + red + white$_4$, or purple

3 $A- dd$ blue + red, or purple

3 $aa\ D-$ blue + white$_2$ + white$_4$, or blue

1 $aa\ dd$ white$_2$ + blue + red, or purple

d. The presence of yellow indicates $bb\ ee$ in the F_2. Therefore, the F_1 is $Bb\ Ee$ and the F_2 is

9 $B- E-$ red + blue, or purple

3 $B- ee$ red + white$_1$, or red

3 $bb\ E-$ yellow + blue, or green

1 $bb\ ee$ yellow + white$_1$, or yellow

41. a. Begin by noting that cross 1 suggests that one gene is involved and that single is dominant to double. Cross 2 supports this conclusion. Now note that in crosses 3 and 4, a 1:1 ratio is seen in the progeny, suggesting that superdouble is an allele of both single and double. Superdouble must be heterozygous, however, and it must be dominant to both single and double. Because the heterozygous superdouble yields both single and double when crossed with the appropriate plant, it cannot be heterozygous for the dominant single allele. Therefore, it must be heterozygous for the recessive double allele. A multiple allelic series has been detected: superdouble > single > double.

For now, assume that only one gene is involved and attempt to rationalize the crosses with the assumptions made above.

CROSS	PARENTS	PROGENY	CONCLUSION
1	$A^S A^S \times A^D A^D$	$A^S A^D$	A^S is dominant to A^D
2	$A^S A^D \times A^S A^D$	$3\ A^S - :1\ A^D A^D$	supports conclusion from cross 1
3	$A^D A^D \times A^{Sd} A^D$	$1\ A^{Sd} A^D:1\ A^D A^D$	A^{Sd} is dominant to A^D
4	$A^S A^S \times A^{Sd} A^D$	$1\ A^{Sd} A^S:1\ A^S A^D$	A^{Sd} is dominant to A^S
5	$A^D A^D \times A^{Sd} A^S$	$1\ A^{Sd} A^D:1\ A^D A^S$	supports conclusion of heterozygous superdouble
6	$A^D A^D \times A^S A^D$	$1\ A^D A^D:1\ A^D A^S$	supports conclusion of heterozygous superdouble

b. While this explanation does rationalize all the crosses, it does not take into account either the female sterility or the origin of the superdouble plant from a double-flowered variety.

A number of genetic mechanisms could be proposed to explain the origin of superdouble from the double-flowered variety. Most of the mechanisms will be discussed in later chapters and so will not be mentioned here. However, it can safely be assumed at this point that, whatever the mechanism, it was aberrant enough to block the proper formation of the complex structure of the female flower. Because of female sterility, no homozygote for superdouble can be observed.

42. a. All the crosses suggest two independently assorting genes. However, that does not mean that there are a total of only two genes governing eye color. In fact, there are four genes controlling eye color that are being studied here. Let

aa = defect in the yellow 1 line

bb = defect in the yellow 2 line

dd = defect in the brown line

ee = defect in the orange line

The genotypes of each line are as follows:

yellow 1: *aa BB DD EE*

yellow 2: *AA bb DD EE*

brown: *AA BB dd EE*

orange: *AA BB DD ee*

b.	P	*aa BB DD EE* × *AA bb DD EE*	yellow 1 × yellow 2
	F$_1$	*Aa Bb DD EE*	red
	F$_2$	9 *A– B– DD EE*	red
		3 *aa B– DD EE*	yellow
		3 *A– bb DD EE*	yellow
		1 *aa bb DD EE*	yellow
	P	*aa BB DD EE* × *AA BB dd EE*	yellow 1 × brown
	F$_1$	*Aa BB Dd EE*	red
	F$_2$	9 *A– BB D– EE*	red
		3 *aa BB D– EE*	yellow
		3 *A– BB dd EE*	brown
		1 *aa BB dd EE*	yellow
	P	*aa BB DD EE* × *AA BB DD ee*	yellow 1 × orange
	F$_1$	*Aa BB DD Ee*	red
	F$_2$	9 *A– BB DD E–*	red
		3 aa *BB DD E–*	yellow
		3 *A– BB DD ee*	orange
		1 *aa BB DD ee*	orange
	P	*AA bb DD EE* × *AA BB dd EE*	yellow 2 × brown
	F$_1$	*AA Bb Dd EE*	red
	F$_2$	9 *AA B– D– EE*	red
		3 *AA bb D– EE*	yellow
		3 *AA B– dd EE*	brown
		1 *AA bb dd EE*	yellow
	P	*AA bb DD EE* × *AA BB DD ee*	yellow 2 × orange
	F$_1$	*AA Bb DD Ee*	red
	F$_2$	9 *AA B– DD E–*	red
		3 *AA bb DD E–*	yellow
		3 *AA B– DD ee*	orange
		1 *AA bb DD ee*	yellow
	P	*AA BB dd EE* × *AA BB DD ee*	brown × orange

F_1 *AA BB Dd Ee*		red
F_2 9 *AA BB D– E–*		red
3 *AA BB dd E–*		brown
3 *AA BB D– ee*		orange
1 *AA BB dd ee*		orange

c. When constructing a biochemical pathway, remember that the earliest gene that is defective in a pathway will determine the phenotype of a doubly defective genotype. Look at the following double-recessive homozygotes from the crosses. Notice that the double defect *dd ee* has the same phenotype as the defect *aa ee*. This suggests that the *E* gene functions earlier than do the *A* and *D* genes. Using this logic, the following table can be constructed:

GENOTYPE	PHENOTYPE	CONCLUSION
aa BB DD ee	orange	*E/e* functions before *A/a*
AA BB dd ee	orange	*E/e* functions before *D/d*
aa bb DD EE	yellow	*B/b* functions before *A/a*
AA bb DD ee	yellow	*B/b* functions before *E/e*
aa BB dd EE	yellow	*A/a* functions before *D/d*
AA bb dd EE	yellow	*B/b* functions before *D/d*

The genes function in the following sequence: *B, E, A, D*. The metabolic path is

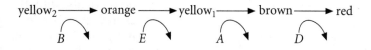

43. a. A trihybrid cross would give a 63:1 ratio. Therefore, there are three *R* loci segregating in this cross.

b.
P R_1R_1	R_2R_2 R_3R_3 × r_1r_1 r_2r_2 r_3r_3	
F_1 R_1r_1	R_2r_2 R_3r_3	
F_2 27	R_1–R_2– R_3–	red
9	R_1–R_2– r_3r_3	red
9	R_1–r_2r_2 R_3–	red
9	$r_1r_1R_2$– R_3–	red
3	R_1–r_2r_2 r_3r_3	red
3	$r_1r_1R_2$– r_3r_3	red
3	$r_1r_1r_2r_2$ R_3–	red
1	$r_1r_1r_2r_2$ r_3r_3	white

c. 1. In order to obtain a 1:1 ratio, only one of the genes can be heterozygous. A representative cross would be R_1r_1 r_2r_2 r_3r_3 × r_1r_1 r_2r_2 r_3r_3.

2. In order to obtain a 3 red:1 white ratio, two alleles must be segregating and they cannot be within the same gene. A representative cross would be $R_1r_1\ r_2r_2\ r_3r_3 \times r_1r_1\ r_2r_2\ r_3r_3$.

3. In order to obtain a 7 red:1 white ratio, three alleles must be segregating, and they cannot be within the same gene. The cross would be $R_1r_1\ r_2r_2\ r_3r_3 \times r_1r_1\ r_2r_2\ r_3r_3$.

d. The formula is $1 - (1/2)^N$, where $N =$ the number of loci that are segregating in the representative crosses above. In F_1 crosses, the formula is $1 - (1/2)^{2N}$, where N is the number of loci.

44. a. and b. The disorder is governed by an autosomal recessive allele, and there are two genes that result in deaf-mutism.

I. *Aa BB × Aa BB, AA Bb × AA Bb*

II. Individuals 1, 3–6:*A– BB*

Individuals 9, 10, 12–15:*AA B–*

Individuals 2, 7:*aa BB*

Individuals 8, 11:*AA bb*

III. All *Aa Bb*

45. a. The first impression from the pedigree is that the gene causing blue sclera and brittle bones is pleiotropic with variable expressivity. If two genes were involved, it would be highly unlikely that all people with brittle bones also had blue sclera.

b. The allele appears to be an autosomal dominant.

c. Both incomplete penetrance and variable expressivity are demonstrated in the pedigree. Individuals II-4, II-14, III-2, and III-14 have descendants with the disorder although they do not themselves express the disorder. Therefore, 4/20 people have the gene, but it is not penetrant in them. That is an 80% penetrance. Of the 16 individuals who have the allele expressed in their phenotype, 9 do not have brittle bones. Usually, expressivity is put in terms of none, variable, and highly variable, rather than expressed as percentages.

46. a. and b. Assuming that both the Brown and the Van Scoy lines were homozygous, the parental cross suggests that hygienic behavior is dominant to nonhygienic. The F_1 cross yielded a 1:2:1 ratio, suggesting a single gene. A consideration of the specific behavior, however, initially causes some puzzlement.

If the behavior is classified as either hygienic or nonhygienic, with removal of dead pupae as the criterion for hygienic behavior, nonhygienic behavior is dominant to hygienic behavior. This supports the conclusion from the parental cross.

If the removal of dead pupae from uncapped compartments is the criterion for hygienic behavior, then hygienic behavior is classically dominant to nonhygienic behavior.

The suggestion from all these data is that there may be two genes acting epistatically. One involves uncapping and one involves removal of dead pupae. If this is true, uncapping and removal of dead pupae, two behaviors that normally go together in the Brown line, have been separated in the F_2 progeny. Those bees that lack uncapping behavior are still able to express removal of dead pupae if environmental conditions are such that they do not need to uncap a compartment first.

Let: U = no uncapping, u = uncapping, R = no removal, r = removal

P $uurr \times UURR$

       (Brown)   (Van Scoy)

F_1 $UuRr$ $\times$ $uurr$

F_2 nonhygenic

      1/4 $UuRr$ 1/4 $uuRr$ uncapping

      1/4 $Uurr$ removal   1/4 $uurr$ hygienic

47. a. Note that the first two crosses are reciprocal and that the male off-spring differ in phenotype between the two crosses. This indicates that the gene is on the X chromosome.

     Also note that the F_1 females in the first two crosses are sickle. This indicates that sickle is dominant to round. The third cross also indicates that oval is dominant to sickle. Therefore, this is a multiple allelic series with oval > sickle > round.

     Let W^o = oval, W^s = sickle, and W^r = round. The three crosses are

Cross 1: $W^s W^s \times W^r Y \longrightarrow W^s W^r : W^s Y$

Cross 2: $W^r W^r \times W^s Y \longrightarrow W^s W^r : W^r Y$

Cross 3: $W^s W^s \times W^o Y \longrightarrow W^o W^s : W^s Y$

b. $W^o W^s \times W^r Y \longrightarrow 1 W^o W^r : 1 W^s W^r : 1 W^o Y : 1 W^s Y$, or

1 female oval:1 female sickle:1 male oval:1 male sickle

48. a. First note that there is a phenotypic difference between males and females, indicating X linkage. This means that the male progeny express the two alleles in the female parent. A beginning statement of the genotypes could be as follows, where H indicates the gene and the numbers 1, 2, and 3 indicate the variants.

P $H^{1+} H^2 \times H^3 Y$

F_1	FEMALES	MALES
	$H^3 H^{1+}$	$H H^{1+} Y$
	$H^3 H^2$	$H^2 Y$

Because both female F_1 progeny obtain allele H^3 from their father, yet only one has a 3-banded pattern, there is obviously a dominance relationship among the alleles.

     The mother indicates that H^{1+} is dominant to H^2. The female progeny must be $H^3 H^{1+}$ and $H^3 H^2$, and they have a 1-banded and 3-banded pattern, respectively. H^3 must be dominant to H^2 because the daughter with that combination of alleles has to be the 3-banded daughter. In other words, there is a multiple-allelic series with $H^{1+} > H^3 > H^2$.

b. The cross is:

P $H^3 H^2 \times H^{1+} Y$

F_1 FEMALES		MALES	
$H^3 H^{1+}$	1-banded	$H^3 Y$	3-banded
$H^2 H^{1+}$	1-banded	$H^2 Y$	2-banded

49. a. The first two crosses indicate that wild type is dominant to both platinum and aleutian. The third cross indicates that two genes are involved rather than one gene with multiple alleles because a 9:3:3:1 ratio is observed.

Let platinum be *a*, aleutian be *b*, and wild type be *A– B–*.

Cross 1: P   *AA BB* × *aa BB*

F₁   *Aa BB*

F₂   3 *A– BB*:1 *aa BB*

Cross 2: P   *AA BB* × *AA b*b

F₁   *AA Bb*

F₂   3 *AA B–*:1 *AA bb*

Cross 3: P   *aa BB* × *AA bb*

F₁   *Aa Bb*

F₂   9 *A– B–*   wild type

3 *A– bb*   aleutian

3 *aa B–*   platinum

1 *aa bb*   sapphire

b.

sapphire × platinum		sapphire × aleutian	
P *aa bb* × *aa BB*		*aa bb* × *AA bb*	
F₁ *aa Bb*	platinum	*Aa bb*	aleutian
F₂ 3 *aa B–*	platinum	3 *A– bb*	aleutian
1 *aa bb*	sapphire	1 *aa bb*	sapphire

50. a. The genotypes are

P   *BB ii* × *bb II*

F₁   *Bb Ii*   hairless

F₂   9 *B– I–*   hairless

3 *B– ii*   straight

3 *bb I–*   hairless

1 *bb ii*   bent

b. In order to solve this problem, first write as much as you can of the progeny genotypes.

4 hairless   *– – I–*

3 straight   *B– ii*

1 bent   *bb ii*

Each parent must have a *b* allele and an *i* allele. The partial genotypes of the parents are

$-b\, -i \times -b\, -i$

At least one parent carries the *B* allele, and at least one parent carries the *I* allele. Assume for a moment that the first parent carries both. The partial genotypes become

$Bb\, Ii \times -b\, -i$

Note that 1/2 the progeny are hairless. This must come from a *Ii* × *ii* cross. Of those progeny with hair, the ratio is 3:1, which must come from a *Bb* × *Bb* cross. The final genotypes are

$Bb\, Ii \times Bb\, ii$

51. a. The first question is whether there are two genes or one gene with three alleles. Note that a black × eyeless cross produces black and brown progeny in one instance but black and eyeless progeny in the second instance. Further note that a black × black cross produces brown, a brown × eyeless cross produces brown, and a brown × black cross produces eyeless. The results are confusing enough that the best way to proceed is by trial and error.

There are two possibilities: one gene or two genes.

One gene. Assume one gene for a moment. Let the gene be *E* and assume the following designations:

E^1 = black

E^2 = brown

E^3 = eyeless

If you next assume, based on the various crosses, that black > brown > eyeless, genotypes in the pedigree become

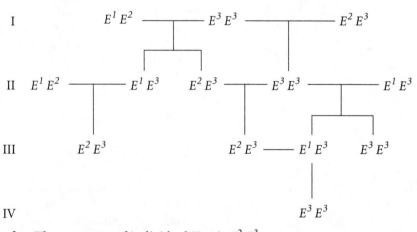

b. The genotype of individual II–3 is $E^2 E^3$.

Two genes. If two genes are assumed, then questions arise regarding whether both are autosomoal or one is X-linked. Eye color appears to be autosomal. The presence or absence of eyes appears to be X-linked. Let

B = black X^E = normal eyes

b = brown X^e = eyeless

The pedigree then is

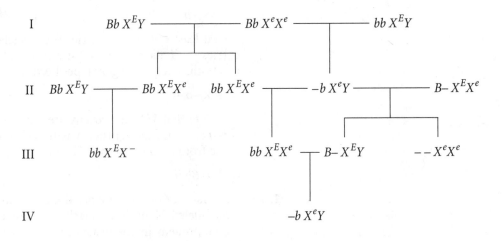

With this interpretation of the pedigree, individual II-3 is $bb\,X^E\,X^e$.

Without further data, it is impossible to choose scientifically between the two possible explanations. A basic "rule" in science is that the more simple answer should be used until such time as data exist that would lead to a rejection of it. In this case, the one-gene explanation is essentially as complex as the two-gene explanation, with one exception. The exception is that eye color is fundamentally different from a lack of eyes. Therefore, the better guess is that two genes are involved.

52. a. and b. Cross 1 indicates that orange is dominant to yellow. Crosses 2–4 indicate that red is dominant to orange, yellow, and white. Crosses 5–7 indicate that there are two genes involved in the production of color. Cross 5 indicates that yellow and white are on different genes. Cross 6 indicates that orange and white are on different genes.

In other words, epistasis is involved and the homozygous recessive white genotype seems to block production of color by a second gene.

Begin by explaining the crosses in the simplest manner possible until such time as it becomes necessary to add complexity. Therefore, assume that orange and yellow are alleles of gene A, with orange dominant.

Cross 1:   P $AA \times aa$

F$_1$ Aa

F$_2$ $3\,A-:aa$

Immediately there is trouble in trying to write the genotypes for cross 2, unless red is a third allele of the same gene. Assume the following dominance relationships: red > orange > yellow. Let the alleles be designated as follows:

red A^R

orange A^O

yellow A^Y

Crosses 1–3 now become

P $A^O A^O \times A^Y A^Y$ $A^R A^R \times A^O A^O$ $A^R A^R \; X \; A^Y A^Y$

F$_1$ $A^O A^Y$ $A^R A^O$ $A^R A^Y$

F$_2$ 3 A^O –:1 $A^Y A^Y$ 3 A^R –:1 $A^O A^O$ 3 A^R –:1 $A^Y A^Y$

Cross 4: To do this cross you must add a second gene. You must also rewrite the above crosses to include the second gene. Let *B* allow color and *b* block color expression, producing white. The first three crosses become

P $A^O A^O BB \times A^Y A^Y BB$ $A^R A^R BB \times A^O A^O BB$ $A^R A^R BB \times A^Y A^Y BB$

F$_1$ $A^O A^Y BB$ $A^R A^O BB$ $A^R A^Y BB$

F$_2$ 3 A^O – BB :1 $A^Y A^Y BB$ 3 A^R – BB :1 $A^O A^O BB$ 3 A^R –:1 $A^Y A^Y BB$

The fourth cross is

P $A^R A^R BB \times A^R A^R bb$

F$_1$ $A^R A^R Bb$

F$_2$ 3$A^R A^R B$–:1$A^R A^R bb$

Cross 5: To do this cross, note that there is no orange appearing. Therefore, the two parents must carry the alleles for red and yellow, and red must be blocked from being expressed.

P $A^Y A^Y BB \times A^R A^R bb$

F$_1$ $A^R A^Y Bb$

F$_2$ 9 A^R – B–   red

       3 A^R – bb white

       3 $A^Y A^Y B$–   yellow

       1 $A^Y A^Y bb$ white

Cross 6: This cross is identical with cross 5 except that orange replaces yellow.

P $A^O A^O BB \times AR AR bb$

F$_1$ $A^R A^O Bb$

F$_2$ 9 A^R – B–   red

       3 A^R – bb white

       3 $A^O A^O B$–   orange

       1 $A^O A^O bb$ white

Cross 7: In this cross, yellow is suppressed by *bb*.

P $A^R A^R BB \times A^Y A^Y bb$

F$_1$ $A^R A^Y Bb$

F$_2$ 9 A^R – B–   red

       3 A^R – bb white

       3 $A^Y A^Y B$–   yellow

       1 $A^Y A^Y b b$ white

▶ TIPS ON PROBLEM
SOLVING

Whenever a cross involving two deviants from normal results in a normal phenotype, more than one gene is involved in producing the phenotype, the normal is dominant to the deviation, and the two parents are abnormal for different genes (see Problem 39).

Whenever the sexes differ with respect to phenotype among the progeny, an X-linked gene is involved (see Problem 10).

If two deviants from normal are crossed and the males are wild type, two genes are involved, the deviation is recessive, and the gene that is deviant in the female is autosomal (see Problem 22).

If the F_1 phenotype differs from that of either parent, two separate genes may be involved. An alternative explanation is either codominance or incomplete dominance. If either codominance or incomplete dominance is involved, then the F_2 progeny will appear in a 1:2:1 ratio. If two genes are involved, then a 9:3:3:1 ratio, or some variant of it, will be observed (see Problems 2 and 3).

Multiple alleles show a 1:2:1 ratio or some modification of that ratio. Multiple genes show a 9:3:3:1 ratio or some modification of that ratio (see Problems 2 and 3).

5

Linkage I: Basic Eukaryotic Chromosome Mapping

▶ IMPORTANT TERMS
AND CONCEPTS

Recombination is the process that generates gametes with gene combinations different from that seen in the parental source. Recombination can be **interchromosomal**, involving genes on nonhomologous chromosomes, or **intrachromosomal**, involving homologous chromosomes.

If the recombination is interchromosomal, an equal frequency of all gamete types results. If the recombination is intrachromosomal, the frequency of gamete types depends on the physical distance between the genes being studied. Genes that are located more than 50 map units from each other are said to be **unlinked**; all gametes occur in equal frequency. Genes that are 50 map units or fewer apart are said to be **linked**; the frequency of parental-type gametes will be greater than 50 percent and the frequency of recombinant-type gametes will be less than 50 percent.

Recombination of linked genes results from a physical exchange of genetic material between homologous chromosomes. The **chiasmata** seen in meiosis I are considered to be the cytological evidence of **crossing-over**, the process that leads to recombination. The percentage of recombination is used as a measure of the physical distance between two genes. It is calculated by the following formula:

$$\frac{(100\%)(\text{number of recombinant progeny})}{\text{total number of progeny}} = \text{number of map units (m.u.)}$$

Linkage maps are formed by combining the results from a series of crosses.

A **three-point testcross** can be done to determine the relative location of three genes simultaneously. To do a three-point testcross, a triple-recessive parent is mated with a triple-heterozygous parent. The contribution of the testcross parent is ignored. This is because the testcross parent must contribute the three recessive alleles to each of the progeny.

The **gene order** in a three-point testcross is determined by comparing the parental-type progeny, which will be most frequent, with the **double-crossover** progeny, which will be the least frequent. The gene that has switched with respect to the other two genes in this comparison must be located between them:

parentals: $D a R, d A r$

double crossovers: $d a R, D A r$

The order is *A D R*; the parental chromosomes are *a D R/A d r*.

Map units in a three-point testcross are determined by identifying progeny that result from a crossover in each region. In the example above, single crossovers in the *A–D* region would be *a d r* and *A D R*. Single crossovers in the *D–R* region would be *a D r* and *A d R*. Because double-crossover progeny result from a crossover in both regions, their frequency is added to the frequency of single crossovers for each region. The formula for calculating the distance between two genes is

$$\frac{(100\%)(\text{number of single crossovers} + \text{number of double crossovers})}{\text{total number of progeny}}$$

$$= \text{number of map units}$$

The double-crossover progeny would be *a d R* and *A D r* in the above example.

The expected types and numbers of progeny can be determined from a linkage map. To determine the expected number of double crossovers, multiply the total number of progeny by the map units in both regions. To determine the expected number of single crossovers, multiply the total number of progeny by the map units in that region and subtract the number of double crossovers from the total.

Interference (I) occurs when a crossover in one region affects the frequency of a crossover in a second region. It is defined as

$$1 - \frac{\text{observed double crossovers}}{\text{expected double crossovers}}$$

If interference exists, it needs to be calculated from data when determining map distances or used to adjust frequencies when calculating the expected progeny types from a linkage map. In the latter situation, the value of I will be given. To calculate the observed number of double crossovers, use the following formula:

observed double crossovers = (1 – I)(expected double crossovers)

Use the **chi square** (χ^2) test to decide whether observations are compatible with the hypothesis that generated the expected values.

Questions for Concept Map 5-1

1. When does meiotic recombination occur in humans?

2. When does meiotic recombination occur in *Neurospora?*

3. When does meiotic recombination occur in male *Drosophila?*

4. What is the name of the cells in humans in which meiotic recombination occurs?

5. What protein structure aids meiotic recombination?

▶ CONCEPT MAP 5-1:
   LINKAGE

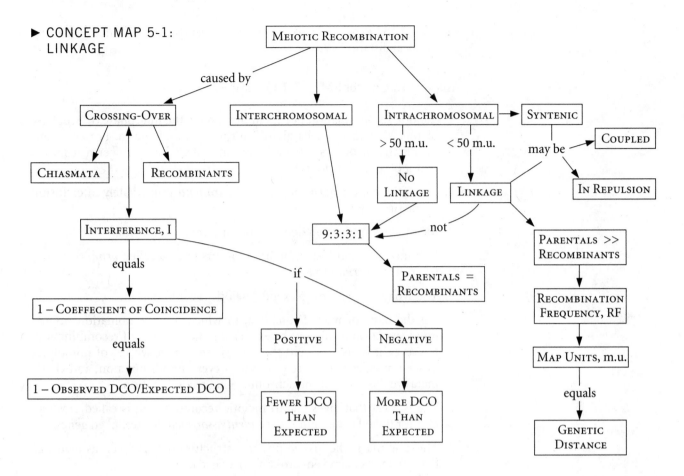

6. If no recombination occurs between homologs, would you expect nondisjunction to occur?

7. What is the process that gives rise to meiotic recombination?

8. What is the visible physical structure that is taken as evidence that the molecular event crossing-over occurred?

9. Define *interference*.

10. Interpret the interferences values of −0.98, 0.0, and +0.98.

11. If genes *A* and *B* show interchromosomal recombination, and the F_1 progeny received *Ab* from one of its parents and *aB* from the other, list the recombinant gametes. What is their frequency relative to parental gametes?

12. Why is a 9:3:3:1 ratio expected with interchromosomal recombination?

13. If two genes are syntenic and unlinked, what is the expected ratio in a dihybrid cross?

14. Can two genes be both coupled and not linked? In repulsion and not linked?

15. Explain how interchromosomal recombination gives the same dihybrid ratio as intrachromosomal recombination of two unlinked (>50 m.u.) genes.

16. Define *RF.*

17. Define m.u.

18. Would you expect a 1:1 correlation between genetic distance and actual physical distance? Why?

Answers to Concept Map 5-1 Questions

1. Meiotic recombination occurs in humans in the interphase or prophase of meiosis I in the cells giving rise to eggs and sperm. For eggs, this means that it occurs during embryonic life, while for sperm it means that it occurs after puberty.

2. Meiotic recombination occurs in *Neurospora* immediately after fusion of the two gametes.

3. Meiotic recombination does not occur in male *Drosophila*.

4. In humans meiotic recombination occurs in cells called *primary oocytes* or *primary spermatocytes*.

5. The synaptonemal complex aids meiotic recombination.

6. On the basis of male *Drosophila*, in which no recombination occurs, nondisjunction would not be expected in the absence of recombination. However, it is also thought that the physical intertwining of homologs, as seen in chiasmata, is required to prevent nondisjunction, and chiasmata are the result of intrachromosomal recombination.

7. The process that gives rise to meiotic recombination is called *crossing-over* for linked genes and *independent assortment* for unlinked genes.

8. The chiasma is the visible physical structure that is taken as evidence that the molecular event crossing-over occurred.

9. Interference is an overall decrease in double crossovers from the expected frequency, based on a genetic map.

10. An interference value of -0.98 indicates that double crossovers occurred at almost twice the expected rate. An interference of 0.0 indicates that double crossovers occurred at the expected rate. An interference of +0.98 indicates that double crossovers occurred at approximately 2% of the expected rate.

11. *Interchromosomal* means "between chromosomes," so the expected recombinant gametes would be *AB and ab* in a 1:1 ratio with the two parental gametes.

12. A 9:3:3:1 ratio is expected with interchromosomal recombination because interchromosomal recombination refers to recombinants that occur with unlinked genes.

13. The expected ratio is 9:3:3:1.

14. Two genes can be both coupled and not linked. They can also be in repulsion and not linked.

15. In both cases, interchromosomal recombination and intrachromosomal recombination of two unlinked (>50 m.u.) genes, the two genes assort independently.

16. *RF* refers to the recombination frequency between two genes, which, because of multiple crossovers, may be an underestimate of the genetic distance.

17. The abbreviation *m.u.* refers to the map units between two genes.

18. There is not an actual 1:1 correlation between genetic distance and actual physical distance. Some regions of a chromosome are "hot spots," where crossing-over occurs at a very high rate, and other regions of the chromosome do not appear to experience crossing-over. In male *Drosophila*, there is no crossing-over at all. Frequently, the rate of crossing-over between two specific genes differs between males and females of a species. For these reasons, a 1:1 correlation is not expected.

▶ SOLUTIONS TO PROBLEMS

Be sure that you have thoroughly read the entire chapter before you attempt any of the problems.

For the following, *CO* is used to designate single recombinants and *DCO* is used to designate double recombinants.

1. The *Aa Bb* progeny are parentals. An expected 90 percent of the progeny are parentals, and they are of two types: *Aa Bb* and *aa bb*. Therefore, 45 percent of the progeny will be *Aa Bb*.

2. P $A\,d/A\,d \times a\,D/a\,D$

    F_1 $A\,d/a\,D$

    F_2 1 $A\,d/A\,d$

        2   $A\,d/a\,D$

        1   $a\,D/a\,D$

3. P $R\,S/r\,s \times R\,S/r\,s$

    gametes $1/2\,(1-0.35)$ $R\,S$

                  $1/2\,(1-0.35)$ $r\,s$

                  $1/2\,(0.35)$ $R\,s$

                  $1/2\,(0.35)$ $r\,S$

F_1	0.1056	$R\,S/R\,S$	0.1138	$r\,s/r\,S$	or	0.6058	$R{-}\,S{-}$
	0.1056	$r\,s/r\,s$	0.1138	$r\,s/R\,s$		0.1056	$rr\,ss$
	0.2113	$R\,S/r\,s$	0.0306	$R\,s/R\,s$		0.1444	$rr\,S{-}$
	0.1138	$R\,S/r\,S$	0.0306	$r\,S/r\,S$		0.1444	$R{-}\,ss$
	0.1138	$R\,S/R\,s$	0.0613	$R\,s/r\,S$			

4. The cross is *Ee Ff* × *ee ff*. If independent assortment exists, the progeny should be in a 1:1:1:1 ratio, which is not observed. Therefore, there is linkage. *E f* and *e F* are recombinants equaling one-third of the progeny. The two genes are 33.3 map units (m.u.) apart.

5. Because only parental types were recovered, the two genes must be quite close to each other, making recombination quite rare.

6. Parental types are the most frequent: 442 *Aa Bb* and 458 *aa bb*. Because one parent was *aa bb*, contributing only *a b* to the offspring, the parental types can be rewritten as 442 *A B/a b* and 458 *a b/a b*. Thus, the female parent was *A B/a b*. The two recombinant types are 46 *A b/a*

b and 54 a B/a b. The frequency of recombination between two genes is

$$\frac{100\%(\text{total number of recombinants})}{\text{total number of progeny}}$$

$$= \frac{100\%(46 + 54)}{(442 + 458 + 46 + 54)} = \frac{100\%(100)}{1000} = 10 \text{ m.u.}$$

should be AaBb

7. **a.** The F_1 is *Aa Bb*, which is crossed to *aa bb*. Among the progeny, $(1/2)(1/2) = 1/4$ will be *aa bb*.

   b. The F_1 is *AB/ab*, which is crossed to *ab/ab*. Half the progeny will be *ab/ab*.

   c. The cross is the same as in part b. The *ab/ab* progeny are a parental type. Parentals occur at a rate of 90%, which means that 45% will be *ab/ab*.

   d. The cross is the same as in part b. The *ab/ab* progeny are a parental type. Parentals occur at a rate of 76%, which means that 38% will be *ab/ab*.

8. Meiosis is occurring in an organism that is *C d/c D*, producing haploid spores ultimately. The parental types are *C d* and *c D*, in equal frequency. The recombinant types are *C D* and *c d*, in equal frequency. Eight map units means 8 percent recombinants. Thus, *C D* and *c d* will each be present at a frequency of 4 percent, and *C d* and *c D* will each be present at a frequency of $(100\% - 8\%)/2 = 46\%$.

   a. 4 percent; **b.** 4 percent; **c.** 46 percent; **d.** 8 percent

9. To solve this problem, you must realize that

   (1) One chiasma involves two of the four chromatids in a homologous pair. Therefore, 16 percent of the meioses having a chiasma will lead to 8 percent recombinants.

   (2) Half the recombinants will be *B r* and half will be *b R*. The answer is b, 4 percent.

10. *Unpacking the Problem*

   a. There is no correct drawing; any will do. Pollen from the tassels is placed on the silks of the females. The seeds are the F_1 corn kernels.

   b. The +'s all look the same because they signify wild type for each gene. The information is given in a specific order, which prevents confusion, at least initially. However, as you work the problem, which may require you to reorder the genes, errors can creep into your work if you do not make sure that you reorder the genes for each genotype in exactly the same way. I strongly suggest that beginning students write the complete genotype, *gre+* instead of +, to avoid confusion.

   c. The phenotype is purple leaves and brown midriff to seeds. In other words, the two colors refer to different parts of the organism.

   d. There is no significance in the original sequence of the data.

   e. A tester is a homozygous recessive for all genes being studied. It is used so that the meiotic products in the organism being tested can be seen directly in the phenotype of the progeny.

f. *Gre* stands for "green leaves." *Sen* stands for "virus-sensitive." *Pla* stands for "plain seed."

g. *Gametes* refers to the gametes of the two pure-breeding parents. *F₁ gametes* refers to the gametes produced by the completely heterozygous F_1 progeny. They indicate whether crossing-over and independent assortment have occurred. In this case, because there is either intrachromosomal or interchromosomal recombination, or both, the data indicate that the three genes are not so tightly linked that zero recombination occurred.

h. The main focus is meiosis I in the F_1 cross, specifically in the complete heterozygote.

i. The gametes from the tester are not shown, because they contribute nothing to the phenotypic differences seen in the progeny.

j. Eight phenotypic classes are expected for three autosomal genes, whether or not they are linked, when all three genes are classically dominant. The general formula for the number of expected phenotypes is 2^n, where n is the number of genes being studied.

k. If the three genes were on separate chromosomes, the expectation is a 1:1:1:1:1:1:1:1 ratio.

l. The four classes of data correspond to the parentals (largest), two groups of single crossovers (intermediate), and double crossovers (smallest).

m. By comparing the parentals with the double crossovers, gene order can be determined. The gene in the middle "flips" with respect to the two flanking genes in the double crossover progeny. In this case, one parental is +++ and one double crossover is +*vb*. This indicates that leaf color is in the middle.

n. If only two of the three genes are linked, the data can still be grouped, but the grouping will differ from that mentioned in *l* above. In this situation, the unlinked gene will show independent assortment with the two linked genes. There will be one class composed of four phenotypes in approximately equal frequency, which combined will total more than half the progeny. A second class will be composed of four phenotypes in approximately equal frequency, and the combined total will be less than half the progeny. For example, if the cross were *ab*/++ *c*/+ × *ab*/*ab* *c*/*c*, then the parental class (more frequent class) would have four components: *abc*, *ab*+, ++*c*, and +++. The recombinant class would be *a*+*c*, *a*++, +*bc*, and +*b*+.

o. *Point* refers to locus. The usage does not imply linkage, but rather a testing for possible linkage. A four-point testcross would look like the following: *a*/+ *b*/+ *c*/+ *d*/+ × *a*/*a* *b*/*b* *c*/*c* *d*/*d*.

p. A *recombinant* refers to an individual who has alleles inherited from two different grandparents, both of whom were the parents of the individual's heterozygous parent. Another way to think about this term is that in the recombinant individual's heterozygous parent, recombination took place among the genes that were inherited from his or her parents. In this case, the recombination took place in the F_1 and the recombinants are among the F_2 progeny.

q. The "recombinant for" columns refer to specific gene pairs and progeny that exhibit recombination between those gene pairs.

r. There are three "recombinant for" columns because three genes can be grouped in three different gene pairs.

s. *R* refers to recombinant progeny, and they are determined by reference back to the parents of their heterozygous parent.

t. Column totals indicate the number of progeny that experience crossing-over between the specific gene pairs. They are used to calculate map units between the two genes.

u. The diagnostic test for linkage is a recombination frequency of less than 50%.

v. A map unit represents 1% crossing-over and is the same as a centimorgan.

w. In the tester, recombination cannot be detected in the gamete contribution to the progeny because the tester is homozygous. The F_1 individuals have genotypes fixed by their parents' homozygous state and, again, recombination cannot be detected in them, simply because their parents were homozygous.

x. Interference, I, = 1 – coefficient of coincidence = 1 – (observed double crossovers/expected double crossovers). The expected double crossovers are equal to *p*(frequency of crossing-over in the first region between genes) × *p*(frequency of crossing-over in the second region between genes) × number of progeny. The probability of crossing-over is equal to map units converted back to percentage.

y. If the three genes are not all linked, then interference cannot be calculated.

z. A great deal of work is required to obtain 10,000 progeny in corn because each seed on a cob represents one progeny. Each cob may contain as many as 200 seeds. While seed characteristics can be assessed at the cob stage, for other characteristics, each seed must separately be planted and assessed after germination and growth. The bookkeeping task is also enormous.

Solution to the Problem

a. The three genes are linked.

b. Comparing the parentals (most frequent) with the double crossovers (least frequent), the gene order is *sen gre pla*. There were 2200 recombinants between *sen* and *gre*, and 1500 between *gre and pla*. The general formula for map units is

m.u. = 100%(number of recombinants)/total number of progeny

Therefore, the map units between *sen* and *gre* = 100%(2200)/10,000 = 22 m.u., and the map units between *gre and pla* = 100% (1500)/10,000 = 15 m.u..

The map is

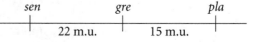

c. I = 1 − observed double crossovers/expected double crossovers

$$= 1 - 132/(0.22)(0.15)(10{,}000) = 1 - 0.4 = 0.6$$

11. a. Gene pairs A/a, B/b, and C/c are linked, and D/d shows no recombination with $A/$a. This is determined by looking at only the two genes you are trying to make a decision about.

Are A/a and B/b linked?

$A\ B = 140 + 305 = 445$

$a\ b = 145 + 310 = 455$

$a\ B = 42 + 6 = 48$

$A\ b = 43 + 9 = 52$

The two genes are 10 m.u. apart.

Are A/a and D/d linked?

$A\ D = 0$

$a\ d = 0$

$A\ d = 43 + 140 + 9 + 305 = 497$

$a\ D = 42 + 145 + 6 + 310 = 503$

The two genes show no recombination = 0 m.u.

Are B/b and D/d linked?

$B\ D = 42 + 6 = 48$

$b\ d = 43 + 9 = 52$

$B\ d = 140 + 305 = 445$

$b\ D = 145 + 310 = 455$

The two genes are 10 m.u. apart.

Are C/c and D/d linked?

$C\ D = 42 + 310 = 350$

$c\ d = 43 + 305 = 348$

$C\ d = 140 + 9 = 149$

$c\ D = 145 + 6 = 151$

The two genes are 30 m.u. apart.

All four genes are linked.

b. and c. Because A/a and D/d show no recombination, first rewrite the progeny omitting D and d (or omitting A and a).

$a\ B\ C$	42
$A\ b\ c$	43
$A\ B\ C$	140
$a\ b\ c$	145
$a\ B\ c$	6
$A\ b\ C$	9

A B c	305
a b C	310
	1000

Note that the progeny now look like those of a typical three-point testcross, with *A B c* and *a b C* the parental types (most frequent) and *a B c* and *A b C* the double recombinants (least frequent). The gene order is *B/b A/a C/c*. This is determined by comparing double recombinants with the parentals; the gene that "switches" in reference with the other two is the gene in the center (*B A c → B a c, b a C → b A C*).

Next, rewrite the progeny again, this time putting the genes in the proper order, and classify the progeny.

B a C	42	CO A–B
b A c	43	CO A–B
B A C	140	CO A–C
b a c	145	CO A–C
B a c	6	DCO
b A C	9	DCO
B A c	305	parental
b a C	310	parental

To construct the map of these genes, use the following formula:

distance between

$$\text{two genes} = \frac{(100\%)(\text{number of single CO} + \text{number of DCO})}{\text{total number of progeny}}$$

For the *A* to *B* distance:

$$\frac{(100\%)(42 + 43 + 6 + 9)}{1000} = \frac{100\%(100)}{1000} = 10 \text{ m.u.}$$

For the *A* to *C* distance:

$$\frac{(100\%)(140 + 145 + 6 + 9)}{1000} = \frac{100\%(300)}{1000} = 30 \text{ m.u.}$$

The map is

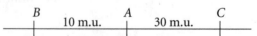

Now it is time to deal with the *D* and *d* alleles. Notice that only two combinations were observed with *A* and *a*: *A d* and *a D*. The parental chromosomes actually were *B (A,d) c/b (a,D) C*, where the parentheses indicate that the order of the genes within is unknown.

d. Interference = 1 – (observed DCO/expected DCO)

$$= 1 - (6 + 9)/[(0.10)(0.30)(1000)]$$

$$= 1 - 15/30 = 0.5$$

12. a. Males must be heterozygous for both genes, and the two must be closely linked: $M\,F/m\,f$.

   b. $m\,f/m\,f$

   c. Sex is determined by the male contribution. The two parental gametes are $M\,F$, determining maleness ($M\,F/m\,f$), and $m\,f$, determining femaleness ($m\,f/m\,f$). Occasional recombination would yield $M\,f$, determining a hermaphrodite ($M\,f/m\,f$), and $m\,F$, determining total sterility ($m\,F/m\,f$).

   d. recombination in the male yielding $M\,f$

   e. Hermaphrodites are rare because the genes are tightly linked.

13. The verbal description indicates the following cross and result:

P $N\!-\!A\!-\times nn\,OO$

F_1 $Nn\,AO \times Nn\,AO$

The results indicate linkage, so the cross and results must be rewritten:

P $N\,A/?\,? \times n\,O/n\,O$

F_1 $N\,A/n\,O \times N\,A/n\,O$

F_2 66% $N\!-\!A\!-$

16% $n\,O/n\,O$

9% $nn\,A\!-$

9% $N\,O/?\,O$

Only one genotype is fully known: 16% $n\,O/n\,O$, a combination of two parental gametes. The frequency of two parental gametes coming together is the frequency of the first times the frequency of the second. Therefore, the frequency of each $n\,O$ gamete is the square root of 0.16, or 0.4. Within an organism the two parental gametes occur in equal frequency. Therefore, the frequency of $N\,A$ is also 0.4. The parental total is 0.8, leaving 0.2 for all recombinants. Therefore, $N\,O$ and $n\,A$ occur at a frequency of 0.1 each. The two genes are 20 m.u. apart.

14. The original cross was

P $P\,L/P\,L \times p\,l/p\,l$

F_1 $P\,L/p\,l \times P\,L/p\,l$

Before proceeding, recognize that crossing-over can occur in both parents and that some crossovers cannot be detected by phenotype. The gametes from each plant are as follows:

parental types: $P\,L, p\,l$

recombinants: $P\,l, p\,L$

The F_2 is as follows:

$P\,L/P\,L$	purple, long	$p\,l/p\,L$	red, long
$P\,L/p\,l$	purple, long	$p\,l/p\,l$	red, round

$P\,L/P\,l$	purple, long	$P\,l/p\,L$	purple, long
$P\,L/p\,L$	purple, long	$P\,l/P\,l$	purple, round
$P\,l/p\,l$	purple, round	$p\,L/p\,L$	red, long

Of these 10 different genotypes, only red, round can be identified unambiguously. It consists of two parental-type gametes and occurs at a frequency of 14.4 percent. The probability of such a genotype can be calculated by multiplying the probability of a gamete from the first parent times the probability of the same gamete from the second parent. Thus, the square root of 14.4 percent will yield the frequency of this one type of parental gamete, or 37.99 percent. Because parental types occur with equal frequency, the parentals are 75.98 percent and the recombinants are 24.02 percent. Therefore, there are approximately 24 m.u. between the two genes.

A more precise calculation involves the use of an advanced statistical technique known as the *method of maximum likelihood.*

15. P $a\,b\,c/a\,b\,c \times a^+b^+c^+/a^+b^+c^+$

    F_1 $a\,b\,c/a^+b^+c^+ \times a\,b\,c/a^+b^+c^+$

    F_2 1364 $a^+-\,b^+-\,c^+-$

            365   $a\,b\,c/a\,b\,c$

             87   $aa\,bb\,c^+-$

             84   $a^+-\,b^+-\,cc$

             47   $aa\,b^+-\,c^+-$

             44   $a^+-\,bb\,cc$

              5   $aa\,b^+-\,cc$

              4   $a^+-\,bb\,c^+-$

Remember that recombination does not occur in the male *Drosophila.*

a. Because you cannot distinguish between $a\,b\,c/a^+\,b^+\,c^+$ and $a^+\,b^+\,c^+/a^+\,b^+\,c^+$, use the frequency of $a\,b\,c/a\,b\,c$ to estimate the frequency of $a^+\,b^+\,c^+$ (parental) gametes from the female.

parentals	730	(2×365)
CO a–b:	91	$(a\,+\,+,\,+\,b\,c = 47 + 44)$
CO b–c:	171	$(a\,b\,+,\,+\,+\,c = 87 + 84)$
DCO:	9	$(a\,+\,c,\,+\,b\,+ = 5 + 4)$
	1001	

    a–b: $100\%(91 + 9)/1001 = 10$ m.u.

    b–c: $100\%(171 + 9)/1001 = 18$ m.u.

b. Coefficient of coincidence = (observed DCO)/(expected DCO)

$$= 9/[(0.1)(0.18)(1001)]$$

$$= 9/18 = 0.5$$

16. a. By comparing the two most frequent classes (parentals: $an\ br^+\ f^+$, an^+ $br\ f$) to the least frequent classes (DCO: $an^+\ br\ f^+$, $an\ br^+\ f$), the gene order can be determined. The gene in the middle switches with respect to the other two (the order is *an f br*). Now the crosses can be written fully.

P   *an f⁺ br⁺/an f⁺ br⁺ × an⁺ f br/an⁺ f br*

F₁   *an⁺ f br/an f⁺ br⁺ × an f br/an f br*

F₂   355   *an f br/an f⁺ br⁺*   parental

339   *an f br/an⁺ f br*   parental

88   *an f br/an⁺ f⁺ r⁺*   CO *an–f*

55   *an f br/an f br*   CO *an–f*

21   *an f br/an⁺ f br⁺*   CO *f–br*

17   *an f br/an f⁺ br*   CO *f–br*

2   *an f br/an⁺ f⁺br*   DCO

2   *an f br/an f br⁺*   DCO

b. *an–f:*   100% (88 + 55 + 2 + 2)/879 = 16.72 m.u.

f–br: 100% (21 + 17 + 2 + 2)/879 = 4.78 m.u.

```
an            f            br
|   16.72 m.u.  |  4.78 m.u.  |
|              |            |
```

c.   Interference = 1 – (observed DCO/expected DCO)

= 1 – 4/(0.1672)(0.0478)(879)

= 1 – 0.569 = 0.431

17. By comparing the most frequent classes (parental: + *v lg*, *b* + +) with the least frequent classes (DCO: + + +, *b v lg*) the gene order can be determined. The gene in the middle switches with respect to the other two, yielding the following sequence: *v b lg*. Now the cross can be written

P   *v b⁺ lg/v⁺ b lg⁺ × v b lg/v b lg*

F₁   305   *v b lg/v b⁺ lg*   parental

275   *v b lg/v⁺ b lg⁺*   parental

128   *v b lg/v⁺ b lg*   CO *b-lg*

112   *v b lg/v b⁺ lg⁺*   CO *b-lg*

74   *v b lg/v⁺ b⁺ lg*   CO *v-b*

66   *v b lg/v b lg⁺*   CO *v-b*

22   *v b lg/v⁺ b⁺ lg⁺*   DCO

18   *v b lg/ v b lg*   DCO

v–b: 100%(74 + 66 + 22 + 18)/1000 = 18.0 m.u.

b–lg: 100%(128 + 112 + 22 + 18)/1000 = 28.0 m.u.

c.c.   = observed DCO/expected DCO

= (22 + 18)/(0.28)(0.18)(1000) = 0.79

18. Let F = fat, L = long tail, and Fl = flagella. The gene sequence is $F\ L\ Fl$ (compare most frequent with least frequent). The cross is

P $\quad F\ L\ Fl/fl\ fl \times fl\ fl/f\ l\ fl$

F_1	398	F	L	Fl/f	l	fl	parental
	370	f	l	fl/f	l	fl	parental
	72	F	L	fl/f	l	fl	CO L–Fl
	67	f	l	Fl/f	l	fl	CO L–Fl
	44	f	L	FL/f	l	fl	CO F–L
	35	F	l	fl/f	l	fl	CO F–L
	9	f	L	fl/f	l	fl	DCO
	5	F	l	Fl/f	l	fl	DCO

L–Fl: $100\%(72 + 67 + 9 + 5)/1000 = 15.3$ m.u.

F–L: $100\%(44 + 35 + 9 + 5)/1000 = 9.3$ m.u.

```
      F              L              Fl
  ____|____9.3 m.u.__|___15.3 m.u.__|____
      |              |              |
```

19. a. and b. The data indicate that the progeny males have a different phenotype than the females. Therefore, all the genes are on the X chromosome. The two most frequent phenotypes in the males indicate the genotypes of the X chromosomes in the female, and the two least frequent phenotypes in the males indicate the gene order. Data from only the males are used to determine map distances. The cross is

P $\quad x\ z\ y^+/x^+\ z^+\ y \times x^+\ z^+\ y^+/Y$

F_1 males

430	$x\ z\ y^+/Y$	parental	
441	$x^+\ z^+\ y/Y$	parental	
39	$x\ z\ y/Y$	CO z–y	
30	$x^+\ z^+\ y^+/Y$	CO z–y	
32	$x^+\ z\ y^+/Y$	CO x–z	
27	$x\ z^+\ y/Y$	CO x–z	
1	$x^+\ z\ y/Y$	DCO	
0	$x\ z^+\ y^+/Y$	DCO	

c. z–y: $100\%(39 + 30 + 1)/1000 = 7.0$ m.u.

x–z: $100\%(32 + 27 + 1)/1000 = 6.0$ m.u.

c.c.   = observed DCO/expected DCO

= $1/[(0.06)(0.07)(1000)] = 0.238$

20. Recall that the parentals are most frequent and the double crossovers are least frequent. Also recall that to determine gene sequence you need to compare parentals with double crossovers: the gene in the center switches with regard to outside markers. Therefore, the gene sequence is

1. *b a c*

2. *b a c*

3. *b a c*

4. *a c b*

5. *a c b*

21. The gene order is *a c b d*.

Recombination between *a* and *c* occurs at a frequency of

$100\%(139 + 3 + 121 + 2)/(669 + 139 + 3 + 121 + 2 + 2{,}280 + 653 + 2{,}215)$

$= 100\%(265/6{,}082) = 4.36\%$.

Recombination between *b* and *c* occurs at a frequency of

$100\%(669 + 3 + 2 + 653)/(669 + 139 + 3 + 121 + 2 + 2{,}280 + 653 + 2{,}215)$

$= 100\%(1{,}327/6{,}082) = 21.82\%$.

Recombination between *b* and *c* occurs at a frequency of

$100\%(8 + 14 + 153 + 141)/(8 + 441 + 90 + 376 + 14 + 153 + 64 + 141)$

$= 100\%(316/1{,}287) = 24.55\%$.

The conflict between the two calculated distances between *b* and *c* is expected because each set of data would not be expected to yield exactly identical results.

Recombination between *b* and *d* occurs at a frequency of

$100\%(8 + 90 + 14 + 64)/(8 + 441 + 90 + 376 + 14 + 153 + 64 + 141)$

$= 100\%(176/1{,}287) = 13.68\%$.

The general map is

```
 a                    c                      b                d
 |      4.4 m.u.       |     21.8–24.6 m.u.   |    13.7 m.u.    |
─┼────────────────────┼──────────────────────┼────────────────┼─
```

22. a. The hypothesis is that the genes are not linked. Therefore, a 1:1:1:1 ratio is expected.

b. χ^2 is

$(54 - 50)^2/50 + (47 - 50)^2/50 + (52 - 50)^2/50 + (47 - 50)^2/50$

$= 0.32 + 0.18 + 0.08 + 0.18 = 0.76$

c. With 3 degrees of freedom, the *p* value is between 0.50 and 0.90.

d. Between 50% and 90% of the time values this extreme from the prediction would be obtained by chance alone.

e. Accept the initial hypothesis.

f. Because the χ^2 value was insignificant, the two genes are assorting independently. The genotypes of all individuals are

P $dp^+ dp^+\ ee \times dpdp\ e^+ e^+$

F_1 $dp^+ dp\ e^+ e$

tester $dpdp\ ee$

progeny	long, ebony	$dp^+ dp\ e\ e$
	long, gray	$dp^+ dp\ e^+ e$
	short, gray	$dp\ dp\ e^+ e$
	short, ebony	$dp\ dp\ e\ e$

23. The cross was $asp\ gal\ rad^+\ aro^+ \times asp^+\ gal^+\ rad\ aro$.

a. The first task is to decide if there is any linkage among the four genes. Arrange the results according to frequencies:

0.136	0.064	0.034	0.016
asp gal + +	*asp gal + rad*	*asp gal aro +*	*asp gal aro rad*
+ + aro rad	*asp + + +*	*asp + aro rad*	*asp + aro +*
asp + + rad	*+ gal aro rad*	*+ gal + +*	*+ gal + rad*
+ gal aro +	*+ + aro +*	*+ + + rad*	*+ + + +*

If all four genes were independently assorting, then all the classes would occur at an equal frequency. This is not observed. Therefore, there is some linkage among these genes.

The two parental classes are present in the highest frequency, as are two other classes: *asp++rad* and *+gal aro+*. Note that *asp* and *gal* are assorting independently of each other, which means that they are unlinked. Also note that *aro* and *rad* are assorting independently of each other and are, therefore, unlinked. Finally, note that *gal* and *rad* are not assorting independently and *asp* and *aro* are not assorting independently. There are two linkage groups, and the original cross was

$gal\ rad^+\ asp\ aro^+ \times gal^+\ rad\ asp^+\ aro$.

Reclassify the data with regard to the *gal-rad* linkage.

PARENTALS		RECOMBINANTS	
0.136	*asp gal + +*	0.064	*asp gal + rad*
0.136	*+ + aro rad*	0.064	*asp + + +*
0.136	*asp + + rad*	0.064	*+ gal aro rad*
0.136	*+ gal aro +*	0.064	*+ + aro +*
0.034	*asp gal aro +*	0.016	*asp gal aro rad*
0.034	*asp + aro rad*	0.016	*asp + aro +*
0.034	*+ gal + +*	0.016	*+ gal + rad*
0.034	*+ + + rad*	0.016	*+ + + +*
0.680		0.320	

The frequency of recombination between *gal* and *rad* is 32%.

Reclassify the data with regard to the *aro-asp* linkage.

PARENTALS		RECOMBINANTS	
0.136	*asp gal + +*	0.034	*asp gal aro +*
0.136	*+ + aro rad*	0.034	*asp + aro rad*
0.136	*asp + + rad*	0.034	*+ gal + +*
0.136	*+ gal aro +*	0.034	*+ + + rad*
0.064	*asp gal + rad*	0.016	*asp gal aro rad*
0.064	*asp + + +*	0.016	*asp + aro +*
0.064	*+ gal aro rad*	0.016	*+ gal + rad*
0.064	*+ + aro +*	0.016	*+ + + +*
0.80		0.200	

The frequency of recombination between *aro* and *asp* is 20%.

b. The map of the four genes is

$$\underset{32\ \text{m.u.}}{\overset{gal \qquad\qquad\qquad rad}{\vdash\!\!-\!\!-\!\!-\!\!-\!\!-\!\!-\!\!-\!\!\dashv}} \qquad \underset{20\ \text{m.u.}}{\overset{aro \qquad\qquad\qquad asp}{\vdash\!\!-\!\!-\!\!-\!\!-\!\!-\!\!-\!\!-\!\!\dashv}}$$

24. a. and b. The wild type is dominant for both traits. First, notice that the F_2 offspring differ with regard to sex. At least one of the genes is X-linked. Second, notice that the two genes are not assorting independently. This means that both genes are on the X chromosome. The cross was

P   *G A/G A* × *g a*/Y

F_1 *G A/g a* females and *G A*/Y males

F_2 females   males

    45% *G A/G A*   45%   *G A*/Y

    45% *g a/G A*   45%   *g a*/Y

    5%   *g A/G A*   5%   *g A*/Y

    5%   *G a/G A*   5%   *G a*/Y

25. a.

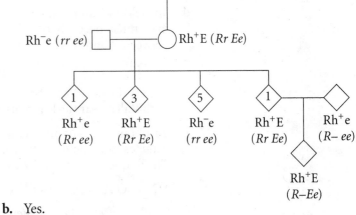

b. Yes.

c. Dominant.

    d. As drawn, the pedigree indicates independent assortment. However, the data also support linkage, with the *R e/r e* individual representing a crossover. The distance between the two genes would be 100%(1/10) = 10 m.u. There is no way to choose between the alternatives without more data.

26. Part a of this problem is solved two ways, once in the standard way, once in a way that emphasizes a more mathematical approach.

    The cross is

P   *PAR/PAR* × par/par

F_1 *PAR*/par × par/par, a three-point testcross

    a. In order to find what proportion will have the Vulcan phenotype for all three characteristics, we must determine the frequency of parentals. Crossing-over occurs 15% of the time between *P* and *A*, which means it does not occur 85% of the time. Crossing-over occurs 20% of the time between *A* and *R*, which means that it does not occur 80% of the time.

    p (no crossover between either gene)

    = *p* (no crossover *P* and *A*)*p*(no crossover *A* and *R*)

    = (0.85)(0.80) = 0.68

        Half the parentals are Vulcan, so the proportion that are completely Vulcan is 1/2(0.68) = 0.34.

    Mathematical method

    Number of parentals = 1 – (single CO individuals – DCO individuals)

    = 1 – {[0.15 + 0.20 – 2(0.15)(0.20)] – (0.15)(0.20)} = 0.68

        Because half the parentals are Earth alleles and half are Vulcan, the frequency of children with all three Vulcan characteristics is 1/2(0.68) = 0.34.

    b. Same as above, 0.34.

    c. To yield Vulcan ears and hearts and Earth adrenals, a crossover must occur in both regions, producing double crossovers. The frequency of Vulcan ears and hearts and Earth adrenals will be half the DCOs, or 1/2(0.15)(0.20) = 0.015.

    d. To yield Vulcan ears and an Earth heart and adrenals, a single crossover must occur between *P* and *A*, and no crossover can occur between *A* and *R*. The frquency will be

    p(CO P–A)*p*(no CO A–R) = (0.15)(0.80) = 0.12

        Of these, 1/2 are *Par* and 1/2 are *pAR*. Therefore, the proportion with Vulcan ears and an Earth heart and adrenals is 0.06.

27. a. To obtain a plant that is abc/abc from selfing of *Abc/aBC*, both gametes must be derived from a single crossover between *A* and *B*. The frequency of the *a b c* gamete is

    1/2 *p*(CO A–B)*p*(no CO B–C) = 1/2 (0.20)(0.70) = 0.07

        Therefore, the frequency of the homozygous plant will be $(0.07)^2 = 0.0049$.

b. The cross is *Abc/aBC* × *abc/abc.*

To calculate the progeny frequencies, note that the parentals are equal to all those that did not experience either a single or a double crossover. Mathematically this can be stated as

parentals = 1 – (single CO *A–B* progeny) – (single CO *B–C* progeny) –

DCO progeny

However, those that experienced only a single crossover are equal in frequency to the probability of a single crossover, minus all those that also experienced the second crossover, or minus DCO. The equation now becomes

parentals = 1 – [p(CO *A–B*) – DCO] – [p(CO *B–C*) – DCO] – DCO

This reduces to

parentals = 1 – p(CO *A–B*) – p(CO *B–C*) + DCO

The progeny are

Abc/abc, parental: 1/2[1 – (CO *A–B*) – (CO *B–C*) + DCO] 1000

= 1/2[1 – 0.20 – 0.30 + (0.20)(0.30)] 1000

= 280

aBC/abc, parental: 1/2[1 – (CO *A–B*) – (CO *B–C*) + DCO] 1000

= 1/2[1 – 0.20 – 0.30 + (0.20)(0.30)] 1000

= 280

ABC/abc, CO *A–B*:   1/2[(CO *A–B*) – DCO] 1000

= 1/2[0.20 – (0.20)(0.30)] 1000 = 70

abc/abc, CO *A–B*:   1/2[(CO *A–B*) – DCO] 1000

= 1/2[0.20 – (0.20)(0.30)] 1000 = 70

Abc/abc, CO *B–C*:   1/2[(CO *B–C*) – DCO] 1000

= 1/2[0.30 – (0.20)(0.30)] 1000 = 120

aBc/abc, CO *B–C*:   1/2[(CO *B–C*) – DCO] 1000

= 1/2[0.30 – (0.20)(0.30)] 1000 = 120

ABc/abc, DCO: 1/2(DCO)(1000)

= 1/2(0.20)(0.30)(1000) = 30

abC/abc, DCO: 1/2(DCO)(1000)

= 1/2(0.20)(0.30)(1000) = 30

c. Interference   = 1 – observed DCO/expected DCO

0.2 = 1 – observed DCO/(0.20)(0.30)

observed DCO = (0.20)(0.30) – (0.20)(0.20)(0.30) = 0.048

Of 1000 progeny, 48 will be DCO. The progeny are

Abc/abc, parental: 1/2[1 – (CO *A–B*) – (CO *B–C*) + DCO] 1000

= 1/2(1 – 0.20 – 0.30 + 0.048)(1000) = 274

aBC/abc, parental: $1/2[1 - (CO\ A\text{–}B) - (CO\ B\text{–}C) + DCO]\ 1000$

$$= 1/2(1 - 0.20 - 0.30 + 0.048)(1000) = 274$$

ABC/abc, CO A–B: $1/2[(CO\ A\text{–}B) - DCO]\ 1000$

$$= 1/2(0.20 - 0.048)(1000) = 76$$

abc/abc, CO A–B: $1/2[(CO\ A\text{–}B) - DCO]\ 1000$

$$= 1/2(0.20 - 0.048)(1000) = 76$$

AbC/abc, CO B–C: $1/2[(CO\ B\text{–}C) - DCO]\ 1000$

$$= 1/2(0.30 - 0.048)(1000) = 126$$

aBc/abc, CO B–C: $1/2[(CO\ B\text{–}C) - DCO]\ 1000$

$$= 1/2(0.30 - 0.048)(1000) = 126$$

ABc/abc, DCO: $1/2(DCO)\ 1000$

$$= 1/2(48) = 24$$

abC/abc, DCO: $1/2(DCO)\ 1000$

$$= 1/2(48) = 24$$

28. Assume there is no linkage. The genotypes should occur with equal frequency, which is the expected value. In each case, there are four genotypes ($n = 4$), which means there are 3 degrees of freedom ($n - 1 = 3$).

$$\chi^2 = \Sigma\ (\text{observed} - \text{expected})^2/\text{expected}$$

(1) $\chi^2 = \dfrac{(310 - 300)^2 + (315 - 300)^2 + (287 - 300)^2 + (288 - 300)^2}{300}$

$$= \frac{(100 + 225 + 169 + 144)}{300} = 2.1266$$

$P > 0.50$, nonsignificant. Therefore, the hypothesis of no linkage cannot be rejected.

(2) $\chi^2 = \dfrac{(36 - 30)^2 + (38 - 30)^2 + (23 - 30)^2 + (23 - 30)^2}{30}$

$$= \frac{36 + 64 + 49 + 49}{30} = 6.6$$

$P > 0.10$, nonsignificant. The hypothesis of no linkage cannot be rejected.

(3) $\chi^2 = \dfrac{(360 - 300)^2 + (380 - 300)^2 + (230 - 300)^2 + (230 - 300)^2}{300}$

$$= \frac{3600 + 6400 + 4900 + 4900}{300} = 66.0$$

$P < 0.005$, significant. The hypothesis of no linkage must be rejected.

$$(4)\ \chi^2 = \frac{(74-60)^2 + (72-60)^2 + (50-60)^2 + (44-60)^2}{60}$$

$$= \frac{196 + 144 + 100 + 256}{60} = 11.60$$

$P < 0.01$, significant. The hypothesis of no linkage must be rejected.

29. The data approximate a 9:3:3:1 ratio, which suggests two genes. Let $A =$ resistance to rust 24, $a =$ susceptibility to rust 24, $B =$ resistance to rust 22, $b =$ susceptibility to rust 22.

    a. P $AA\ bb$ (770B) $\times aa\ BB$ (Bombay)

    F_1 $Aa\ Bb \times Aa\ Bb$

    F_2 184 $A–B–$

    63    $A–bb$

    58    $aa\ B–$

    $\underline{\ \ 15\ \ }$ $aa\ bb$

    320

    b. Expect:

    $(320)(9/16) = 180$ $A–B–$

    $(320)(3/16) = 60$ $A–bb$

    $(320)(3/16) = 60$ $aa\ B–$

    $(320)(1/16) = 20$ $aa\ bb$

    $$\chi^2 = \frac{(184-180)^2}{180} + \frac{(63-60)^2}{60} + \frac{(58-60)^2}{60} + \frac{(15-20)^2}{20}$$

    $$= \frac{16}{180} + \frac{9}{60} + \frac{4}{60} + \frac{25}{20} = 1.555$$

    P (3 df) > 0.58, nonsignificant. A P value is the probability that the result would be observed by chance alone. Therefore, the hypothesis of two independently assorting genes cannot be rejected.

30. **a.** Both disorders must be recessive to yield the patterns of inheritance that are observed. Notice that only males are affected, strongly suggesting X linkage for both disorders. In the first pedigree there is a 100% correlation between the presence or absence of both disorders, indicating close linkage. In the second pedigree, the presence and absence of both disorders are inversely correlated, again indicating linkage. In the first pedigree, the two defective alleles must be cis within the heterozygous females to show 100% linkage in the affected males, while in the second pedigree the two defective alleles must be trans within the heterozygous females.

b. and c. Let *a* stand for the allele giving rise to steroid sulfatase deficiency (vertical bar) and *b* stand for the allele giving rise to
ornithine transcarbamylase deficiency (horizontal bar). Crossing-
over cannot be detected without attaching genotypes to the pedigrees. When this is done, it can be seen that crossing-over need not
occur in either of the pedigrees to give rise to the observations.

First Pedigree

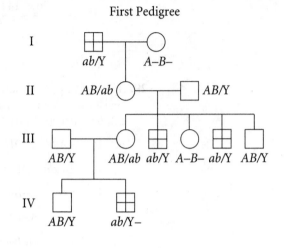

Second Pedigree

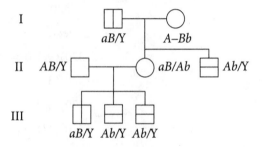

31. a. Blue sclerotics appears to be an autosomal dominant disorder.
Hemophilia appears to be an X-linked recessive disorder.

b. If the individuals in the pedigree are numbered as generations I
through IV and the individuals in each generation are numbered
clockwise, starting from the top right-hand portion of the pedigree,
their genotypes are

I: *bb Hh, Bb HY*

II: *Bb HY, Bb HY, bb HY, Bb H–, bb HY, Bb Hh, Bb H–, bb H–*

III: *bb H–, Bb H–, bb hY, bb HY, Bb HY, Bb H–, Bb HY, Bb hY, Bb
H–, bb HY, Bb H– bb HY, Bb H–, Bb HY, Bb hY, bb HY, bb
HY, bb H–, bb HY, bb HY, Bb H–, Bb HY, Bb hY*

IV: *bb H–, Bb H–, Bb H–, bb Hh, bb Hh, bb HY, bb HH, bb HY, bb
Hh, bb H–, bb H–, bb HY, bb HY, bb H–, bb HY, bb HY, Bb HY,
bb HY, bb HH, bb HY, bb HY, bb HH, bb H–, bb H–, bb H–, bb
HY, bb HY, bb HY, bb Hh, Bb H–, Bb HY, bb HY, Bb HY, bb H–*

c. There is no evidence of linkage between these two disorders.
Because of the modes of inheritance for these two genes, no linkage
would be expected.

d. The two genes exhibit independent assortment.

e. No individual could be considered intrachromosomally recombinant. However, a number show interchromosomal recombination: all individuals in generation III that have both disorders.

32. a. Note that only males are affected by both disorders. This suggests that both are X-linked recessive disorders.

b. Individual II-2 must have inherited both disorders in a trans configuration. Therefore, individual III-2 represents recombination.

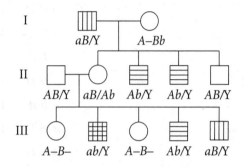

c. Because the genes are on the X chromosome, this is intrachromosomal recombination. However, the progeny size is too small to give a reliable estimate of crossing-over.

33. If *h* = hemophilia and *b* = colorblindness, the genotypes for individuals in the pedigree can be written as

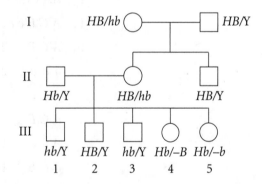

The mother of the two women in question would produce the following gametes:

0.45   *H B*   0.05   *H b*

0.45   *h b*   0.05   *h B*

Woman III-4 can be either *H b/H B* (0.45 chance) or *H b/h B* (0.05 chance), because she received *B* from her mother. If she is *H b/h B* [0.05/(0.45 + 0.05) = 0.10 chance], she will produce the parental and recombinant gametes with the same probabilities as her mother. Thus, her child has a 45 percent chance of receiving *h B*, a 5 percent chance of receiving *h b*, and a 50 percent chance of receiving a Y from his father. The probability that her child will be a hemophiliac son is (0.1)(0.50)(0.5) = 0.025 = 2.5 percent.

Woman III-5 can be either *H b/H b* (0.05 chance) or *H b/h b* (0.45

chance), because she received *b* from her mother. If she is *H b/h b* [0.45/(0.45 + 0.05) = 0.90 chance], she has a 50 percent chance of passing *h* to her child, and there is a 50 percent chance that the child will be male. The probability that she will have a son with hemophilia is (0.9)(0.5)(0.5) = 0.225 = 22.5 percent.

34. a. Cross 1 reduces to

P   *AA BB DD* × *aa bb dd*

F_1 *A B D/a b d* × *a b d/a b d*, correct order

F_2 *A B D*   316   parental

   a b d   314   parental

   A B d   31   CO B–D

   a b D   39   CO B–D

   A b d   130   CO A–B

   a B D   140   CO A–B

   A b D   17   DCO

   a B d   13   DCO

A–B: 100%(130 + 140 + 17 + 13)/1000 = 30 m.u.

B–D: 100%(31 + 39 + 17 + 13)/1000 = 10 m.u.

Cross 2 reduces to

P   *AA CC EE* × *aa cc ee*

F_1 *A C E/a c e* × *a c e/a c e*, correct order

F_2 *A C E*   243   parental

   a c e   237   parental

   A c e   62   CO A–C

   a C E   58   CO A–C

   A C e   155   CO C–E

   a c E   165   CO C–E

   a C e   46   DCO

   A c E   34   DCO

A–C: 100% (62 + 58 + 46 + 34)/1000 = 20 m.u.

C–E: 100% (155 + 165 + 46 + 34)/1000 = 40 m.u.

The map can be put together in one way that accommodates all the data:

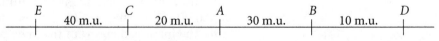

E		C		A		B		D
	40 m.u.		20 m.u.		30 m.u.		10 m.u.	

b. Interference (I) = 1 – [(observed DCO)/(expected DCO)]

For cross 1: I = 1 – {30/[(0.30)(0.10)(1000)]}

$$= 1 - 1 = 0, \text{ no interference.}$$

For cross 2: $\text{I} = 1 - \{80/[(0.20)(0.40)(1000)]\}$

$$= 1 - 1 = 0, \text{ no interference.}$$

35. a–c. The cross is *GG yy* × *gg YY*. If the two are unlinked, a 9:3:3:1 ratio should be observed in the progeny. The proper way to decide on linkage is to conduct a chi-square analysis of the data. The chi-square value indicates that it is highly unlikely that the two genes are unlinked.

In the absence of chi-square analysis, the best way to decide whether the data indicate linkage is to realize that a 9:3:3:1 ratio would predict 562.5:187.5:187.5:62.5 and that the data are far from the predicted ratio. Because the observations do not match the expected, assuming independent assortment, then linkage is a viable alternative.

Assuming linkage, the cross becomes

P   *Gy/Gy* × *gY/gY*

F$_1$ *Gy/gY* × *Gy/gY*

F$_2$ data as in problem

In this situation, it is necessary to realize that the phenotype yellow, glabrous represents the double homozygous recessive. To obtain a double homozygous recessive, both parents must donate a recombinant gamete. In other words, the frequency of

$$(1/2 \text{ recombinants})^2 \times 1000 = 15$$

Let 1/2 the recombinants = *x*, and divide each side of the equation by 1000. Then the equation becomes

$$x^2 = 0.015$$

$$x = 0.122$$

If 1/2 the recombinants is 0.122, then the recombinants total 0.24, or 24% of the progeny. The map distance between the two genes is 24 m.u.

36. The three crosses are

Cross 1:   P   *LH/LH* × *lh/lh*

F$_1$ *LH/lh*

Cross 2:   P   *Lh/Lh* × *lH/lH*

F$_1$ *Lh/lH*

Cross 3:   P   *LH/lh* × *Lh/lH*

a.   Each parent must donate *lh*. Therefore,

$p(lh/lh) = p(lh \text{ from first parent})p(lh \text{ from second parent})$

$= p(1/2 \text{ parentals})p(1/2 \text{ recombinants}).$

Because there are 16 map units between the two genes, 16% of the progeny are recombinant and 84% are parental. There are two parental and two recombinant types, and only half of each are *lh*. Therefore, the probability of *lh/lh* = 1/2(0.84) 1/2(0.16) = 0.0336.

b. There are two ways to obtain each type of gamete *Lh and lh*. Therefore,

$$p(Lh/lh) = p(lh \text{ from first parent})p(Lh \text{ from second parent}) +$$

$$p(Lh \text{ from first parent})p(lh \text{ from second parent})$$

$$= 1/2(0.84)1/2(0.84) + 1/2(0.16)1/2(0.16)$$

$$= 0.176 + 0.006 = 0.182$$

37. a. and b. Again, the best way to determine whether there is linkage is through chi-square analysis, which indicates that it is highly unlikely that the three genes assort independently. To determine linkage by simple inspection, look at gene pairs. Because this is a testcross, independent assortment predicts a 1:1:1:1 ratio.

Comparing shrunken and white, the frequency of all possibilities is

+ +   113 + 4

s wh   116 + 2

+ wh   2708 + 626

s   +   2538 + 601

There is not independent assortment between shrunken and white, which means that there is linkage.

Comparing shrunken and waxy, the frequencies are

+ +   626 + 4

s wa   601 + 2

+ wa   2708 + 113

s   +   2538 + 116

There is not independent assortment between shrunken and waxy, which means that there is linkage.

Comparing white and waxy, the frequencies are

 + +   2538 + 4

wh wa   2708 + 2

wh +   626 + 116

 + wa   601 + 113

There is not independent assortment between waxy and white, which means that there is linkage.

Because all three genes are linked, the parentals are clearly + s +/wh + wa and wh s wa/wh s wa (compare most frequent, parentals, to least frequent, double crossovers, to obtain the gene order). The cross can be written as

P + s +/wh + wa × wh s wa/wh s wa

F_1 as in problem

Crossovers between white and shrunken and shrunken and waxy are

113   601

116   626

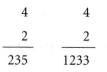

$$\frac{\begin{array}{c}4\\2\end{array}}{235}\qquad \frac{\begin{array}{c}4\\2\end{array}}{1233}$$

Dividing by the total number of progeny and multiplying by 100% yields the following map:

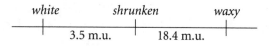

white   shrunken   waxy

3.5 m.u.   18.4 m.u.

c. Interference

= 1 – observed double crossovers/expected double crossovers

= 1 – 6/(0.035)(0.184)(6,708)

= 1 – 0.139

= 0.86

38. The cross is *Ab/aB* × *Ab/aB*. The *A* and *B* genes are 10 m.u. apart. Therefore, the gametes should occur in the following frequencies:

0.45 *Ab*

0.45 *aB*

0.05 *ab*

0.05 *AB*

Construct a Punnett square, and sum the values for each progeny phenotype. These values are

0.5025 *A–B–*

0.0025 *aabb*

0.2475 *A–bb*

0.2475 *aaB–*

The *z*-value is calculated using the following formula:

$z = (A–B–)(aabb)/(A–bb)(aaB–)$

Substituting the values from the Punnett square, this becomes

$z = (0.5025)(0.0025)/(0.2475)(0.2475) = 0.020$

Using the table of *z*-values, 0.020 is equivalent to 10 m.u.

In a three-point testcross, the gene order is determined by comparing the parental-type progeny, which will be most frequent, with the double-crossover progeny, which will be least frequent. The gene that has switched with respect to the other two genes in this comparison must be located between them:

parentals: *D a R, d A r*

double crossovers: *d a R, D A r*

► TIPS ON PROBLEM
SOLVING

The order is $A\ D\ R$; the parental chromosomes are $a\ D\ R/A\ d\ r$ (see Problem 10).

In three-point testcrosses, if the data indicate that the progeny males have a different phenotype than the females, the genes are on the X chromosome. The two most frequent phenotypes in the males indicate the genotypes of the X chromosomes in the female, and the two least frequent phenotypes in the males indicate the gene order (see Problem 19). To estimate the frequency of crossing-over, utilize the data from the males only (see Problem 19).

If, in a three-point cross, the data indicate that the genes are autosomal, it is frequently necessary to focus on the homozygous recessive progeny exclusively (see Problem 15). Also, if the cross is not a testcross, it may be necessary to use the homozygote recessive frequency to estimate the frequency of crossing-over (see Problem 13).

6

Linkage II: Special Eukaryotic Chromosome Mapping Techniques

▶ IMPORTANT TERMS
AND CONCEPTS

Multiple crossovers in larger intervals result in an underestimate of map distance between two genes. The relationship between real map distance and the **recombinant frequency**, RF, is not linear. The **mapping function** provides a closer approximation of the real relationship.

The **Poisson distribution** describes the frequency of 0, 1, 2, . . . n crossovers, given the average number of crossovers. The general expression is

$$f(i) = e^{-m}m^i/i!$$

where e is the base of natural logarithms, m is the mean number of events, and i is an integer that ranges from 0 to n.

Meiotic events in diploid organisms are studied by means of an estimation based on the observation of random meiotic products. An indirect calculation of the recombinant frequency is conducted using the Poisson distribution. In contrast, the four products of a single meiotic event can be studied directly through **tetrad analysis** in certain haploid organisms. This allows for a direct calculation of the recombinant frequency, again using the Poisson distribution.

Linear tetrad analysis is frequently conducted using *Neurospora*. Recall that meiosis, followed immediately by mitosis, results in eight meiotic products.

A monohybrid cross in which no crossing-over occurs between the gene studied and the centromere results in **first-division segregation** (M_I) of alleles and a 4:4 pattern of the progeny. (See the first figure on page 128.)

Any deviation from this pattern is the result of **second-division segregation** (M_{II}) of alleles, and signals crossing-over. One example of a second-division segregation pattern is shown in the second figure on page 128.

Because each crossover produces two recombinant and two nonrecombinant products, the frequency of recombinant asci must be adjusted to reflect the true RF value. The distance of a locus from the centromere is therefore

$$RF = \frac{1/2 \text{ number of recombinant asci}}{\text{total number of asci}}$$

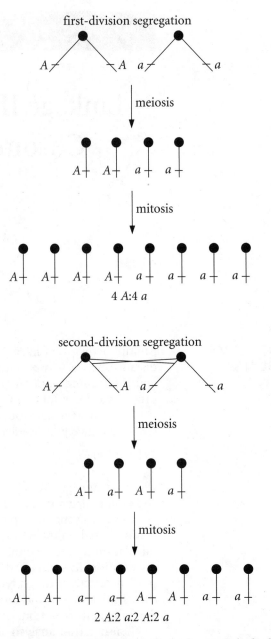

Dihybrid crosses may involve two unlinked or two linked genes. As will become apparent below, the relative frequency of asci types allows the determination of linkage or nonlinkage.

First-division segregation of two unlinked loci results in two types of equally frequent 4:4 patterns, **parental ditype**, PD, and **nonparental ditype**, NPD. Parental ditype means that there are two types of spores and each spore is parental in allelic content. Nonparental ditype means that each spore is recombinant and there are two types of spores. For example,

Parentals: $A\,B \times a\,b$

PD: $A\,B,\ A\,B,\ A\,B,\ A\,B,\ a\,b, a\,b,\ a\,b,\ a\,b$

            $4\,A\,B{:}4\,a\,b$

NPD: $A\,b,\ A\,b,\ A\,b,\ A\,b,\ a\,B,\ a\,B,\ a\,B,\ a\,B$

            $4\,A\,b{:}4\,a\,B$

Any deviation from the two 4:4 patterns for two unlinked loci results from second-division segregation, signaling crossing-over between a gene and the centromere. One example is *A b, A b, A B, A B, a b, a b, a B, a B,* or 2:2:2:2. The asci that contain the allelic combinations *A B, a b, A b,* and *a B* are called **tetratypes**, T, indicating that there are four different types of meiotic products. A tetratype in unlinked loci indicates that crossing-over has occurred between a gene and its centromere. Whether or not crossing-over occurs, the number of parental ditypes equals the number of nonparental ditypes, and the NPD/T ratio is between 1/4 and infinity when the two genes are not linked. The RF between a gene and its centromere is calculated separately for each gene, using the formula

$$RF = \frac{1/2 \text{ number of recombinant asci}}{\text{total number of asci}}$$

If two loci are linked, the number of parental ditypes significantly exceeds the number of nonparental ditypes. The NPD/T ratio is between 0 and 1/4. No crossing-over leads to a 4:4 pattern, while a single crossover between a gene and its centromere results in a tetratype. Two crossovers result in a parental ditype, a tetratype, or a nonparental ditype, depending on which strands are involved. The RF between a gene and its centromere is calculated separately for each gene, using the formula

$$RF = \frac{1/2 \text{ number of recombinant asci}}{\text{total number of asci}}$$

The RF between the two linked genes is calculated using the following formula:

$$RF = \frac{100\% \; [1/2T + NPD]}{\text{total number of asci}}$$

Because two crossovers can lead to a parental ditype, this formula results in an underestimate of the RF.

These relationships are summarized below:

UNLINKED	LINKED
PD = NPD	PD >> NPD
NPD/T = 1/4 to infinity	NPD/T = 0 to 1/4
RF = 50%	RF < 50%

Mitotic crossing-over leading to recombination and **mitotic nondisjunction** can result in **mitotic segregation** of alleles following mitosis. Mitotic segregation has been extensively studied in fungal cells. A **heterokaryon** generated by hyphal fusion in *Aspergillus* consists of two nuclei in a common cytoplasm. Phenotypic variegation indicates mitotic segregation in the heterokaryons.

Human chromosomes have been mapped using **somatic cell hybridization.** Fusion of two nuclei, one human and the other frequently mouse, results in a gradual and random loss of human chromosomes. The presence or absence of human enzymes is correlated with the presence or absence of human chromosomes in several cell lines.

► CONCEPT MAP 6-1:
   MAPPING

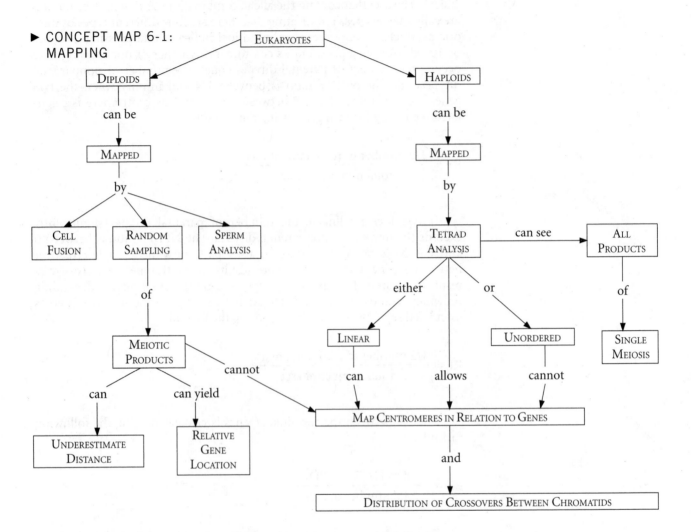

Questions for Concept Map 6-1

1. What are three ways eukaryotes can be mapped?

2. What is meant by *random sampling of meiotic products?*

3. If meiotic products are only sampled, is the sample always a correct reflection of what occurred?

4. With random sampling of meiotic products, can a double crossover be detected between two genes if there are not markers between them?

5. If double crossovers are not detected with diploids, how does this affect the map distance between the two genes?

6. What is the formula for calculating map distance between two genes in diploids?

7. What information can be obtained for two linked genes in diploids?

8. What is tetrad analysis?

9. If one is looking at a tetrad, why are there eight ascospores?

10. What is the biological difference between linear and unordered tetrad analysis?

11. What types of tetrads are seen?

12. What does a PD tetrad indicate?

13. What does an NPD tetrad indicate?

14. What does a T tetrad indicate?

15. What is the formula for calculating map distance between two genes that form linear tetrads?

16. What does it mean to *map centromeres in relation to genes*?

17. If haploids are haploid, how is it that recombination can occur within them?

18. What is cell fusion?

19. What is sperm analysis?

20. Does first-division segregation occur in diploids?

Answers to Concept Map 6-1 Questions

1. Three ways eukaryotes can be mapped are random sampling of meiotic products, cell fusion, and sperm analysis.

2. *Random sampling of meiotic products* means that all outcomes of a single meiosis cannot be detected. In females, only one of every four meiotic products becomes an egg. In males, all meiotic products from thousands of meioses are pooled and released at one time. One is never able to look at four outcomes and know that they are from the same progenitor cell.

3. Because meiotic products are sampled, by chance alone the sample can contain a result that does not accurately reflect what occurred, just as a family can have six boys in a row by chance alone.

4. With random sampling of meiotic products, a double crossover cannot be detected between two genes if there are not markers between them.

5. If double crossovers are not detected with diploids, this results in an underestimate of the map distance between the two genes.

6. The formula for calculating map distance between two genes in diploids is $100\% \times$ number of recombinants/total number of progeny.

7. Only the map distance can be obtained for two linked genes in diploids.

8. Tetrad analysis is the meiotic mapping of haploid organisms that form tetrads. This allows one to see the complete set of outcomes from a single meiotic cell.

9. As part of the life cycle of organisms that form tetrads, immediately following meiosis there is one mitotic division, yielding eight products, or ascospores.

10. The biological difference between linear and unordered tetrad analysis

is that in organisms that form linear tetrads, meiosis is constrained physically so that the alignment of chromosomes during both meiotic divisions is set. For those organisms that form unordered tetrads, meiosis is not physically constrained.

11. Three types of tetrads are seen: parental ditypes (PD), nonparental ditypes (NPD), and tetratypes (T).

12. A PD tetrad indicates that no crossover occurred.

13. An NPD tetrad indicates that two crossovers occurred.

14. A T tetrad indicates that one crossover occurred.

15. The formula for calculating map distance between two genes that form linear tetrads is 100%(1/2T + NPD)/total number of asci.

16. Literally the map distance between a gene and its centromere is determined.

17. Recombination can occur within haploids because they have a transiently diploid stage.

18. Cell fusion is the joining of two mitotic cells, usually because of experimental manipulation. Such studies for mapping purposes fuse cells from two species. Chromosome elimination then allows mapping.

19. Sperm analysis can be done at the single-cell level. It allows the detection of DNA variants.

20. First-division segregation occurs in diploids.

▶ SOLUTIONS TO PROBLEMS

Be sure that you have thoroughly read the entire chapter before you attempt any of the problems.

1. a. +, +, *al–2, al–2*, +, +, *al–2, al–2*

    al–2, al–2, +, +, *al–2, al–2*, +, +

   b. The 8 percent value can be used to calculate the distance between the gene and the centromere. That distance is 1/2 the percentage of second-division segregation, or 4 percent.

2. a. *arg–6 al–2* and + +

   b. *arg–6* +, *arg–6 al–2*, + +, and + *al–2*

   c. *arg–6* + and + *al–2*

3. The formula for this problem is $f(i) = e^{-m}m^i/i!$ where $m = 2$ and $i = 0, 1,$ and 2.

   a. $f(0) = e^{-2}2^0/0! = e^{-2} = 0.135$

   b. $f(1) = e^{-2}2^1/1! = e^{-2}(2) = 0.27$

   c. $f(2) = e^{-2}2^2/2! = e^{-2}(2) = 0.27$

4. a. A region 1 crossover will yield the following chromosomes:

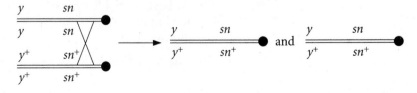

Mitosis can yield two equally likely results: $y\ sn/y\ sn$ and $y^+\ sn^+/y^+$ sn^+ or both daughter cells $y\ sn/y^+\ sn^+$. In the first case, one yellow singed spot will exist in a brown unsinged fly. In the second case, the resulting cells will be wild type.

b. A region 2 crossover will yield the following chromosomes:

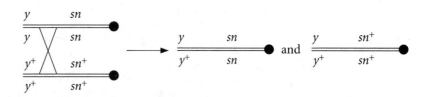

Mitosis can yield the following combinations:

$y\ sn/y^+\ sn^+$ and $y^+\ sn/y\ sn^+$ wild type

or

$y\ sn/\ y\ sn^+$ and $y^+\ sn/y^+\ sn^+$ yellow, unsinged spot

5. This problem is analogous to meiosis in organisms that form linear tetrads. Let red $= R$ and blue $= r$. Then meiosis is occurring in an organism that is Rr (but there are 4 "alleles" because the "chromosomes" are at the "two-chromatids-per-chromosome" stage), and the "alleles" are at loci far from the centromere. The patterns, their frequencies, and the division of segregation are given below. Notice that the probabilities change as each ball/allele is selected. This occurs when there is sampling without replacement.

$1/2\ R$
- $1/3\ R - 1/1\ r - 1/1\ r\ =\ 1/6\ RRrr$ first division
- $2/3\ r$
  - $1/2\ R - 1/1\ r\ =\ 1/6\ RrRr$ second division
  - $1/2\ r - 1/1\ R\ =\ 1/6\ RrrR$ second division

$1/2\ r$
- $1/3\ r - 1/1\ R - 1/1\ R\ =\ 1/6\ rrRR$ first division
- $2/3\ R$
  - $1/2\ R - 1/1\ r\ =\ 1/6\ rRRr$ second division
  - $1/2\ r - 1/1\ R\ =\ 1/6\ rRrR$ second division

These results indicate one-third first-division segregation and two-thirds second-division segregation.

6. a. The formula is RF $= 1/2(1 - e^{-m})$, where recombination $= 0.2$. Therefore, $e^{-m} = 1 - 0.4 = 0.6$, and $m = 0.51$ (from e^{-m} tables). Because an m value of $1.0 = 50$ map units (m.u.), 0.51×50 m.u. $= 25.5$ m.u.

b. The problem is the interpretation of 45 m.u. That could represent two loci approximately 45 m.u. apart, or it could represent two unlinked loci. A χ^2 test is needed for decision making. Hypothesis: no linkage, resulting in a 1:1:1:1 ratio.

$$\chi^2 = \frac{(58-50)^2 + (52-50)^2 + (47-50)^2 + (43-50)^2}{50}$$

$$= \frac{(64+4+9+49)}{50} = 2.52$$

With 3 degrees of freedom, the probability is greater than 10% that the genes are not linked. Therefore, the hypothesis of no linkage can be accepted.

7. Rewrite the column headings to note what is missing from the media, and then count the different types of patterns. Growth will occur if the wild-type gene is present or if the medium supplies whatever gene product is missing:

–LEU	–NIC	–AD	–ARG	NUMBER
+	+	–	–	6
–	–	+	+	4
–	+	–	+	5
+	–	+	–	4
+	–	–	–	1
+	–	–	+	0
–	+	+	–	0

The first four categories have a 1:1:1:1 pattern, indicating independent assortment of two chromosomes. Furthermore, the pattern of growth on two of the media types and no growth on two of the media types indicates that two genes are located on each of the two chromosomes, rather than one gene on one chromosome and three genes on the second.

The two missing categories indicate which genes are linked.

Remember that the cross is

$arg^- \ ad^- \ nic^+ \ leu^+ \times arg^+ \ ad^+ \ nic^- \ leu^-$.

Growth is not seen in the (–Nic and –Ad) or (–Leu and –Arg) media simultaneously, which means that *nic* is linked to *ad* and *leu* is linked to *arg*. Both *nic* and *ad* assort independently with *leu* and *arg*, suggesting that *nic* is not linked to *leu* and *arg* and that *ad* is not linked to *leu* and *arg*.

a. The parents were

$$\underline{ad^- \ nic^+} \ \underline{leu^+ \ arg^-} \times \underline{ad^+ \ nic^-} \ \underline{leu^- \ arg^+}$$

b. Culture 16 resulted from a crossover between *ad* and *nic*. The reciprocal did not show up in the small sample.

8. Recall that larger RF values are less accurate measures of map distance than smaller RF values. Also recall that multiple crossovers reduce the calculated map distance. Therefore, an RF of 36% is an underestimate of the true map distance between waxy and shrunken.

The formula that is needed to calculate the proportion of meiocytes with different numbers of crossovers is $f(i) = e^{-m}m^i/i!$, where $i = 0, 1$, and 2, e is the base of natural logarithms, and m is the mean number of crossovers.

To calculate m, use the formula RF $= (1 - e^{-m})/2$.

$0.36 = (1 - e^{-m})/2$

$e^{-m} = 0.28$

$m = 1.27$

a. The probability of no crossover is

$f(0) = e^{-1.27}(1.27)^0/0! = e^{-1.27} = 0.28$

b. The probability of one crossover is

$f(1) = e^{-1.27}(1.27)^1/1! = 0.35$

c. The probability of two crossovers is

$f(2) = e^{-1.27}(1.27)^2/2! = 0.22$

d. The probability of at least one crossover is

p (at least 1 CO) $= 1 - p$ (0 CO) $= 1 - 0.28 = 0.72$

9. To work these problems, it is first necessary to draw the chromosomes and explore the consequences of crossing-over in different regions. The genes can be assumed to be coupled or in repulsion.

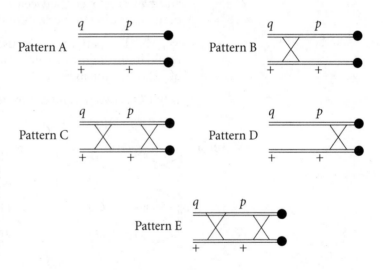

With no crossovers, the pattern is $p\ q$, $p\ q$, $+\ +$, $+\ +$, a parental ditype. Both genes show an M_I pattern.

With one CO between q and the centromere, the pattern is $p\ q$, $p\ +$, $+\ q$, $+\ +$. This is a T, with gene p showing M_I and gene q showing M_{II} segregation.

With one CO between p and the centromere and one CO between p and q, the pattern is $p\ q$, $+\ q$, $p\ +$, $+\ +$. This is a T, with M_{II} for p and M_I for q. (Note: a 4-strand double would also give the same result.)

With one CO between p and the centromere, the pattern is $p\ q$, $++$, This is a PD, with both genes showing M_{II} segregation.

The pattern is $p\ +$, $+\ +$, $p\ q$, $+\ q$. This is a T, with both genes showing M_{II} segregation.

a. $M_I M_I$, PD is pattern A, no crossovers. The probability is

p(no CO from centromere to p)p(no CO from centromere to q)

$= (0.88)(0.80) = 0.704$

b. $M_I M_I$, NPD requires two crossovers between q and the centromere, which cannot occur according to the rules of the problem. The probability is 0.

c. $M_I M_{II}$, T is pattern B. The probability is

p(no CO from p to centromere)p(CO between q and centromere)

$= (0.88)(0.2) = 0.176$

d. $M_{II} M_I$, T is pattern C. Because there are two ways to achieve this result, in the following calculation there is an adjustment by a factor of 1/2. The probability is

$1/2p$(CO between p and centromere)p(CO between q and centromere)

$= (0.5)(0.12)(0.2) = 0.012$

e. $M_{II} M_{II}$, PD is pattern D. The probability is

p(CO between p and centromere)p(no CO from q to centromere)

$= (0.12)(0.8) = 0.096$

f. $M_{II} M_{II}$, NPD requires one crossover between p and the centromere and two crossovers between q and the centromere, which cannot occur according to the rules of the problem. The probability is 0.

g. $M_{II} M_{II}$, T is pattern E. Because there are two ways to achieve this result, in the following calculation there is an adjustment by a factor of 1/2. The probability is

$1/2p$(CO between p and centromere)p(CO between q and centromere)

$= (0.5)(0.12)(0.2) = 0.012$

10. a.

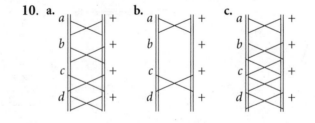

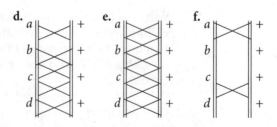

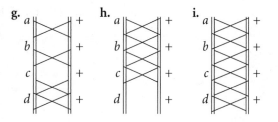

11. **Unpacking the Problem**

a. Fungi are generally haploid.

b. There are four pairs, or eight ascospores, in each ascus. One member of each pair is presented in the data.

c. A mating type in fungi is analogous to sex in humans, in that the mating types of two organisms must differ in order to have a mating that produces progeny. Mating type is determined experimentally simply by seeing if progeny result from specific crosses.

d. The mating types *A* and *a* do not indicate dominance and recessiveness. They simply symbolize the mating-type difference.

e. *Arg–1* indicates that the organism requires arginine for growth. Testing for the genotype involves isolating nutritional mutants and then seeing if arginine supplementation will allow for growth.

f. *Arg–1*$^+$ indicates that the organism is wild-type and does not require supplemental arginine for growth.

g. *Wild type* refers to the dominant form of an organism in a population, whether it be in the laboratory or in the "wild."

h. *Mutant* means that, for the trait being studied, an organism differs from the wild type.

i. The actual function of the alleles in this problem does not matter in solving the problem.

j. *Linear tetrad analysis* refers to the fact that the ascospores in each ascus are in a line within the ascus that reflects the order in which the two meiotic divisions occurred to produce them. By tracking traits and correlating them with position, it is possible to detect crossing-over events that occurred at the tetrad (four-strand, homologous pairing) stage prior to the two meiotic divisions.

k. Linear tetrad analysis allows for the mapping of centromeres in relation to genes, which cannot be done with unordered tetrad analysis.

l. A cross is made in *Neurospora* by placing the two organisms in the same test tube or Petri dish and allowing them to grow. Gametes develop and fertilization, followed by meiosis, mitosis, and ascus formation, occurs. The asci are isolated, and the ascospores are dissected out of them with the aid of a microscope. The ascus has an octad, or eight spores, within it, and the spores are arranged in four (tetrad) pairs.

m. Meiosis occurs immediately following fertilization in *Neurospora*.

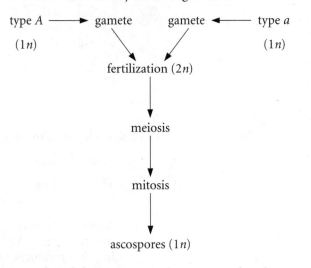

n. Meiosis produced the ascospores that were analyzed.

o. The cross is $A\ arg–1 \times a\ arg–1^{+}$.

p. Although there are eight ascospores, they occur in pairs. Each pair represents one chromatid of the originally paired chromosomes. By convention, both members of a pair are represented by a single genotype.

q. The seven classes represent the seven types of outcomes. The specific outcomes can be classified as follows:

Class	1	2	3	4	5	6	7
Outcome	PD	NPD	T	T	PD	NPD	T
A/a	I	I	I	II	II	II	II
+/arg	I	I	II	I	II	II	II

where PD = parental ditype, NPD = nonparental ditype, T = tetratype, I = first-division segregation, and II = second-division segregation.

Other classes can be detected, but they indicate the same underlying process. For example, the following three asci are equivalent.

1	2	3
A arg	*A +*	*A +*
A +	*A arg*	*A arg*
a arg	*a arg*	*a +*
a +	*a +*	*a arg*

In the first ascus, a crossover occurred between chromatids 2 and 3, while in the second ascus it occurred between chromatids 1 and 3, and in the third ascus the crossover was between chromatids 1 and 4. A fourth equivalent ascus would contain a crossover between chromatids 2 and 3. All four indicate a crossover between the second gene and its centromere and all are tetratypes.

r. This is exemplified in the answer to q above.

s. The class is identical with class 1 in the problem, but inverted.

t. *Linkage arrangement* refers to the relative positions of the two genes and the centromere along the length of the chromosome.

u. A genetic interval refers to the region between two loci, whose size is measured in map units.

v. It is not known whether the two loci are on separate chromosomes or are on the same chromosome.

w. Recall that there are eight ascospores per ascus. By inspection, the frequency of recombinant A arg–1^+ ascospores is $4(125) + 2(100) + 2(36) + 4(4) + 2(6) = 800$. There is also the reciprocal recombinant genotype a arg–1.

x. Class 1 is parental; class 2 is a nonparental ditype. Because they occur at equal frequencies, the two genes are not linked.

Solution to the Problem

a. The cross is A $arg \times a$ arg^+. Use the classification of asci in part q above. First decide if the two genes are linked by using the formula PD>>NPD, when the genes are linked, while PD = NPD when they are not linked. PD = 127 + 2 = 129 and NPD = 125 + 4 = 129, which means that the two genes are not linked. Alternatively,

$$RF = 100\%(1/2\ T + NPD)/\text{total asci}$$

$$= 100\%[(1/2)(100 + 36 + 6) + (125 + 4)]/400 = 50\%.$$

Next calculate the distance between each gene and its centromere using the formula RF = 100%(1/2 number of recombinant asci)/(total number of asci).

$$A-\text{centromere} = 100\%(1/2)(36 + 2 + 4 + 6)/400$$

$$= 100\%(24/400) = 6\ \text{m.u.}$$

$$arg^+ - \text{centromere} = 100\%(1/2)(100 + 2 + 4 + 6)/400$$

$$= 100\%(14/400) = 14\ \text{m.u.}$$

b. Class 6 can be obtained if a single crossover occurred between chromatids 2 and 3 between each gene and its centromere.

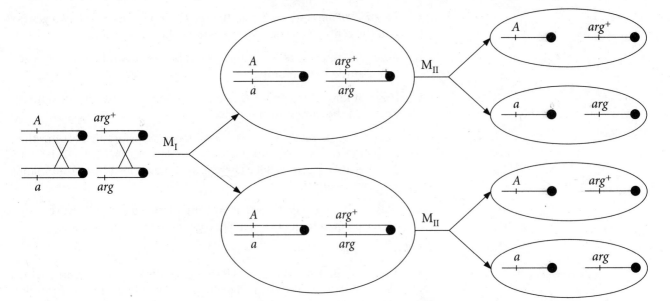

12. a. The cross is $t\ m^+ \times t^+\ m$. The six asci types can be classified as follows:

Class	PD	T	PD	T	NPD	T
t	I	I	II	II	I	II
m	I	II	II	I	I	II
Number	260	76	4	54	1	5

where PD = parental ditype, NPD = nonparental ditype, T = tetratype, I = first-division segregation, and II = second division segregation.

First decide whether the two genes are linked. The formula to use is PD >> NPD with linkage and PD = NPD without linkage. In this case, 264 PD >> 1 NPD. Alternatively,

RF = 100%(1/2T + NPD)/total asci = 100%[1/2(76 + 54 + 5) + 1]/400 = 17.125%. Therefore, the genes are linked.

Next, calculate the distance between the two genes. The formula to use is t–m = 100%(1/2T +NPD)/total.

t–m = 100[1/2(76 + 54 + 5) + 1]/400 = 100%(67.5 + 1)/400 = 17.1 m.u.

The distance between each gene and the centromere is calculated using the following formula:

RF = 100%(1/2 number of recombinant asci)/total number of asci.

In this case,

t–centromere = 100%[1/2(4 + 54 + 5)]/400 = 7.9 m.u.

m – centromere = 100%[1/2(76 + 4 + 5)]/400 = 10.6 m.u.

The map is

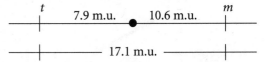

b.

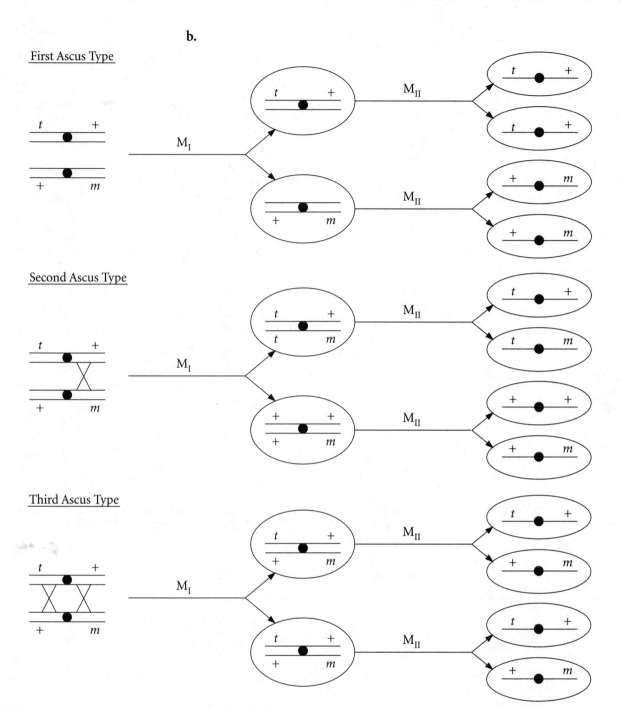

Fourth Ascus Type

Fifth Ascus Type

Sixth Ascus Type

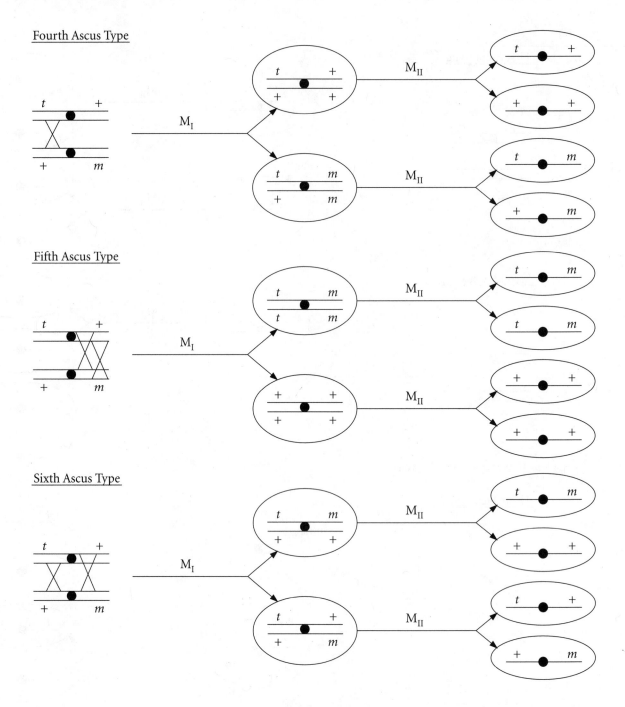

c. Only spores that are ++ will grow on minimal medium. Look at class 2. One-fourth of the spores will grow in minimal medium. The 76 asci represent 19% of the total number of asci. Therefore, of 1000 randomly chosen spores, 76/400 will come from class 2 (or 19% × 1000 = 190), and 1/4 (or 1/4 × 190 = 48) of these will be able to grow in minimal medium. In a similar manner, the contribution of the other three classes containing ++ asci can be determined, as follows:

100%(1000)[1/4(76/400) + 1/4(54/400) + 1/2(1/400) + 1/4(5/400)]

= 85.6

Of 1000 randomly chosen spores, 86 would grow in minimal medium.

13. Before beginning this problem, classify all asci as PD, NPD, or T and determine whether there is M_I or M_{II} segregation for each gene:

	ASCI TYPE						
	1	2	3	4	5	6	7
TYPE	PD	NPD	T	T	PD	NPD	T
gene a (M)	I	I	I	II	II	II	II
gene b (M)	I	I	II	I	II	II	II

If RF = 100% (1/2T + NPD)/total < 50% there is linkage. If the RF value is 50%, there is no linkage. The distance between a gene and its centromere = 100%(1/2)(M_{II})/total. The distance between two genes = 100%(1/2T + NPD)/total.

Cross 1: PD = NPD and RF = 50%; the genes are not linked.

a–centromere: 100%(1/2)(0)/100 = 0 m.u. Gene a is very close to the centromere.

b–centromere: 100%(1/2)(32)/100 = 16 m.u.

Cross 2: PD >> NPD and RF = 8.5%; the genes are linked.

a–b: 100%[1/2(15) + 1]/100 = 8.5 m.u.

a–centromere: 100%(1/2)(0)/100 = 0 m.u. Gene a is very close to the centromere.

b–centromere: 100%(1/2)(15)/100 = 7.5 m.u.

The data contradict because of asci that show an M_I pattern for b even though crossing-over must have occurred between b and the centromere. Therefore, the estimate of 8.5 m.u. between the two genes is the better estimate.

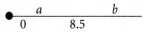

Cross 3: PD >> NPD and RF = 23%; the genes are linked.

a–b: 100%[1/2(40) + 3]/100 = 23 m.u.

a–centromere: 100%(1/2)(2)/100 = 1 m.u.

b–centromere: 100%(1/2)(40 + 2)/100 = 21 m.u.

Again, the data are contradictory for the same reasons as in *Cross 2*.

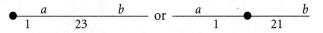

The first diagram is the better interpretation of the data.

Cross 4: PD >> NPD and RF = 11%; the genes are linked.

a–b: 100%[1/2(20) + 1]/100 = 11 m.u.

a–centromere: 100%(1/2)(10)/100 = 5 m.u.

b–centromere: 100%(1/2)(18 + 8 + 1)/100 = 13.5 m.u. Again, the *b*-centromere distance is underestimated.

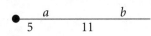

Cross 5: PD = NPD and RF = 49%; the genes are not linked.

a–centromere: 100%(1/2)(22 + 8 + 10 + 20)/99 = 30.3 m.u.

b–centromere: 100%(1/2)(24 + 8 + 10 + 20)/99 = 31.3 m.u.

For values this large, in tetrad analysis, the genes are considered unlinked to their centromeres.

Cross 6: PD >> NPD and RF = 4%; the genes are linked.

a–b: 100%[1/2(1 + 3 + 4) + 0]/100 = 4 m.u.

a–centromere: 100%(1/2)(3 + 61 + 4)/100 = 34 m.u.

b–centromere: 100%(1/2)(1 + 61 + 4)/100 = 33 m.u.

The *b*–centromere distance is underestimated. Genes *a* and *b* are more than 50 m.u. from the centromere and are 4 m.u. apart.

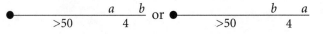

Cross 7: PD >> NPD and RF = 2.5%; the genes are linked.

a–b: 100%[1/2(3 + 2) + 0]/100 = 2.5 m.u.

a–centromere: 100%(1/2)(2)/100 = 1 m.u.

b–centromere: 100%(1/2)(3)/100 = 1.5 m.u.

Cross 8: PD = NPD and RF = 50%; the genes are not linked.

a–centromere: 100%(1/2)(22 + 12 + 11 + 22)/100 = 33.5 m.u.

b–centromere: 100%(1/2)(20 + 12 + 11 + 22)/100 = 32.5 m.u.

Same as cross 5.

Cross 9: PD >> NPD and RF = 16%; the genes are linked.

a–b: 100%[1/2(10 + 18 + 2) + 1]/100 = 16 m.u.

a–centromere: 100%(1/2)(18 + 1 + 2)/100 = 10.5 m.u.

b–centromere: 100%(1/2)(10 + 1 + 2)/100 = 6.5 m.u.

Cross 10: PD = NPD and RF = 49.5%; the genes are not linked.

a–centromere: 100%(1/2)(60 + 1 + 2 + 5)/100 = 34 m.u.

b–centromere: $100\%(1/2)(2 + 1 + 2 + 5)/100 = 5$ m.u.

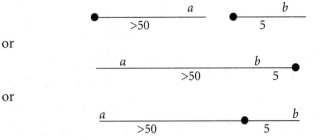

or

or

Cross 11: PD = NPD and RF = 49%; the genes are not linked.

a–centromere: $100\%(1/2)(0)/100 = 0$ m.u.

b–centromere: $100\%(1/2)(0)/100 = 0$ m.u.

14. The cross is $a +/+ b$, with the two genes unlinked. Each gene has a given probability of a crossover between it and the centromere. Remember that map units = $100\%[1/2(\text{number crossover asci})]/\text{total asci}$. Therefore, if there are 5 m.u. between a gene and its centromere, there will be 10 percent crossover, or second-division asci (M_{II}), and 90 percent non-crossover, or first-division asci (M_I).

A 4:4 pattern exists with no crossing-over. Thus, if neither gene shows a crossover, two equally likely patterns are possible: $a\,b$, $a\,b$, ++, ++ (NPD) and $a +$, $a +$, $+ b$, $+ b$ (PD).

If one gene (assume gene a) experiences a crossover, the pattern for it is a, +, a, + or a, +, +, a. Combined with a 4:4 pattern for gene b, this would result in four equally likely outcomes: $a\,b$, $+ b$, $a +$, ++ (T); or $a\,b$, $+ b$, ++, $a +$ (T); or $+ b$, $a\,b$, ++, $a +$ (T); or $a\,b$, $+ b$, ++, $a +$ (T). Note that patterns 2 and 4 are identical.

If both genes experience crossovers, the pattern is 1/4 PD: 1/2 T:1/4 NPD (you should check to see if you can produce this ratio).

The following table presents the probabilities for both genes:

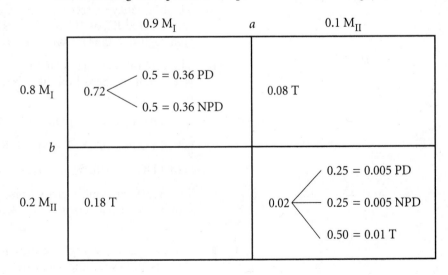

To understand the table, look at cell 1. No crossovers occur between either gene and its centromere (p = 0.8 × 0.9). The pattern for a is thus a

$a + +$, and the pattern for b is $b\ b + +$. The b gene can line up in either orientation with the a gene: $a\ b,\ a\ b,\ ++,\ ++$ or $a +,\ a +,\ + b,\ + b$. Therefore, there is a 1:1 chance of $a +,\ a +,\ + b,\ + b$ (PD) and $a\ b,\ a\ b,\ ++,\ ++$ (NPD). The other cells can be interpreted in a similar manner.

a.   36.5% PD: the total of PD from the table

b.   36.5% NPD: the total of NPD from the table

c.   27% T: the total of T from the table

d.   50% recombinants: 1/2 T + NPD

e.   25% ++

15. The frequency of recombinants is equal to NPD + 1/2T. The uncorrected map distance is based on RF = (NPD + 1/2T)/total. The corrected map distance = 50(T + 6NPD)/total.

Cross 1:

recombinant frequency $= 4\% + 1/2(45\%) = 26.5\%$

uncorrected map distance $= [4\% + 1/2(45\%)]/100\% = 26.5$ m.u.

corrected map distance $= 50[45\% + 6(4\%)]/100\% = 34.5$ m.u.

Cross 2:

recombinant frequency $= 2\% + 1/2(34\%) = 19\%$

uncorrected map distance $= [2\% + 1/2(34\%)]/100\% = 19$ m.u.

corrected map distance $= 50[34\% + 6(2\%)]/100\% = 29$ m.u.

Cross 3:

recombinant frequency $= 5\% + 1/2(50\%) = 30\%$

uncorrected map distance $= [5\% + 1/2(50\%)]/100\% = 30$ m.u.

corrected map distance $= 50[50\% + 6(5\%)]/100\% = 40$ m.u.

16. First, classify the asci: 138 T, 12 NPD, and 150 PD, for a total of 300 asci.

a.   The frequency of recombinant asci is 50 percent, which leads to an uncorrected RF of $100\%(12 + 69)/300 = 27$ m.u. The corrected RF is $m = -\ln(1 - 0.54) = -\ln(0.46) = 0.63$, which is 31.5 m.u.

b.   The general formula is DCOs = 4(NPD). Therefore, DCOs = 4(12) = 48. Half of them look like tetratypes, and half look like parental ditypes. Therefore

actual 0 crossovers = PD − 12 = 150 − 12 = 138, or 46%

actual 1 crossovers = T − 24 = 138 − 24 = 114, or 38%

actual DCOs = 48, or 16%

c.   To correct for double crossovers, the general formula for the mean number of crossovers is $m = $ T + 6NPD = $[138 + 6(12)]/300 = 0.70$, which is 35 m.u.

17. a.   The cross is $arg^- \times arg^-$. Because arg^+ progeny result, more than one locus is involved, and each deviation from wild type is recessive. The cross can be rewritten

$arg\text{-}1^+\ arg\text{-}2^- \times arg\text{-}1^-\ arg\text{-}2^+$

A 4:0 ascus is a PD ascus, because all spores require arginine. The 3:1 ascus must represent a T ascus. The 2:2 ascus is an NPD ascus.

b. The data support independent assortment of the two genes. PD = NPD = 40, and NPD/T = 40/20 = 2.00

18. Because PD = NPD, the *his-?* is not linked to *ad-3*. The T-type ascospores require one crossover between the gene and the centromere. Because only 10 tetrads were analyzed, *his-2* would be expected to produce one-tenth of a T ascus [10(0.01) = 0.1], *his-3* would be expected to produce one T ascus [10(0.1) = 1], and *his-4* would be expected to produce four T asci [10(0.4) = 4]. Only *his-4* is located far enough from its centromere to result in 60 percent T asci. Therefore, *his-?* is *his-4*.

19. Because the two mutants, when crossed, result in some black spores, two separate genes are involved, and both deviations from wild type are recessive. Let mutant 1 = $w\ t^+$, mutant 2 = $w^+\ t$, and wild type = $w^+\ t^+$. The cross involving the two mutants is

P $w\ t^+$ (white) $\times w^+\ t$ (tan)

Asci types are

4 black:4 white = 4 $w^+\ t^+$:4 $w\ t$ (NPD)

4 tan:4 white = 4 $w^+\ t$:4 $w\ t^+$ (PD)

4 white:2 black:2 tan = 2 $w\ t^+$ (white):2 $w\ t$ (white): 2 $w^+\ t^+$ (black):2 $w^+\ t$ (tan)

Notice that there are two types of white, indicating that there is an epistatic relationship between *w* and *t*. The *w* allele blocks expression of the t^+ allele.

20. At meiosis I, normal segregation is between homologous chromosomes. The two attached-X chromosomes segregate together because they structurally are the equivalent of a single chromosome. This produces one daughter cell with two copies of X (disomic) and one daughter cell with no copies of X (nullisomic):

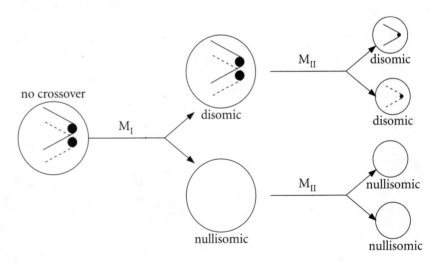

EGG	SPERM	PROGENY
XX	X	XXX, nonviable
XX	Y	XXY, viable female
O	X	XO, sterile male
O	Y	OY, nonviable

a. A single crossover between *a* and *b*, involving strands 1 and 2 or involving strands 3 and 4, does not result in recombination. However, a 1-3 (illustrated below), 1-4, 2-3, or 2-4 crossover between *a* and *b* results in recombination.

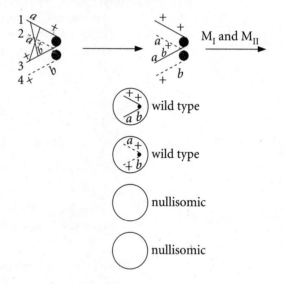

A single crossover between *b* and the centromere, involving strands 1 and 2 or involving strands 3 and 4, does not result in recombination. A 1-3 (illustrated below) or 2-4 crossover between *b* and the centromere also does not result in recombination.

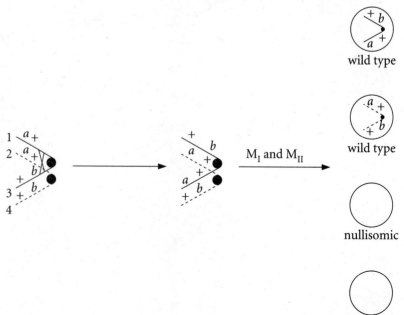

However, a 2-3 (illustrated below) or a 1-4 single crossover between *b* and the centromere does result in recombination.

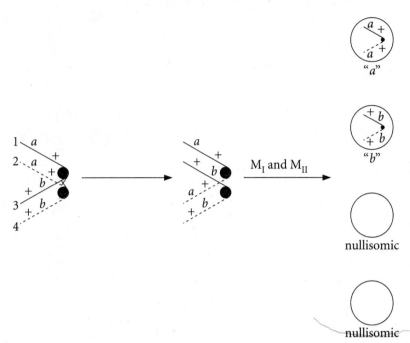

b. With double crossovers, the following are produced:

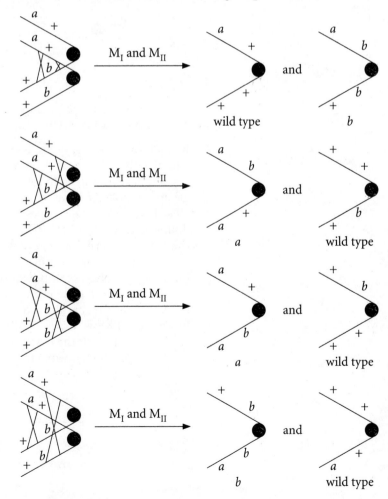

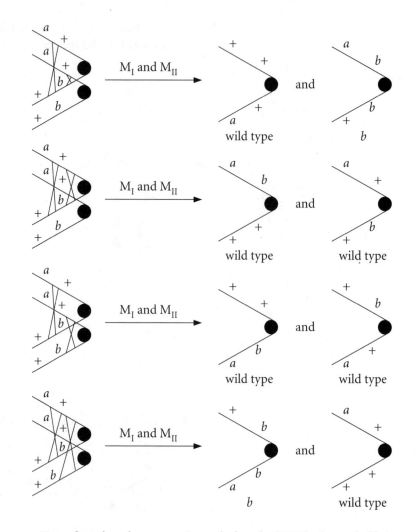

Note that the phenotype in each female (XXY) shows half the meiotic products. Thus, each female can be regarded as a half-tetrad.

21. First classify the asci types. Number 1 is a tetratype (T), number 2 is a parental ditype (PD), and number 3 is a nonparental ditype (NPD).

Recall from the textbook (page 167) that tetratypes arise from single and double crossovers. A SCO yields 100% T asci, and a DCO yields 50% T asci. Parental ditypes arise from no crossovers (100% PD) and double crossovers (25% PD). Nonparental ditypes arise from 25% of the double recombination events.

Also recall the formula RF = 100%(1/2T + NPD)/total asci. In other words, the recombination frequency is a function of 1/2 the tetratype asci. If a gene is 5 map units from the centromere or another gene, that means that the crossover frequency is actually 10%. In order to understand this, draw two homologous chromosomes with a single crossover. Only half the chromatids are involved.

Consider the following map derived from the problem:

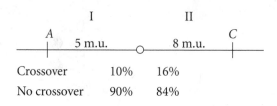

Now consider the following combination of events and their frequencies:

(CO in I)(no CO in II) = $0.10 \times 0.84 = 0.084$, all T

(no CO in I)(CO in II) = $0.90 \times 0.16 = 0.144$, all T

(CO in I)(CO in II) = $0.10 \times 0.16 = 0.016$, with 1/4 PD:1/2 T:1/4 NPD

(no CO in I)(no CO in II) = $0.90 \times 0.84 = 0.756$, all PD

Total the frequency of each asci type:

PD = $0.756 + 0.004 = 0.760$

T = $0.084 + 0.144 + 0.008 = 0.236$

NPD = 0.004

Number 1 is a T ascus, which occurs at a frequency of 0.236.
Number 2 is a PD ascus, which occurs at a frequency of 0.760.
Number 3 is a NPD ascus, which occurs at a frequency of 0.004.

22. First classify the asci types and determine whether they are M_I or M_{II} ascii for each gene:

TYPE	un	cyh
1. T	II	I
2. T	I	II
3. PD	I	I
4. PD	II	II
5. NPD	II	II
6. T	II	II

Notice that PD (47 + 2) >> NPD (2) and RF = 26.5%. The two genes are linked.

a. $un - cyh = 100\% \ (1/2T + NPD)/\text{total asci}$

$$= \frac{100\% \ [1/2 \ (15 + 29 + 5) + 2]}{15 + 29 + 47 + 2 + 2 + 5}$$

$$= 100\% \ [1/2 \ (49) + 2]/ \ 100 = 26.5 \ \text{m.u.}$$

$un - \text{centromere} = 100\% \ (1/2)(M_{II} \ \text{asci})/\text{total asci}$

$$= 100\% \ (1/2)(15 + 2 + 2 + 5)/100 = 12 \ \text{m.u.}$$

$cyh - \text{centromere} = 100\% \ (1/2)(M_{II} \ \text{asci})/\text{total asci}$

$$= 100\% \ (1/2)(29 + 2 + 2 + 5)/100$$

$$= 19.0 \ \text{m.u.}$$

The final map is

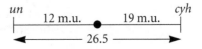

b.

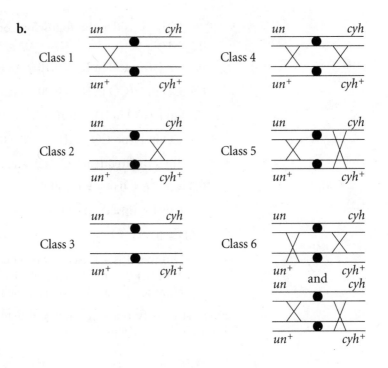

c. Figure 6-14 in the text can be used as a guide to determine the missing tetrads if the number of crossovers is limited to two. Classify the asci types as to the number of crossovers involved using the table from the text:

TYPE	un	cyh	C O
1. T	II	I	SCO
2. T	I	II	SCO
3. PD	I	I	NCO
4. PD	II	II	DCO
5. NPD	II	II	DCO
6. T	II	II	DCO
7. NPD	I	I	DCO

Seven asci types are presented here because of the seven drawings in part b.

The above table indicates that the seven basic asci types that result from 0, 1, or 2 crossovers are present, although the reciprocals have not been listed. For example, the reciprocal of type 1 is

un^+ cyh^+

un cyh^+

un^+ cyh^+

un cyh

Other missing asci types, such as those given below, result from three or more crossovers.

un^+ cyh^+ un cyh

un cyh un^+ cyh^+

un cyh un^+ cyh^+

un^+ cyh^+ un cyh

The missing asci types probably did not occur because of the low frequency of three or more crossovers and the relatively low number of asci that were studied.

d. Classes 4, 5, and 6 all involve double crossovers. Class 4 involves a two-strand double crossover, with both crossovers occurring between the centromere and each gene. There is only one way to achieve this result. Class 5 involves a four-strand double crossover, one between the centromere and *un* and one between *un* and *cyh*. There is only one way of achieving this result. Class 6 involves a three-strand double crossover, with one between *un* and the centromere and one between *cyh* and the centromere. There are two ways of achieving this result. Therefore, class 6 should occur at twice the frequency of either of the other two classes.

23. The chromosome map is

The cross is $+ cys2 \times leu3 +$. First classify the asci types:

TYPE

a. NPD

b. PD

c. T

d. T

e. NPD

f. PD

g. T

Because *leu3* segregates with the centromere, there are only two possible outcomes:

1. If a crossover occurs between *cys2* and the centromere, only tetratypes occur and they are at a frequency of 16%.

2. If no crossover occurs in that region, only parental ditypes occur, and they are at a frequency of 84%.

    a. The ascus is a NPD with M_I segregation for both loci. This can happen only when a double crossover occurs. The problem states that double crossovers should not be considered. Therefore, the answer is 0%.

    b. The ascus is a PD with M_I segregation for both loci. This occurs only with no crossovers. The frequency is 84%.

c.   The ascus is a T, with a crossover between *cys2* and the centromere. The frequency is 16%.

d–g. All four ascus types show M$_{II}$ segregation for *leu3*, which cannot occur. The probability of each type is 0%.

24. The cross is

P   *g C s/g C s × G c S/G c S*

F$_1$ *g C s/G c S* wild type

In order to achieve the mutants that were observed, mitotic crossing-over had to occur.

The first set of mutants were *gg C– ss* and *G– cc S–*. A single
crossover can yield these results:

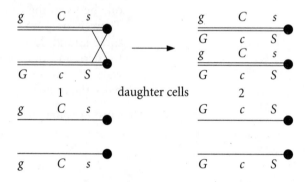

The second set of mutants were *gg C– S–* and *G– cc S–*. A single
crossover can yield these results:

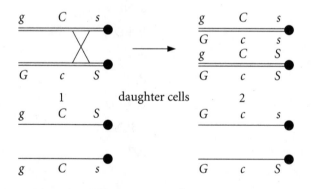

The third mutants were *gg C– S–*. A single crossover can yield these
results:

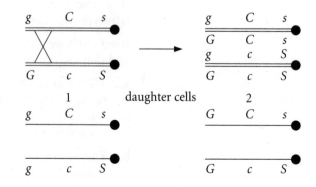

25. Recall that white diploid sectors will have two copies of the chromosome and two copies of each gene. The original chromosomes were

ad col phe pu sm w/ad$^+$ col$^+$ phe$^+$ pu$^+$ sm$^+$ w$^+$ (gene order is alphabetical).

a. The sectors must be *ww*. To be *ww*, mitotic crossing-over must have occurred. The following diagram illustrates this for alleles *w/w$^+$* only:

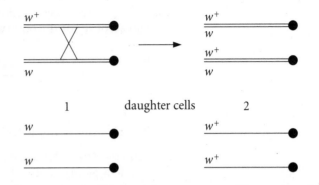

b. Note that any gene distal to the crossover will remain with the *w* allele if no other crossovers occur, while any gene proximal to the crossover will be separated from the *w* allele if no other crossovers occur. This means that the closer a proximal gene is to the *w* allele, the less likely it will be separated from *w* and the more likely it will be homozygous recessive.

Using this logic, *pu* (with *ww* in 100% of the sectors) is distal to *w*. The *col* allele is never observed with *w*, suggesting that it is proximal to it. In other words, *col* lies very close to the centromere. However, as will be discussed below, on which side of the centromere *col* is located cannot be determined. The other genes are between *w* and the centromere: *ad* (with *ww* in 95% of the sectors), *sm* (with *ww* in 65% of the sectors), and *phe* (with *ww* in 41% of the sectors). The final gene order is

pu — w — ad — sm — phe — col — centromere

c. The relative map distances can be calculated directly from the cosegregation of phenotypes. If two genes are found together X% of the time, then crossing-over occurs between them (100 − X)% of the time.

w − pu: 100% − 100% = 0 relative map units. This finding indicates that the relative order determined above may not be exactly correct, because *w* and *pu* are so tightly linked. It is possible that the gene order is

w — pu — ad — sm — phe — col — centromere

Without further information, no choice can be made between the two possibilities.

w − ad: 100% − 95% = 5 relative map units

w − sm: 100% − 65% = 35 relative map units

w − phe: 100% − 41% = 59 relative map units

The *w −* centromere distance cannot be calculated precisely.

d. The phenotype associated with homozygous *col* was never observed.

This means that *col* was very far from *w*. If *col* were between *w* and the centromere, then some of the sectors should have had a *col* phenotype unless *col* is so close to the centromere that no crossing-over occurs between them. If *col* were on the other side of the centromere from *w*, but close to the centromere so that no mitotic crossing-over occurred, then it would never be observed with *w*. Therefore, *col* is very close to the centromere but could be on either arm.

26. Let *gg* = green, *Gg* = yellowish, and *GG* = yellow. Mitotic crossing-over can account for the observations:

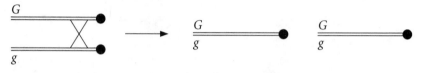

The resulting daughter cells would be *GG* (yellow) and *gg* (green).

27. Ignoring the white allele for a moment, the 1:1:1:1 ratio indicates independent assortment of two units, or chromosomes. The *a* and *c* alleles are found in equal proportions with *b* and *b*⁺, indicating that *b* assorts independently of *a* and *c*. If *w* were linked on either chromosome, then the white allele would be found more frequently with one gene combination than another. Because it is not, *w* is not linked to either chromosome.

28. Remember that the segregants are yellow and diploid (80 percent *yy* *r*⁺–, 20 percent *yy rr*). To get *yy* segregants, crossing-over must occur between *y* and the centromere. There are three possible arrangements:

1. yellow — centromere — ribo

2. centromere — yellow — ribo

3. centromere — ribo — yellow

If 1 is correct, after crossing-over the chromosomes would be

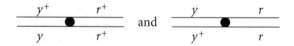

Without a crossover between *r* and the centromere, no *rr* segregants would be observed. Because the two genes are far away from each other, a 1:1 ratio of ribo-requiring and non-ribo-requiring (wild type for ribo) would be expected.

If 2 is correct, after crossing-over the chromosomes would be

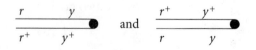

Unless the two genes are more than 50 m.u. apart, most of the *yy* segregants would be ribo-requiring.

If 3 is correct, crossing-over between *r* and the centromere would give ribo-requiring segregants, and crossing-over between *y* and the centromere would give wild type for ribo. Therefore, both types of segregants would be observed with only one crossover. Furthermore, the 80:20 ratio suggests that crossing-over occurs more frequently between *y* and the centromere.

29. a. All *fpa/fpa* diploid segregants experienced a crossover between *fpa* and the centromere. The data indicate that *pro* and *paba* are linked to *fpa*. The different frequencies are a measure of the distances involved. Because there are no progeny requiring only proline, *pro* is closer to the centromere than is *paba*.

b.

```
          pro              paba      fpa
●─────────┼────────────────┼─────────┼─────────
```

pro–paba: 100%(110)/154 = 71.4 relative m.u.

paba–fpa: 100%(35)/154 = 22.7 relative m.u.

pro–centromere: 100%(9)/154 = 5.8 relative m.u.

c. *pro paba fpa*

30. a. To work this problem, you must first realize that resistance to fluorophenylalanine is recessive and not wild-type. Thus, selection was for *fpa* or *fpa/fpa*, depending on whether cells were haploid or diploid.

Note that all the haploid colonies were *fpa leu ribo*, which suggests linkage between these three genes because neither *leu* nor *ribo* was selected. There were approximately equal numbers of colonies that were either *ad w* or *ad*+ *w*+, which suggests that these two genes are linked and independently assorting from the *fpa*-bearing chromosome.

b. You are given information regarding only one chromosome for the colonies that appeared among the diploids. That chromosome is the *fpa*-bearing chromosome. The *phenotypes* of the colonies are

24 fpa leu+ ribo+

17 fpa leu+ ribo

14 fpa leu ribo

──

55

Because the colonies are diploid, mitotic crossing-over had to occur. All colonies represent a mitotic crossover between *fpa* and the centromere. This is diagrammed below, with the assumption that the gene order is centromere–*leu*–*ribo*–*fpa*. If order existed other than the one assumed, more complicated crossovers would be required to yield one or more of the phenotypes.

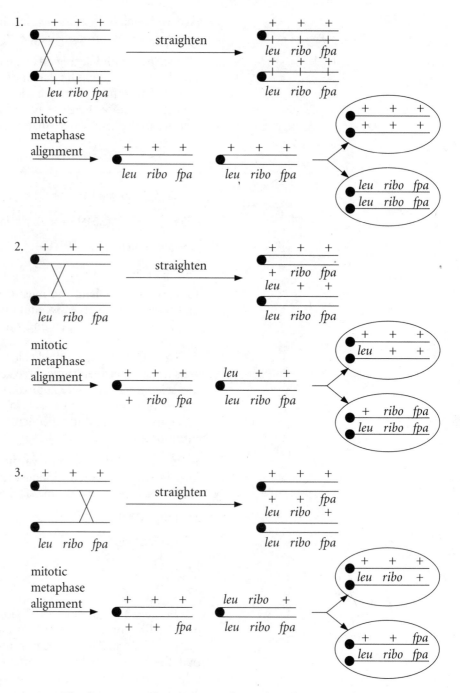

The frequency of each type reflects the relative size for each region.

▶ COMMENTS ON THIS
MATERIAL

The material covered in Chapter 6 of *An Introduction to Genetic Analysis*, 6th ed., will challenge the very best of students. Students experience more difficulty with this chapter than with any other chapter in the textbook. One reason for this may be insufficient mathematical skills, a difficulty that cannot be solved here. Another reason this material is so difficult for so many is that it requires the integration of all that you have learned up to this point. It requires a thorough understanding of the foregoing material and systematic application of the problem-solving techniques discussed earlier. As you are struggling with some of these problems, it may help to know that you may have to spend well over an hour in solving a specific problem in this chapter. Persistence, coupled with seeking help when you are truly stuck, does pay off in the end.

▶ SOME TOUGH ADVICE ON
HOW TO STUDY GENETICS

It is my sincere hope you have the habit and the methods of critical analysis that have been presented in both your textbook and this *Companion*. At this point in the text, the material becomes much more descriptive. You now have two main tasks: (1) to memorize the presented material and (2) to integrate the new material with all that has come before. The first task, memorization, should be relatively easy. It requires a skill that you have been perfecting since long before you entered kindergarten.

The second task, integration, is quite difficult for many students. For example, in Chapter 11 you will cover the structure of DNA, in Chapter 16 you will put that DNA into chromosomes, and in Chapter 20 you will study recombination at the molecular level. All these major topics must be integrated with one another and with what you have learned about recombination at the level of transmission genetics.

If at this point your grasp of transmission genetics is weak, go back to Chapter 2 and learn it. You may be able to memorize the following material without a firm foundation in transmission genetics, and you may be able to do very well on tests covering the new material; but, ultimately, that new material will carry little meaning for you unless it can be applied to transmission genetics.

Gene Mutation

A gene and its function are not detectable without **variants** to reveal the existence of that gene. Variants arise by two mechanisms: (1) **gene mutation**, which is also known as **point mutation**, and (2) **chromosome mutation.**

The standard form of a gene is the **wild type.** Any deviation from the standard form is a **forward mutation.** Any change from the forward mutation back toward the standard form is called variously a **reverse mutation**, a **reversion**, or a **back mutation.**

A **somatic mutation** occurs in nonreproductive cells and leads to a **clone** of cells that differs genetically from the rest of the organism. This results in the organism's being classed as a **mosaic,** because it has two or more differing cell lines. Somatic mutations can be inherited if the organism undergoes vegetative reproduction.

A **germinal mutation** occurs in reproductive tissue. Germinal mutations can be inherited.

Morphological mutations cause a change in form. **Lethal** mutations result in death. **Conditional** mutations are expressed under the **restrictive** condition and are not expressed under the **permissive** condition. **Biochemical** mutations result in a change in metabolism. Organisms that are nutritionally self-sufficient given a standard set of growth conditions are **prototrophic.** Mutants that require additional supplementation are **auxotrophic. Resistance mutations** confer the ability to grow in the presence of a specific inhibitor such as a poison.

The **mutational rate** is the number of mutational events per unit time. The **mutational frequency** is the number of mutant individuals per total number of organisms.

Selective systems aid in the detection of rare mutations.

Evolution occurs through mutations, although most mutations are deleterious.

Mutations can be induced by a number of differing techniques, biochemical and physical. Natural and induced mutations are random in direction, in the gene being affected, and in the cell in which they occur.

Questions for Concept Map 7-1

1. What does it mean to say that an organism is in balance with the environment?

2. Does an organism affect the environment? How? Give some examples.

▶ CONCEPT MAP 7-1:
MUTATIONS

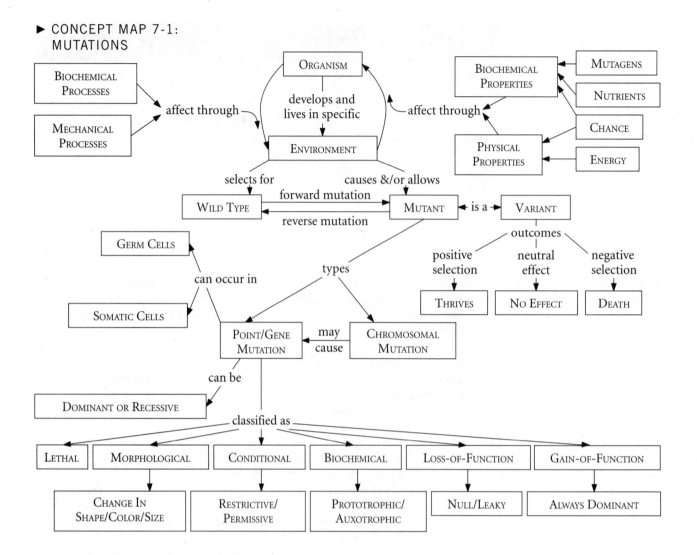

3. Does the environment affect an organism? How? Give some examples.

4. What role does developmental noise play in the adaptation of a specific individual?

5. Can developmental noise lead to positive selection? Negative selection?

6. Can the outcome of developmental noise be neutral?

7. Is the wild type in a specific environment the best-adapted form of an organism?

8. What does it mean to say that mutations occur randomly? Which set of experiments demonstrated this?

9. Can a forward mutation eventually become the wild type?

10. What is the role of time in mutations?

11. Does a reverse mutation restore the wild-type phenotype exactly, or does it restore the underlying DNA sequence?

12. Can a reverse mutation restore the phenotype without restoring the original DNA sequence?

13. Which do you think occur more frequently: forward or reverse mutations? Why?

14. How can a mutation have no effect?

15. What is the definition of *selection?*

16. How does a chromosomal mutation lead to a point mutation?

17. Are somatic mutations inherited?

18. Do germ mutations affect the individual in which they occur?

19. Give three examples in humans of a morphological mutation.

20. Will lethal mutations in somatic tissue always lead to the immediate death of an individual in which it occurs?

21. Will a lethal mutation in a germ cell always lead to the immediate death of an organism developing from that cell?

22. Define *restrictive* and *permissive conditional mutations.*

23. Humans and guinea pigs cannot make vitamin C. How is it obtained by these two species?

24. Because humans and guinea pigs must obtain vitamin C from the environment, are the two species auxotrophic for vitamin C?

25. Are loss-of-function mutations generally dominant or recessive?

26. Explain why gain-of-function mutations are always dominant.

Answers to Concept Map 7-1 Questions

1. An organism, to survive, needs to be adapted to its environment. That means that its nutritional needs must be met by the environment and that its reproductive activities cannot exceed what the environment can support without deterioration of the environment. Furthermore, the mechanical and biochemical effects of the organism cannot destroy any aspect of the environment. If the balance between organism and environment is lost, then the organism must either change or die eventually. Complete destruction of the environment may also occur. Right now, humans are not in balance with the environment. They are in the process of destroying the environment through mechanical and biochemical effects due to reproduction beyond what the environment can maintain.

2. An organism affects the environment through its nutritional needs and its waste products. Humans are also destroying vast regions of the earth through overplanting, clear-cutting, and construction.

3. The environment places limits on population size through the nutrients available. It also may contain mutagens that lead to organismal change and possibly death. Other aspects of the environment that affect organisms are the amount of water available, temperature, and presence or absence of shelter.

4. Developmental noise is specific to individuals in the sense that it produces noninherited variability that allows the individual to be either more or less adapted to the environment. Differences produced by developmental noise can ultimately affect the reproductive potential of an individual, which in turn may shape the adaptation of a species to a specific environment.

5. Developmental noise leads to both positive and negative selection.

6. The outcome of developmental noise can be neutral.

7. The wild type in a specific environment is not necessarily the best-adapted form of an organism. This is because the wild type is a product of a specific time period in the history of an organism and refers simply to the characteristics of the most numerous individuals. If the wild type is not the best-adapted form, then the characteristics of what constitutes the wild type will change over time so that the organism will be in better balance with the environment. However, because the environment also changes over time, this adaptation can never be perfect and, in instances of radical environmental change, can lead to an elimination of the organism from the environment.

8. Mutations occur randomly in that a change in any gene is random and the type of change is also random. That is, organisms do not mutate in a manner required for survival in a particular environment. Rather, random mutations occur, and the environment selects for those that aid the organism and against those that are detrimental to the organism. This was demonstrated in the replica plating experiments by Lederberg and Lederberg.

9. A forward mutation can eventually become the wild type.

10. What is considered wild type or mutant is a function of the particular time at which a species is observed.

11. Whether a reverse mutation restores the wild-type phenotype exactly or restores the underlying DNA sequence is a function of the level at which it is being assessed.

12. A reverse mutation can restore the phenotype without restoring the original DNA sequence.

13. Forward mutations occur much more frequently than reverse mutations. A forward mutation can go in many different directions, while a reverse mutation must go back to the original situation.

14. A mutation can have no effect if it occurs in such a way as to result in the same amino acid in a protein (knowledge of the genetic code is required to understand this), or if it inserts an amino acid that has characteristics similar to the original amino acid, or of it occurs in a nonimportant region of a protein.

15. *Selection* is a function of the characteristics of the environment that allow or do not allow the survival and reproduction of a specific variant.

16. A chromosomal mutation can lead to a point mutation if the break point occurs within a gene.

17. Somatic mutations are inherited by organisms that have vegetative reproduction.

18. Germ mutations do not affect the individual in which they occur.

19. Three examples in humans of a morphological mutation are extra fingers, shortened limbs, and cleft lip.

20. Lethal mutations in somatic tissue do not always lead to the immediate death of an individual in which they occur, because the immune system

may destroy the cells possessing it, or the cell and its progeny can lead simply to lowered level of function, that is, disease.

21. A lethal mutation in a germ cell will always lead to death of an organism developing from that cell, but the timing depends on when the gene is active. That is, a lethal mutation may not be expressed until adulthood or it may be expressed at the embryonic stage.

22. *Restrictive conditional mutations* are those that allow expression of a mutant phenotype only under certain environmental conditions, while the wild type is expressed under permissive environmental conditions.

23. Vitamin C is obtained by humans and guinea pigs through diet.

24. Humans and guinea pigs are not auxotrophic for vitamin C, because that is the wild type of both species.

25. Loss-of-function mutations are generally recessive.

26. Gain-of-function mutations are always dominant because they do something that the species was unable to do before they occurred.

Questions for Concept Map 7-2

1. Does this concept map refer to a population or a species?
2. Name five functions included within the category of "most genes."
3. What is distinctive about cell cycle regulatory genes? What do they do?
4. What is the site of action of mutagens?
5. What constitutes DNA damage?
6. When does DNA repair occur in the cell cycle?
7. How does DNA repair result in the wild type?
8. What would block the occurrence of DNA repair?
9. Is DNA repair under the control of organismal genes?
10. If a repair system is mutated, how would this affect an organism's ability to survive?
11. If a repair system is inhibited by medicines, how would this affect the ability to survive?
12. What is the difference between a repair system and the immune system?
13. If the immune system is impaired, how will this affect the ability of an organism to survive?
14. How does aging relate to the repair system? The immune system?
15. What is one underlying factor in aging?
16. Distinguish genetically between an inherited form of a specific type of malignancy and a randomly produced form.
17. What is the difference between a dominant and a recessive mutation?
18. What is another term for *invasive?*
19. All cancer cells have changed biochemical properties that are reflected in their cell membranes. How does this connect to an immune system failure?

► CONCEPT MAP 7-2:
  MUTATION EFFECTS

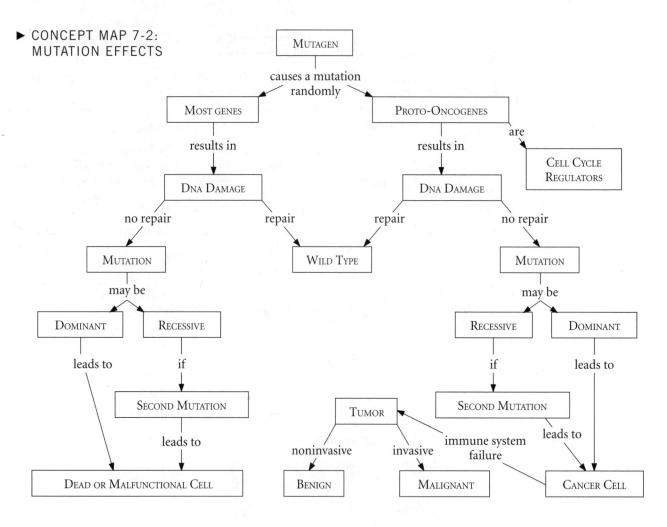

20. How does the immune system become impaired?

21. AIDS is characterized by a high rate of malignancy. Explain why this occurs.

Answers to Concept Map 7-2 Questions

1. This concept map refers to an individual more than to populations or species.

2. Five functions included within the category of "most genes" could be lung function, kidney function, digestion in the small intestine, heart function, movement of a specific muscle, or many other functions carried out through genes.

3. Cell cycle regulatory genes determine whether a cell reproduces and at what rate reproduction occurs.

4. Mutagens ultimately affect the DNA of an organism.

5. DNA damage can be breakage, biochemical modification, gain or loss of DNA segments, rearrangement of DNA segments, or the substitution of one component for another.

6. DNA repair occurs during interphase.

7. DNA repair can result in the wild type by returning the alteration to what was. If a break occurred, for instance, the break can be repaired. If one component (base) was substituted for another, then the wrong component can be removed and replaced by the right one.

8. DNA repair might not take place if the damage is in a gene controlling repair, or if the cell cycle has been speeded up to the point where there is no time for repair before mitosis.

9. DNA repair is under the control of organismal genes.

10. If a repair system is mutated, then improper or no repair would occur. This will drastically shorten the life span of an organism.

11. If a repair system is inhibited by medicines, then the organism is at a high risk for DNA damage, consequent mutation, and the loss of function through disease or the development of tumors.

12. A repair system fixes DNA damage, and the immune system destroys cells that have become damaged to the point that they are not recognized as self.

13. If the immune system is impaired, the ability of an organism to survive is drastically reduced.

14. The process of aging is, in part, the accumulation of unrepaired DNA damage, which can also occur in the repair and immune systems. DNA damage in these systems can lead to a cascade effect whereby mutations can occur with increasing frequency while the immune system less effectively eliminates cells that have changed to such a degree that they no longer are self (that is, they are tumors).

15. An underlying factor in aging is the gradual accumulation of unrepaired DNA damage.

16. An inherited form of a specific type of malignancy is a recessive or dominant mutation passed from parent to child. If recessive, then any cell in the specific organ has the potential to become malignant through a mutation of the wild-type allele. If dominant, then all cells will eventually express the malignant phenotype. Noninherited malignancy requires that two alleles of a cell be mutated in the same direction at some point in time, if recessive. If dominant, then only one allele must experience the mutation.

17. A dominant mutation requires only one allele for expression, while a recessive mutation requires that it be the only allele present for expression to occur.

18. Another term for *invasive* is *metastatic.*

19. Because all cancer cells have changed biochemical properties that are reflected in their cell membranes, they should be recognized as non-self by the immune system and destroyed. Therefore, malignancy requires both mutation and an immune system failure.

20. The immune system becomes impaired through accumulated DNA damage and metabolic inhibition (such as by the use of cortisone).

21. The normal organism is continually experiencing DNA damage and continually makes errors during replication. Such faulty DNA is repaired, leads to cell death, or changes the cells to the point where they

are non-self and are destroyed by the immune system. AIDS is charac-terized by a high rate of malignancy because the AIDS virus specifically destroys the immune system, and therefore the ability to destroy non-self cells is lost.

► SOLUTIONS TO
PROBLEMS

Be sure that you have thoroughly read the entire chapter before you attempt any of the problems.

1. The petal will now be *Ww*, or blue, either in whole or in part, depend-ing on the timing of the reversion.

2. Grow spores in the absence of leucine and in the presence of an antibi-otic that will kill only proliferating cells. Use filtration enrichment to obtain spores that were unable to grow because they were *leu⁻*.

3. Plate the cells on medium lacking proline. Nearly all colonies will come from revertants. The remainder will be second-site suppressors.

4. Streak the yeast on minimal medium plus arginine. When colonies appear, replica-plate them onto minimal medium. The absence of growth in minimal medium will identify the arginine-requiring mutants.

5. Assume that you are working with 20 specific nutrients that were added to the minimal medium. Group the nutrients, with five to a group. (The choice of how many nutrients are included in each group is completely arbitrary.) Test each auxotroph against each group. When an auxotroph grows in one of the groups, test the auxotroph separately against each member of the group. A flow sheet would look something like the following:

TEST 1	GROWTH	TEST 2	GROWTH
group A (1 to 5)	–	11	–
group B (6 to 10)	–	12	–
group C (11 to 15)	+ ⟶	13	+
group D (16 to 20)	–	14	–
		15	–

These results tell you that nutrient 13 was required for growth.

If a mutant cannot be identified to have a requirement for a specific nutrient, then it may be a double or multiple mutant. Alternatively, it could require an unidentified component of the complete medium.

6. You need to apply the Poisson distribution (Chapter 6) to answer this problem. Mutants were *not* observed (zero class) on 37 plates out of 100. The formula is e^{-un} = number in zero class/total number. Or $e^{-u \times 10^6} = 37/100$, and $u = -\ln(0.37) \times 10^{-6} = 1/10^{-6}$ cell divisions.

7. There are many ways to carry out this experiment. Using *Drosophila*, you could raise flies with (experimental) and without (control) caffeine added to the diet. You could do the same with mice, rats, cats, or dogs, or with mammalian cells in culture. Alternatively, you could inject a

solution with or without caffeine into an organism. Finally, you could set up more elaborate tests such as the ClB test in *Drosophila*.

8. Stain pollen grains, which are haploid, from a homozygous *Wx* parent with iodine. Look for red pollen grains, indicating mutations to *wx*, under a microscope.

9. The most straightforward explanation is that a mutation from wild type to black occurred in the germ line of the male wild-type mouse. Thus, he was a gonadal mosaic of wild-type and black germ cells.

10. An X-linked disorder cannot be passed from father to son. Because the gene for hemophilia must have come from the mother, the nuclear power plant cannot be held responsible.

     It is possible that the achondroplastic gene mutation was caused by exposure to radiation.

11. The mutation rate needs to be corrected for achondroplastic parents and put on a "per gamete" basis, which requires subtracting the two cases for which a parent was achondroplastic and the parents involved in producing those children:

$$(10 - 2)/[2 \times (94{,}075 - 2)] = 4.25 \times 10^{-5} \text{ gametes}$$

You do not have to worry about revertants in this problem because the problem asks for the net mutation frequency to achondroplasia.

12. The commission was looking for induced recessive X-linked lethal mutations, which would show up as a shift in the sex ratio. A shift in the sex ratio is the first indication that a population has sustained lethal genetic damage. Other recessive mutations might have occurred, of course, but they would not be homozygous and therefore would go undetected. All dominant mutations would be immediately visible, unless they were lethal. If they were lethal, there would be lowered fertility, an increase in detected abortions, or both, but the sex ratio would not shift as dramatically.

13. a. reddish all over

    b. reddish all over

    c. many small, red spots

    d. a few large, red spots

    e. like part c, but with fewer reddish patches

    f. like part d, but with fewer reddish patches

    g. some large spots and many small spots

14. *Unpacking the Problem*

a.

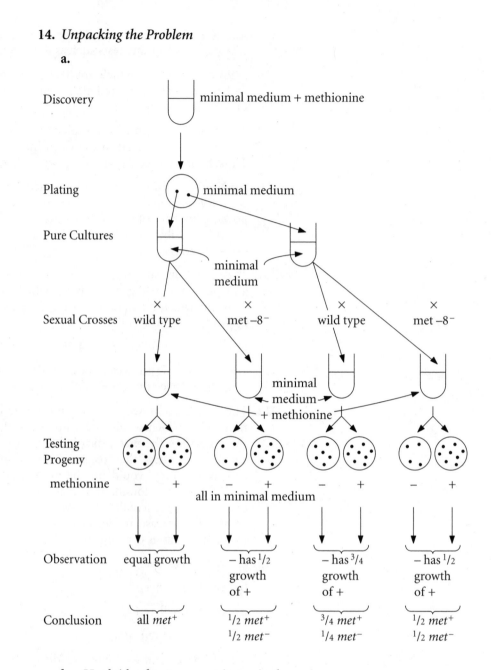

b. Haploid refers to possessing a single genome.

Auxotrophic means that an organism requires dietary provision of some substance that normally is not required by members of its species.

Methionine is an amino acid.

Asexual spores are a mode of propagation used by some species in which the spores are derived from an organism without a genetic contribution from another organism. Of necessity, the spores have the same number of genomes as the organism from which they are derived.

Prototrophic means that an organism does not have any special dietary requirements beyond those normal for the species.

A colony is a collection of cells or organisms all derived mitotically from a single cell or organism and all possessing the same genotype.

A mutation is the process that generates alternative forms of genes, and it results in an inherited difference between parent and progeny.

c. The "*8*" in *met-8* refers to the eighth locus found that leads to a methionine requirement. It is unnecessary to know the specifics of the mutation in order to work the problem.

d. The following crosses were made in this problem:

Cross 1: prototroph 1 × wild type

Cross 2: prototroph 1 × backcross to *met-8*

Cross 3: prototroph 2 × wild type

Cross 4: prototroph 2 × backcross to *met-8*

e. Use *met-8*★ to indicate the prototroph derived from the *met-8* strain.

Cross 1: met-8★ 1 × *met-8*$^+$

Cross 2: met-8★ 1 × *met-8*

Cross 3: met-8★ 2 × *met-8*$^+$

Cross 4: met-8★ 2 × *met-8*

f. In this organism, asexual spores give rise to an organism that is capable of forming sexual spores following a mating. Therefore, the original mutation occurred in somatic tissue that subsequently gave rise to germinal tissue.

g. Because the trait being selected is the ability to grow in the absence of methionine, a reversion is being studied.

h. Only two revertants were observed because reversion occurs at a much lower frequency than forward mutation.

i. The millions of asexual spores did not grow because they required methionine and the medium used did not contain methionine.

j. A low percentage of the millions of spores that did not grow would be expected to have other mutations that rendered them incapable of growth. In addition, a low percentage would be expected to have chromosome abnormalities that would lead to death.

k. The wild type used in this experiment was prototrophic, by definition; that is, *wild type* refers to the norm for a species, which means prototrophic.

l. It is highly unlikely that visual inspection could distinguish between wild type and prototrophic revertants.

m. One way to select for a *met-8* mutation is to grow a large number of spores on a medium that lacks methionine. Filtration will separate those spores capable of growth from those incapable of growth. Once spores have been isolated that are incapable of growth in a medium lacking methionine, they can be tested for a methionine mutation by plating them on medium containing methionine. If they are capable of growth on this second medium, they are methionine auxotrophs.

n. The starting auxotrophic spores were haploids. Both mitotic crossing-over and haploidization require diploids. Therefore, it is unlikely that either process is involved with producing the observed results.

o.   *Cross 1:* met-8 × wild type → 1 met-8:1 wild type

   Cross 2: met-8 × met -8 → all met-8

   Cross 3: wild type × wild type → all wild type

p.   While the analysis could have been conducted using tetrad analysis, it is more likely that random selection of progeny was used.

q.

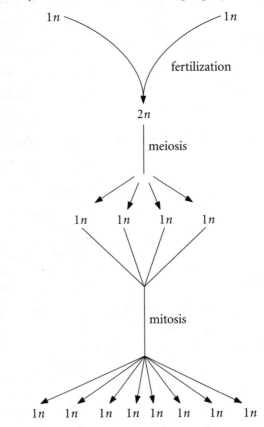

r.   If a 3:1 ratio is obtained in haploids, then two genes must be segregating.

Solution to the Problem

a. and b. The pattern of growth for prototroph 1 suggests that it is a reversion of the original mutation. When crossed with wild type, a reversion would be expected to produce all met^+ progeny, and when backcrossed it would be expected to produce a 1:1 ratio. Let the reversion be symbolized by *met-8*★ The crosses are

P   *met-8*★ × *met-8*$^+$ P *met-8*★ × *met-8*

F$_1$ 1/2 *met-8*★ prototrophic   F$_1$ 1/2 *met-8*★ prototrophic

   1/2 *met-8*$^+$ prototrophic   1/2 *met-8* auxotrophic

   The pattern of growth for prototroph 2 suggests that a suppressor at another locus is responsible for its prototrophic growth. Let *met -s* symbolize this suppressor. Then the crosses are

P   *met-8 met-s* × *met-8*$^+$ *met-s*$^+$ P *met-8 met-s* × *met-8 met-s*$^+$

F$_1$ 1/4 *met-8 met-s*   prototrophic   F$_1$ 1/4 *met-8 met-s* prototrophic

1/4 *met-8⁺ met-s⁺* prototrophic   1/4 *met-8 met-s⁺* auxotrophic

1/4 *met-8⁺ met-s* prototrophic   1/4 *met-8 met-s* prototrophic

1/4 *met-8 met-s⁺* auxotrophic   1/4 *met-8 met-s⁺* auxotrophic

15. The most likely explanations for the five observed colony types are

   a. White colonies: the cells are parental in phenotype. That is, the cells of most colonies are *ad⁺/ad*.

   b. (1) Diploid red, round cells: the *ad⁺* gene was inactivated by mutation, deleted, or mutated to the *ad* allele.

   (2) Diploid red, elongated cells: the chromosome carrying round and white was broken and these two genes were lost. Mitotic crossing-over is a more likely alternative.

   c. (1) No growth originally but grew in a complete medium: mutation for *met⁺*, *nic⁺*, or both.

   (2) No growth originally, nor in a complete medium: loss of some function, through mutation of either an allele or the entire chromosome, that was essential for viability. Alternatively, a dominant lethal mutation occurred.

16. The five mutations can be characterized as follows:

   Mutant 1: An auxotrophic mutation permitting growth only on complete medium

   Mutant 2: A non-nutritional temperature-sensitive mutation

   Mutant 3: A leaky auxotrophic mutation

   Mutant 4: A leaky, non-nutritional temperature-sensitive mutation

   Mutant 5: A temperature-sensitive auxotrophic mutation

   Other possibilities exist, for example, an auxotrophic mutation could also be characterized as a biochemical mutation.

17. **a.** *A* is 13.5 × 2, or 27, map units from *B*.

   A is 17.9 × 2, or 35.8, map units from *C*.

   B is 5.0 × 2, or 10, map units from *C*.

   The map is

   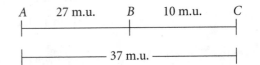

   b. *Cross 4:* 2-1 × B: Because the map units indicate that 2-1 is the same distance as A from B, the 2-1 spore is type A. The genotypes are

   $A B⁺ × A⁺ B → A⁺ B⁺$

   Cross 5: 2-2 × B: One explanation is that a double crossover occurred, as diagrammed below:

   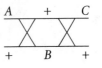

This would lead to an expected 0.027 frequency of recombinants, half of which (0.013) would be trp^+. That is,

$$A\ B^+\ C \times A^+\ B\ C^+ \rightarrow A^+\ B^+\ C^+$$

An alternative explanation is that the spore is neither A nor C. This is a new mutation that arose at a site that is 1.3 map units from B. The genotypes are

$$X\ B^+ \times X^+\ B \rightarrow X^+\ B^+$$

Cross 6: 2-3 × B: The map units indicate that 2-3 is C. The genotypes are

$$C\ B^+ \times C^+\ B \rightarrow C^+\ B^+$$

Cross 7: 2-4 × B: The map units indicate that 2-4 is A. The genotypes are

$$A\ B^+ \times A^+\ B \rightarrow A^+\ B^+$$

Cross 8: 2-5 × B: The map units indicate that 2-5 is also C. The genotypes are

$$C\ B+ \times C^+B \rightarrow C^+\ B^+$$

18. a. The experiment was designed to detect the number of genes that are involved with locomotion.

b. To solve this, look for a lack of complementation, indicating that mutations are in the same gene. There are a total of four genes detected that are involved in locomotion. The genes are as follows:

GENE	MUTANTS
A	1, 5
B	2, 6, 8, 10
C	3, 4
D	7, 11, 12

c. Mutant 1: $AA\ B^+B^+\ C^+C^+\ D^+D^+$

Mutant 2: $A^+A^+\ BB\ C^+C^+\ D+D^+$

Mutant 5: $AA\ B^+B^+\ C^+C^+\ D^+D^+$

Hybrid 1/2: $AA\ B^+B^+\ C^+C^+\ D^+D^+ \times A^+A^+\ BB\ C^+C^+\ D^+D^+ \rightarrow AA^+\ B^+B\ C^+C^+\ D^+D^+$

Hybrid 1/5: $AA\ B^+B^+\ C^+C^+\ D^+D^+ \times AA\ B^+B^+\ C^+C^+\ D^+D^+ \rightarrow AA\ B^+B^+\ C^+C^+\ D^+D^+$

In hybrid 1/2, there is complementation because the mutations are in different genes. In hybrid 1/5, the mutations are in the same gene, and complementation cannot occur.

19. a. If the nonsense mutation is homozygous, then complete translation will occur only with T^s. The traditional definition of a recessive allele is that it is seen in the phenotype when it is the only type of allele present. By this definition, T^+ is recessive to the dominant T^s.

b. In a trihybrid cross, the white phenotype will occur only when one or both of the enzymes is homozygous for a nonsense mutation and the wild-type form of T is also homozygous. The frequencies and phenotypes seen are

27	$A-$	$B- T^{s}-$	purple
9	$a^n a^n$	$B- T^{s}-$	purple
9	$A-$	$b^n b^n T^s-$	purple
9	$A-$	$B- T^+ T^+$	purple
3	$a^n a^n$	$b^n b^n T^s-$	purple
3	$a^n a^n$	$B- T^+ T^+$	white
3	$A-$	$b^n b^n T^+ T^+$	white
1	$a^n a^n$	$b^n b^n T^+ T^+$	white

or 57 purple to 7 white.

20. Let the new mutation be designated m. Cross the homozygote with different lines of tomatoes, each of which has mutations on different chromosomes. That is, if you let the known mutations be designated as a, you will be crossing $+ m/+ m$ with $a +/a +$. All progeny are $+ m/a +$. Next, let these F_1 plants self and look for independent assortment between a and m. If independent assortment is seen, the two genes are not linked. If there is not independent assortment, the genes are linked.

As an example of a specific cross, assume that the new mutation is on chromosome 1. You could use a plant that is mottled, dwarf, peach, oblate, normal (nonwooly), necrotic, compound, beaked, and multiloculed to cross with the new mutation in a background of homozygote wild type. The new mutation will show independent assortment with all genes except those within 50 map units of it.

As for the very practical questions asked in these problems, such as how much field space is needed, they cannot be answered from the text. You can imagine, however, that a rather large field space would be required, and at least two growing seasons would be needed to be confident that the new mutation had been mapped correctly.

21. The cross is $arg^r \times arg^+$, where arg^r is the revertant. However, it might also be $arg^- su \times arg^+ su^+$, where su^+ has no effect on the arg gene.

a. If the revertant is a precise reversal of the original change that produced the mutant allele, 100% of the progeny would be arginine-independent.

b. If a suppressor mutation on a different chromosome is involved, then the cross is $arg^- su \times arg^+ su^+$. Independent assortment would lead to the following:

1 $arg^- su$ arginine-independent

1 $arg^+ su^+$ arginine-independent

1 $arg^- su^+$ arginine-dependent

1 $arg^+ su$ arginine-independent

c. If a suppressor mutation 10 map units from the arg locus occurred, then the cross is $arg^- su \times arg^+ su^+$ but it is now necessary to write the diploid intermediate as arg^- su $/arg^+$ su^+. The two parental types would occur 90% of the time, and the two recombinant types would occur 10% of the time. The progeny would be

45% arg^- su arginine-independent

45% arg^+ su^+ arginine-independent

5% arg^- su^+ arginine-dependent

5% arg^+ su arginine-independent

22. a. The results show that the new mutants have the following epistatic relationships: $w^- > p^- > y^- > b^- > o^-$.

 b. There are a total of 10 heterokaryon pairs possible. All would have the phenotype of the wild type because of complementation.

 c. The cross is o^- p^- × o^+ p^+. Remember that haploid fungi immediately enter meiosis after gamete fusion, producing haploid progeny. Because the two genes are 16 map units apart, the recombinants will total 16% of the progeny and the parentals will be 84%. The genotypes and phenotypes of the progeny are given below.

42% o^- p^- pink

42% o^+ p^+ red

8% o^- p^+ orange

8% o^+ p^- pink

8

Chromosome Mutation I: Changes in Chromosome Structure

▶ IMPORTANT TERMS
  AND CONCEPTS

Chromosome mutations are changes in the genome involving chromosome parts, whole chromosomes, or whole chromosome sets. They are also known as **chromosome aberrations.** They can be detected either by genetic tests or by viewing the chromosomes under the microscope.

Cytogenetics is the combined study of cells (cytology) and genetics.

Chromosomes can be distinguished by size, centromere position, nucleolar organizers, satellites, and staining patterns.

Specific autosomal chromosomes within a genome are numbered. The larger the chromosome, the smaller the number. If two chromosomes are of the same size, the one with the more centrally positioned centromere has the lower number.

The **primary constriction** is the centromere. The **secondary constriction** is the **nucleolar organizing region**, or NO.

Metacentric chromosomes have the centromere in the middle of the chromosome. **Acrocentric** chromosomes have the centromere off center. **Telocentric** chromosomes have the centromere at the end of the chromosome. **Acentric** chromosomes do not have a centromere. **Dicentric** chromosomes have two centromeres.

A **telomere** is the end of a chromosome. A **satellite** is a small piece of chromosome distal to the nucleolar organizing region.

Heterochromatin is a densely staining region thought to be genetically inert. **Constitutive** heterochromatin does not vary from cell to cell within a species. **Facultative** heterochromatin varies with cell type within a species. **Euchromatin** stains very lightly and is thought to be genetically active.

Banding patterns along the length of chromosomes can occur naturally in some species, such as *Drosophila*, or be induced by various treatments in other species, such as humans.

Endomitosis is a process of chromosome replication not followed by cell division. It leads to an increase in the total number of chromosomes in a cell. In humans, a normal feature of liver and other cells is one round of endomitosis, leading to 92 chromosomes. In *Drosophila* salivary gland cells, endomitosis results in **polytene chromosomes.**

Chromosomal rearrangements consist of deletions, duplications, inversions, and translocations.

A **deletion** is the loss of a chromosome segment. A **terminal** deletion results from one chromosome break. An **interstitial** deletion results from two chromosome breaks; the region between the two breaks is lost when chromosome repair occurs. Deletions frequently result in **pseudodominance**, the expression of a recessive gene when present in a single copy. A deletion results in an unpaired loop during synapsis. Homozygous deletions usually are lethal.

A **duplication** is the presence of more than one copy of a chromosome segment on one chromosome. Adjacent duplicated segments occur in **tandem sequence** with respect to each other (abcdabcd) or they may occur in **reverse order** with respect to each other (abcddcba). Duplications, like deletions, can disturb the genetic balance of the genome, resulting in abnormal development or function. They also supply additional genetic material capable of evolving new functions. Duplications result in an unpaired loop during synapsis.

Inversions result from two chromosome breaks, with a subsequent "flipping" of the middle segment with respect to the two ends, followed by chromosome repair. **Paracentric** inversions do not involve the centromere. Crossing-over in the inverted region of a paracentric inversion leads to an acentric fragment and a dicentric chromosome, which then enters the **breakage-fusion-bridge cycle**, resulting in duplications and deletions. **Pericentric** inversions involve the centromere. Crossing-over in the inverted region of a pericentric inversion leads to duplications and deletions. Inversion heterozygosity results in a reduction of viable recombinant gametes. Inversion heterozygotes have a paired loop during synapsis.

A **translocation** is the movement of a segment of a chromosome to a new location. A **reciprocal translocation** occurs when there is one break in each of two chromosomes followed by an exchange of the acentric fragments. New linkage relations are created by translocations. When nonhomologous chromosomes are involved, pairing at synapsis in the heterozygote results in a cross configuration involving four chromosomes. Synapsis in the homozygote does not result in the cross structure. Translocation heterozygosity leads to greatly reduced fertility because segregation usually results in both duplications and deletions for entire chromosomes.

Position effect is the alteration in a gene's functioning caused by a change in the gene's location.

Questions for Concept Map 8-1

1. If a chromosome at the one-chromatid stage breaks and the acentric piece is lost, what happens to the other portion that has a centromere?

2. What causes chromosome breakage?

3. Is the information that is gained, lost, or rearranged "good" information, or is it information that is not normally found in healthy cells?

4. What is position effect?

5. Could position effect also occur with a deletion?

6. Why can there be no reversion of a deletion?

7. Is any other mutation besides a deletion characterized by no reversion?

8. What is *pseudodominance*?

9. Do duplications have to be in tandem?

► CONCEPT MAP 8-1:
  CHROMOSOME
  MUTATIONS

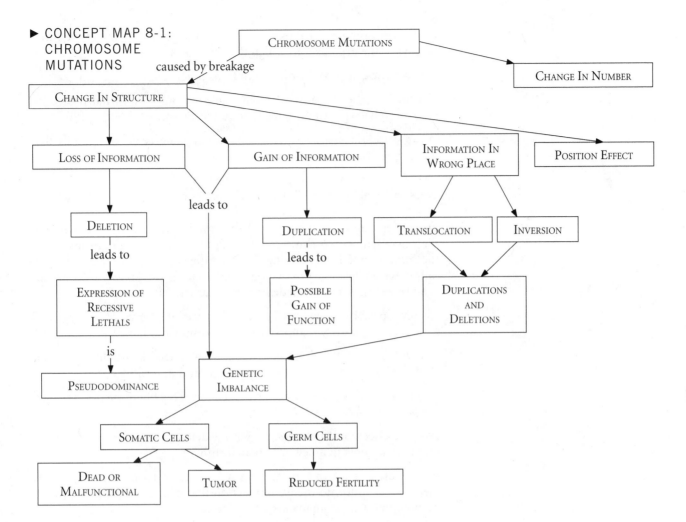

10. How many chromosome breaks are required for an inversion? A translocation?

11. Is the size of the genetic material that is gained, lost, or moved of any significance?

12. Is the information encoded within the genetic material that is gained, lost, or moved of any significance?

13. How does a gain of function arise from a duplication?

14. What is meant by *genetic imbalance?*

15. Which would be more unbalancing, the loss of a chromosome or the gain of a chromosome?

16. Which would be more unbalancing, the loss of a large chromosome or the loss of a small chromosome?

17. Which would be more unbalancing, the gain of a large chromosome or the gain of a small chromosome?

18. In humans, which chromosome is larger, the X or number 21?

19. Can humans tolerate the loss or gain of chromosome 21 better than we can tolerate the loss or gain of an X chromosome?

20. Which process partially protects against the unbalancing effects of a loss or gain of the X chromosome?

21. Do chromosome mutations that occur in the germ cells affect the functioning of the organism in which the germ cells reside?

22. Are somatic cell chromosome mutations inherited?

23. If a tumor arises because of a chromosome mutation, what protective system in the organism has failed to function properly?

24. What would the characteristics of a human pedigree be for a family that has a chromosome mutation in it?

Answers to Concept Map 8-1 Questions

1. During replication the broken end behaves as if it were sticky and fuses with the newly synthesized chromatid that also has a broken end. At anaphase, the chromosome forms a bridge between the two poles, ultimately breaks, only to repeat the same process again and again. This is known as the breakage-fusion-bridge cycle, first described by Barbara McClintock.

2. Chromosome breakage can be caused by various forms of energy (UV, X rays, etc.), chemicals, and viruses, and simply by chance.

3. The information that is gained, lost, or rearranged is usually "good," or wild-type, information that is normally found in healthy cells.

4. The location of a gene affects its expression. This phenomenon is called the *position effect*.

5. Position effect could also occur with a deletion because the two ends that fuse are now in a new microenvironment.

6. A deletion is a missing piece, and it is extremely unlikely that a new piece will be put into the exact location that contains exactly the information that was originally lost.

7. Although chromosome rearrangements theoretically revert to wild type, it is highly unlikely. Likewise, it would be highly unlikely for a monosomy to revert, although it is theoretically possible.

8. *Pseudodominance* is the expression of a recessive within an organism that is heterozygous. It requires either a loss of the dominant allele or a mutation of the dominant allele. Loss can occur through nondisjunction or mitotic crossing-over.

9. Duplications do not have to be in tandem.

10. Two chromosome breaks are required for an inversion or a translocation.

11. The size of the genetic material that is gained, lost, or moved can be of major significance.

12. The information encoded within the genetic material that is gained, lost, or moved is of major significance.

13. A gain of function arises from a duplication through subsequent mutation of one copy.

14. *Genetic imbalance* means that an organism does not possess the right number of all the genes that the wild type has.

15. The loss of a chromosome is more unbalancing than the gain of a chromosome. In addition, recessive alleles can be uncovered by a loss.

16. The loss of a large chromosome produces more genetic imbalance than the loss of a small chromosome because, presumably, the larger the chromosome, the greater the amount of genetic material.

17. The gain of a large chromosome would result in greater genetic imbalance than the gain of a small chromosome.

18. In humans, the X is larger than chromosome 21.

19. The loss or gain of an X chromosome is tolerated by humans better than the loss or gain of chromosome 21.

20. X-inactivation partially protects against the unbalancing effects of a loss or gain of the X chromosome.

21. Chromosome mutations that occur in the germ cells do not affect the functioning of the organism in which the germ cells reside.

22. Somatic cell chromosome mutations are inherited in species that have vegetative reproduction.

23. If a tumor arises because of a chromosome mutation, the immune system in the organism has failed to function properly.

24. A human pedigree of a family that has a chromosomal mutation in it would have a large number of spontaneous abortions over several generations.

Questions for Concept Map 8-2

1. What is speciation?

2. What is partial reproductive isolation?

3. Why do heterozygotes for a chromosomal mutation have reduced fertility? Be specific.

4. What are the specific recombinational changes associated with translocations? Inversions?

5. Inversions are sometimes called *recombination suppressors*. What is actually suppressed?

6. What is the name of the process that leads to an exchange of genetic material between homologous chromosomes? Nonhomologous chromosomes? Sister chromatids?

7. Draw a Robertsonian translocation and take the resulting chromosome or chromosomes through meiosis. Do the same for a reciprocal translocation, a paracentric inversion, and a pericentric inversion.

8. Using a pericentric inversion, show the meiotic outcome of a crossover in the noninverted region and the inverted region.

9. What is a dicentric chromosome?

10. What happens to dicentric chromosomes at anaphase?

Answers to Concept Map 8-2 Questions

1. Speciation is the evolution of a new species.

► CONCEPT MAP 8-2:
TRANSLOCATIONS AND
INVERSIONS

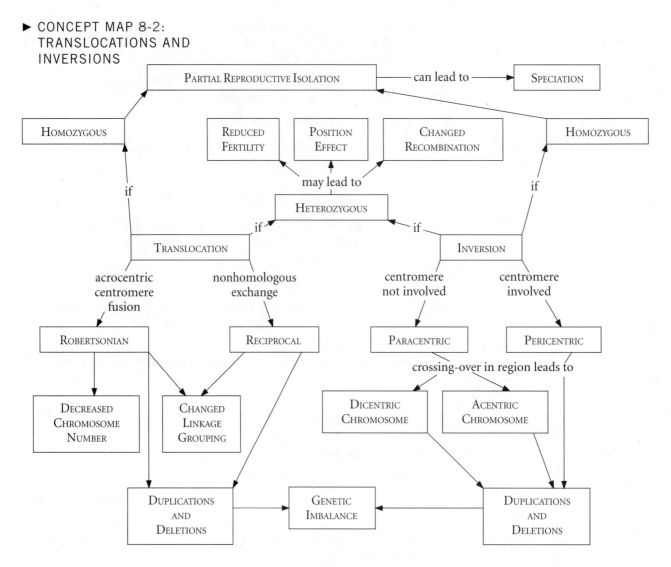

2. Partial reproductive isolation is the reduced fertility that comes from the crossing of two breeds of a species. It can be mechanical, behavioral, geographic, or chromosomal as the result of inversions and translocations.

3. Heterozygotes for a chromosome mutation have reduced fertility because meiotic pairing can produce unbalanced gametes.

4. There are no specific recombinational changes that result from translocations. The reproductive problems are due to alignment on the metaphase plate that can lead to a reduction in some categories of progeny. Inversions cause specific recombinational changes because a single crossover within an inverted region leads to unbalanced gametes.

5. Inversions are called *recombination suppressors* because the products of crossing-over within an inverted region are not viable and, therefore, not observed in the progeny.

6. The exchange of genetic material between homologous chromosomes is crossing-over. The exchange of genetic material between nonhomologous chromosomes in called *translocation*. The exchange of genetic material between sister chromatids is called *sister chromatid exchange*.

7.   Robertsonian Translocation

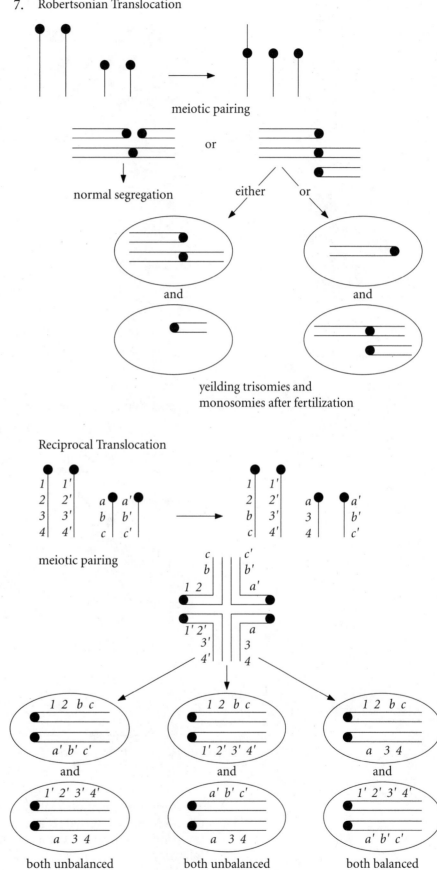

meiotic pairing

or

normal segregation

either   or

and   and

yeilding trisomies and
monosomies after fertilization

Reciprocal Translocation

meiotic pairing

and   and   and

both unbalanced   both unbalanced   both balanced

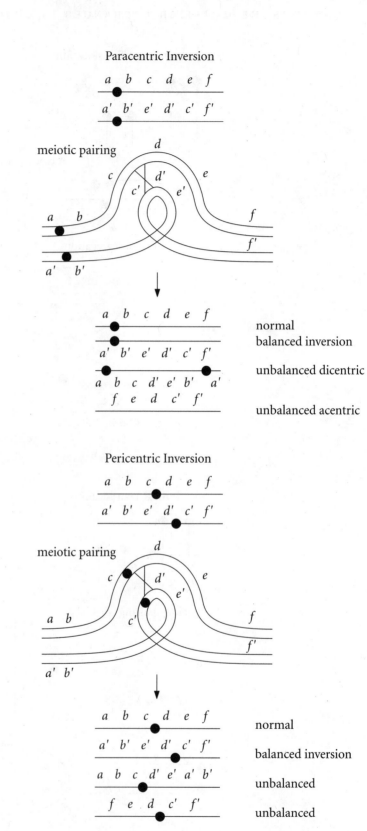

Paracentric Inversion

meiotic pairing

normal
balanced inversion

unbalanced dicentric

unbalanced acentric

Pericentric Inversion

meiotic pairing

normal

balanced inversion

unbalanced

unbalanced

8. The answer to Question 7, above, showed the meiotic outcome of a crossover in the inverted region of a chromosome with a pericentric inversion. A crossover in the noninverted region would yield the following:

Pericentric Inversion, crossover in noninverted region

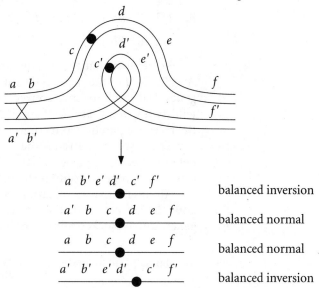

a b' e' d' c' f'   balanced inversion

a' b c d e f   balanced normal

a b c d e f   balanced normal

a' b' e' d' c' f'   balanced inversion

9. A dicentric chromosome is a chromosome with two centromeres.

10. At anaphase, the chromosome forms a bridge between the two mitotic poles, and then both daughter cells enter the breakage-fusion-bridge cycle.

▶ SOLUTIONS TO PROBLEMS

Be sure that you have thoroughly read the entire chapter before you attempt any of the problems.

1. **a.** Deletions lead to a shorter chromosome with missing bands, if banded, an unpaired loop during homologous pairing, and the expression of hemizygous recessive alleles.

   b. Duplications lead to a longer chromosome with repeated bands, if banded, and an unpaired loop during homologous pairing; there may be disturbed development.

   c. Inversions can be detected by banding, and they show the typical twisted homologous pairing for heterozygotes. Pericentric inversions result in a change in the p:q ratio. No crossover products are seen for genes within the inversion in the heterozygote, and linked genes show a decrease in RF.

   d. Reciprocal translocations can be detected by banding. They show the typical cross structure during homologous pairing, lead to new linkage groups, and show altered linkage relationships. The heterozygote has a high rate of unbalanced gamete production.

2. **a.** In heterozygotes, the products of crossing-over will be nonviable 25 percent of the time. Thus, the RF will be about three-fourths of the normal value, or 27 percent.

   b. The homozygote will have no trouble with crossing-over, so the RF will be about 36 percent.

3. P   *A–B–C– D– E–F–* × *aa bb cc dd ee ff*

F$_1$ 1/2 *Aa Bb Cc Dd Ee Ff*

1/2 *Aa Bb Cc dd ee Ff*

Remember that these genes are linked. Because all progeny flies are *A– B– C– F–*, the wild type must have been homozygous for these genes. Half the progeny received *D E*, which had to have come from the wild-type parent, and half received *d e*, which presumably came from the recessive parent. No recombinants were seen.

The best explanation is that the wild-type fly was heterozygous for a deletion spanning genes *D* and *E*. With this explanation, the second class of F$_1$ progeny would be hemizygous for *d* and *e*, or *Aa Bb Cc de Ff*.

An alternative explanation is that the wild-type fly was heterozygous *D E/d e* and that a heterozygous inversion spanned these two genes, which blocked all recombination products from being seen.

4. **a.** paracentric inversion

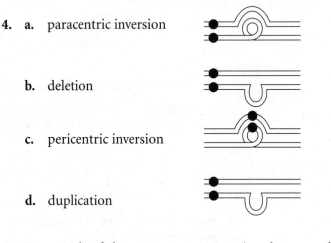

b. deletion

c. pericentric inversion

d. duplication

5. From each of the statements concerning the rare cells, you should be able to draw the following conclusions:

STATEMENT	CONCLUSION
require leucine	*leu*$^+$ lost
do not mate	one mating type lost
will not grow at 37°C	*un*$^+$ lost
cross only with *a* type	*a* lost
only nucleus 1 recovered	deletion occurred in nucleus 2

Because *ad-3A*$^+$ and *nic*$^+$ function are required by the heterokaryon, these genes must have been retained. Therefore, the most reasonable explanation is that a deletion occurred in the left arm of the large chromosome in nucleus 2 and that *leu*$^+$, mating type *a*, and *un*$^+$ were lost.

6. The single waltzing female that arose from a cross between waltzers and normals is expressing a recessive gene when it is present in one copy. That means that the normal allele must have been deleted.

When she was mated to a waltzing male, all the progeny were waltzers. This further supports the conclusion that there was a deletion

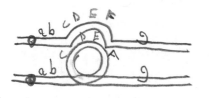

of the normal allele. Had it been present, half the progeny would have been normal.

When she was mated to a normal homozygous male, all the progeny were normal. This indicates that her waltzing was due to a recessive allele and does not represent a dominant variant for waltzing. If a deletion occurred, half these offspring are heterozygous normal and half are hemizygous normal.

When some of these offspring are mated, the progeny are normal. The offspring are w^+w (normal) or $w^+w^\star$ (normal, with deletion). Without a deletion in the normal female, one-fourth of their progeny would be waltzers. With a deletion, progeny homozygous for a deletion would be expected to die. Progeny that are $ww^\star$ would be expected to be waltzers. These waltzers were not observed. Because this is the only discrepancy, it must be assumed that not enough crosses were performed.

The cytological data also support the conclusion that a deletion occurred, because the abnormal waltzing female and her abnormal progeny had one member of a chromosome pair that was abnormally short.

7. The colonies that would not revert most likely had a deletion within the *ad-3B* gene. If the gene had not been deleted, at least some reversion would have been seen. Because the gene was deleted, however, there was nothing there for the mutagens to work on. These colonies could grow only with adenine supplementation.

8. Compare deletions 1 and 2: allele *b* is more to the left than alleles *a* and *c*. The order is *b* (*a c*), where the parentheses indicate that the order is unknown.

    Compare deletions 2 and 3: allele *e* is more to the right than (*a c*). The order is *b* (*a c*) *e*.

    Compare deletions 3 and 4: allele *a* is more to the left than *c* and *e*, and *d* is more to the right than *e*. The order is *b a c e d*.

    Compare deletions 4 and 5: allele *f* is more to the right than *d*. The order is *b a c e d f*.

ALLELE	BAND
b	1
a	2
c	3
e	4
d	5
f	6

9. **a. and b.** When a deletion is crossed with a recessive point mutation in the same gene, the recessive point mutation is expressed. When the deletion and the point mutation are in different genes, wild type is observed. This is exactly like the results seen in a cross between two allelic variants (non-wild-type progeny) and a cross between organisms that have defects in two different genes (wild-type progeny).

MUTANT	DEFECT
1	deletion of at least part of genes *h* and *i*
2	deletion of at least part of genes *k* and *l*
3	deletion of at least part of gene *m*
4	deletion of at least part of genes *k*, *l*, and *m*
5	deletion not within the *h* through *m* genes, or a recessive point mutation

10. a. Eighteen map units (m.u.) were either deleted or inverted. A large inversion would result in semisterility, whereas a large deletion would most likely be lethal. Thus, an inversion is more likely.

b. Normal   Inversion

Normal: a b c d e f, P1 ... sm

Inversion: a d c b e f, P1 sm

Alternatively, *sm* could be located external to the *e* locus on both the normal and the inversion chromosomes.

c. The semisterility is the result of crossing-over in the inverted region. All products of crossing-over would have both duplications and deletions.

11. The testcross is *P B Q/p B q × p b q/p b q*.

a. To obtain normal eye shape, the bar allele must be deleted. Actually, bar eye is the result of a tandem duplication of a normal allele rather than a variant allele that results in bar eye. Thus, to delete the extra copy of the gene, synapsis must occur out of register between the two chromosomes:

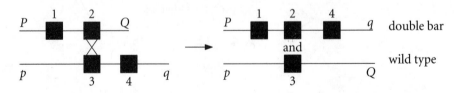

b. In the diagram above, the flanking markers are *P q* and *p Q* after crossing-over. If gene 1 had paired with gene 4, then the wild-type phenotype would have been associated with *P q*, and the double bar would have been with *p Q*.

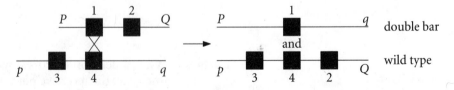

12. a. Classical dominance states that one copy of the dominant allele is sufficient for the dominant phenotype. In this case, the female progeny have one copy of the dominant allele, yet they do not have the dominant phenotype. To explain these results, it must be assumed that both the dominant and the variant alleles have a product and that the phenotype is the result of the ratio of one product to the other.

b. The female parents are v^+v^-/v^+v^-. The ratio of the two alleles is 1:1, which results in the wild-type phenotype. The male parents are v^-/Y. No dominant alleles are present, and the males are vermilion.

c. The male progeny are v^+v^-/Y, and the ratio of the two alleles is 1:1, yielding a wild-type phenotype identical with that of their female parents. The female progeny are v^+v^-/v^-. The ratio of dominant alleles to vermilion is 1:2. Thus, they have the vermilion phenotype.

13. The most likely explanation is that one or both break points were located within essential genes, leading to a lethal recessive mutation.

14. a. Single crossovers lead to tetratypes. Also, only half the tetratype asci are crossover products, so the 10 m.u. must be multiplied by 2 to yield the right frequencies.

$un3^+\ ad3^+$	$un3^+\ ad3^+$	$un3^+\ ad3^+$
$un3^+\ ad3^+$	$un3^+\ ad3^+$	$un3^+\ ad3^+$
$un3^+\ ad3^+$	$un3^+\ ad3$	$un3^+\ ad3$
$un3^+\ ad3^+$	$un3^+\ ad3$	$un3^+\ ad3$
$un3\ ad3$	$un3\ ad3^+$	$un3\ ad3$
$un3\ ad3$	$un3\ ad3^+$	$un3\ ad3$
$un3\ ad3$	$un3\ ad3$	$un3\ ad3^+$
$un3\ ad3$	$un3\ ad3$	$un3\ ad3^+$
80%	10%	10%

b. The aborted spores result from an inversion in the wild type. Crossing-over led to nonviable spores because they were unbalanced. This could be tested by selecting $un3\ ad3$ double mutants from the wild type and then crossing them with the $un3^+\ ad3^+$ inverted strain. The $un3$ to $ad3$ distance should be altered.

15. a.

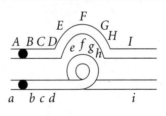

b.

1 A B C d h G F E D C B A   dicentric

2 I H g f e i   acentric

3 a b c D E F G H I   viable

4 a b c d h g f e i

c.

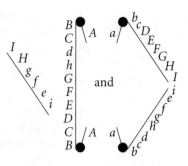

d. The chromosomes numbered 3 and 4 will give rise to viable progeny. The genotypes of those progeny will be *Aa Bb Cc DD EE FF GG HH II* and *Aa Bb Cc Dd Ee Ff Gg Hh Ii*.

16. a. This is a paracentric inversion.

b.

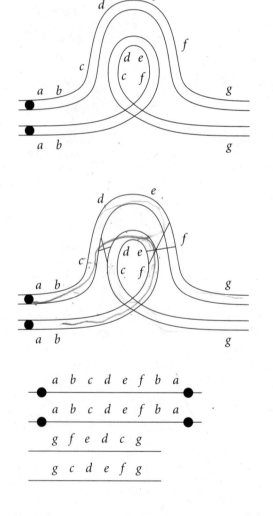

c.

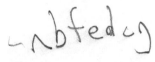

abcdefba

abfedcba

gfedcg

gcdefg

d.

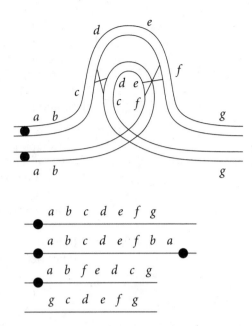

e.   The strand not involved in the three-strand double crossover will be normal.

17. a.   The Sumatra chromosome is a pericentric inversion compared with the Borneo chromosome.

b.

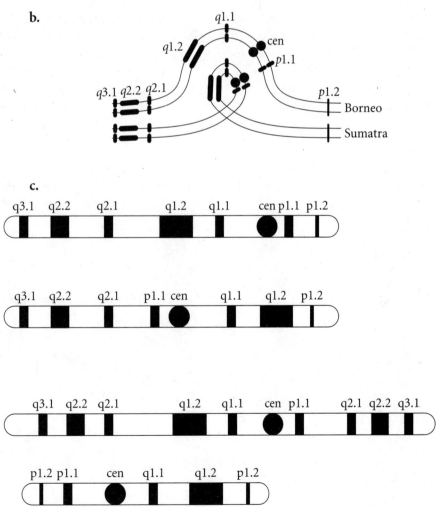

c.

d. Recall that all single crossovers within the inverted region will lead to 50% viable parental types and 50% extremely unbalanced gametes that will most likely be nonviable. In other words, if 30% of the meioses have a crossover in this region, 15% of the gametes will not lead to viable progeny. That means that 85% of the gametes should produce viable progeny.

18. a. Two crosses show 28 map units between the loci for body color and eye shape in a testcross of the F_1: California × California and Chile × Chile. The third type of cross, California × Chile, leads to only 4 map units between the two genes when the hybrid is testcrossed. This indicates that the genetic distance has decreased by 24 map units, or 100%(24/28) = 85.7%. A deletion cannot be used to explain this finding, nor can a translocation. Most likely the two lines are inverted with respect to each other for 85.7% of the distance between the two genes.

b.

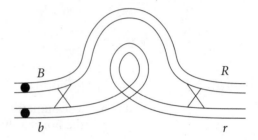

A single crossover in either region would result in 4% crossing-over between B and R. The products are

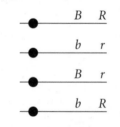

19. a. To construct the maps, look at three genes at a time. For example, the Okanagan sequence for a, b, and c is

a to b: 12 m.u.

a to c: 14 m.u.

b to c: 2 m.u.

The only possible sequence is $a\ b\ c$

Okanagan sequence: Spain sequence:

b. Diagram the heterozygote during homologous pairing. Crossing-over can occur between *c* and *f* and between *d* and *b*, as indicated in the following figure.

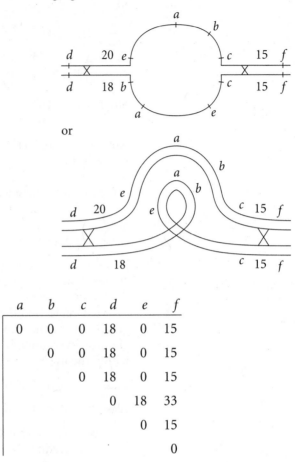

	a	*b*	*c*	*d*	*e*	*f*
a	0	0	0	18	0	15
b		0	0	18	0	15
c			0	18	0	15
d				0	18	33
e					0	15
f						0

The only recombination events that will lead to viable progeny occur in two regions: (1) the 18 map units between *d* and the breakpoint and (2) the 15 map units between *c* and *f*. Therefore, the recombination frequency between *d* and any gene in the inverted region is 18%. The *c-f* recombination frequency is obviously 15%. Finally, the *d-f* recombination frequency is 18% + 15%, or 33%.

20. a. The aberrant plant is semisterile, which suggests an inversion. Since the *d–f* and *y–p* frequencies of recombination in the aberrant plant are normal, the inversion must involve *b* through *x*.

b. To obtain recombinant progeny when an inversion is involved, either a double crossover occurred within the inverted region or single crossovers occurred between *f* and the inversion, which occurred someplace between *f* and *y*.

21. The cross is

P   *cc bzbz wxwx shsh dd* × *CC BzBz WxWx ShSh DD*

F₁   *Cc Bzbz Wxwx Shsh Dd*

Backcross   *Cc Bzbz Wxwx Shsh Dd* × *cc bzbz wxwx shsh dd*

a. The total number of progeny is 1000. Classify the progeny as to where a crossover occurred for each type. Then, total the number of crossovers between each pair of genes. Calculate the observed map units (m.u.).

REGION	# CO	M.U. OBSERVED	M.U. EXPECTED
C–Bz	103	10.3	12
Bz–Wx	13	1.3	8
Wx–Sh	13	1.3	10
Sh–D	186	18.6	20

Notice that a reduction of map units, or crossing-over, is seen in two intervals. Results like this are suggestive of an inversion. The inversion most likely involves the *Bz*, *Wx*, and *Sh* genes in the wild-type plant as compared with the standard stock.

Further notice that all those instances in which crossing-over occurred in the proposed inverted region involved a double crossover. This is the expected pattern.

b. A number of possible classes are missing: four single-crossover classes resulting from crossing-over in the inverted region, eight double-crossover classes involving the inverted region and the non-inverted region, and triple crossovers and higher. The 10 classes detected were the only classes that were viable. They involved a single crossover outside the inverted region or a double crossover within the inverted region.

c. Class 1   parental; increased due to nonviability of some crossovers

Class 2   parental; increased due to nonviability of some crossovers

Class 3   crossing-over between *C* and *Bz*; approximately expected frequency

Class 4   crossing-over between *C* and *Bz*; approximately expected frequency

Class 5   crossing-over between *Sh* and *D*; approximately expected frequency

Class 6   crossing-over between *Sh* and *D*; approximately expected frequency

Class 7   double crossover between *C* and *Bz* and between *Sh* and *D*; approximately expected frequency

Class 8   double crossover between *C* and *Bz* and between *Sh* and *D*; approximately expected frequency

Class 9   double crossover between *Bz* and *Wx* and between *Wx* and *Sh*; approximately expected frequency

Class 10   double crossover between *Bz* and *Wx* and between *Wx* and *Sh*; approximately expected frequency

d. Cytological verification could be obtained by looking at chromosomes during meiotic pairing. Genetic verification could be achieved by crossing two homozygotes for the inversion wildtype.

22. a. and b. The F_1 females are *y cv v f B^+ car/y^+ cv^+ v^+f^+ B^+ car^+*. These are crossed with *y cv v f B car*/Y males.

Class 1:   parental

Class 2:   parental

Class 3:   DCO *y–cv* and *B–car*

Class 4:   reciprocal of class 3

Class 5:   DCO *cv–v* and *v–f*

Class 6:   reciprocal of class 5

Class 7:   DCO *cv–v* and *f–car*

Class 8:   reciprocal of class 7

Class 9:   DCO *v–cv* and *v–f*

Class 10: reciprocal of class 9

Class 11: This class is identical with the male parent's X chromosome and could not have come from the female parent. Thus, the male sperm must have donated it to the offspring. In *Drosophila*, sex is determined by the ratio of X chromosomes to the number of sets of autosomes. The ratio in males is 1X:2A, where A stands for the autosomes contributed by one parent (the ratio in females is 2X:2A). Thus, this class of males must have arisen from the union of an X-bearing sperm with an egg that was the product of nondisjunction for X and contained only autosomes.

Because all nonparental progeny are double crossovers, there must be an inversion spanning the genes that were studied.

c.   Class 11 should have only one sex chromosome, which could be checked cytologically.

23. The inversion results in no viable crossover products from heterozygous females. When *Cu pr*/*Cu pr* females are crossed with irradiated wild-type males, all female progeny will be heterozygous for the inversion and for any recessive lethal mutation induced by irradiation. They will have curled wings and wild-type eyes (*Cu pr*/*Cu*$^+$ *pr*$^+$?). Each female will carry a different mutation (indicated by "?"), if any were induced. Cross the females individually with a homozygous *Cu pr* male to generate groups of flies with the same mutation. Then, cross the normal-eyed progeny among themselves (*Cu pr*/*Cu*$^+$ *pr*$^+$?). This results in

1/4 *Cu*$^+$ *pr*+ ?/*Cu*$^+$ *pr*$^+$?   normal wings and eyes

1/2 *Cu pr*/*Cu*$^+$ *pr*$^+$ curled wings and normal eyes

1/4 *Cu pr*/*Cu pr*   curled wings and purple eyes

If a lethal mutation had been induced, the class with normal wings and eyes would be missing.

24.

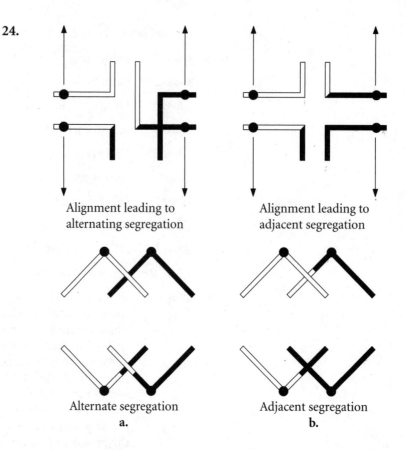

Alignment leading to   Alignment leading to
alternating segregation   adjacent segregation

Alternate segregation   Adjacent segregation
         a.   **b.**

25. If the *a* and *b* genes are on separate chromosomes, independent assortment should occur, giving equal frequencies of *a b*, *a⁺ b⁺*, *a b⁺*, and *a⁺ b*. This was not observed; instead, the two genes are behaving as if they were linked, with 10 m.u. between them. This behavior is indicative of a reciprocal translocation in one of the parents, most likely the wild type from nature.

At meiosis prior to progeny formation, the chromosomes would look like the following figure (centromere not included because its position has no effect on the results):

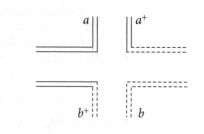

Only alternate segregation avoids duplications and deletions for many genes. Therefore, the majority of the progeny would be parental *a b* and *a⁺b⁺*. The *a b⁺* and *a⁺ b* progeny would result from crossing-over between either gene and the breakpoint locus.

26. Notice that the males have the male parent phenotype and the females have the female parent phenotype. This suggests a translocation of the *Cy* chromosome to the Y chromosome:

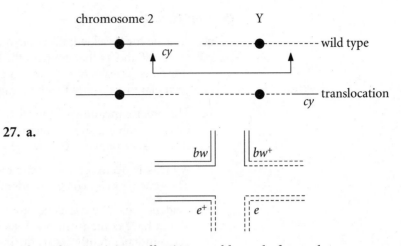

27. a.

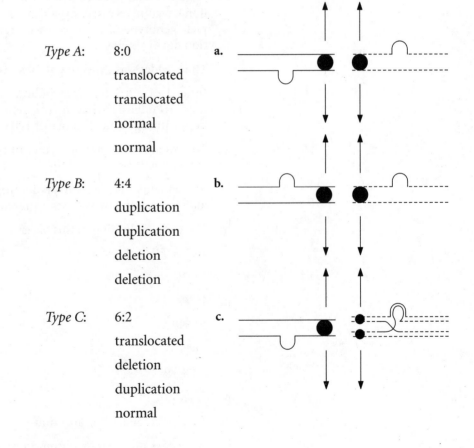

b. The surviving offspring would result from alternate segregation and would be either *bw e* (brown eye, ebony body) or *bw⁺e⁺* (wild type), in a 1:1 ratio.

28. The size of the insertional translocation will determine whether the translocated region or the rest of the region will dominate homologous pairing. Below, it is assumed that the insertion is quite small in relation to the rest of both chromosomes.

   The following asci types will be seen:

Type A: 8:0   **a.**
            translocated
            translocated
            normal
            normal

Type B: 4:4   **b.**
            duplication
            duplication
            deletion
            deletion

Type C: 6:2   **c.**
            translocated
            deletion
            duplication
            normal

29. *Unpacking the Problem*

  a. A "gene for tassel length" means that there is a gene with at least two alleles (*T* and *t*) that controls the length of the tassel. A "gene for rust resistance" means that there is a gene which determines whether the corn plant is resistant to a rust infection or not (*R* and *r*).

  b. The precise meaning of the allelic symbols for the two genes is irrelevant to solving the problem because what is being investigated is the distance between the two genes.

  c. A locus is occupied by a gene pair on homologous chromosomes. The gene pair can consist of identical or different alleles.

  d. Evidence that the two genes are normally on separate chromosomes would have come from previous experiments showing that the two genes independently assort during meiosis.

  e. Routine crosses could consist of F_1 crosses, F_2 crosses, backcrosses, and testcrosses.

  f. The genotype *Tt Rr* is a double heterozygous, or dihybrid, or F_1 genotype.

  g. The pollen parent is the male that contributes to the pollen tube nucleus, the endosperm nucleus, and the progeny.

  h. Testcrosses are crosses that involve a genotypically unknown and a homozygous recessive organism. They are used to reveal the complete genotype of the unknown organism and to study recombination during meiosis.

  i. The breeder was expecting to observe 1 *Tt Rr* :1 *Tt rr* :1 *tt Rr* :1 *tt rr*.

  j. Instead of a 1:1:1:1 ratio indicating independent assortment, the testcross indicated that the two genes were linked in a cis configuration, with a genetic distance of $100\%(3 + 5)/210 = 3.8$ map units.

  k. The equality and predominance of the first two classes indicate that the parentals were *TR / tr*.

  l. The equality and lack of predominance of the second two classes indicate that the two classes represent recombinants.

  m. The gametes leading to this observation were

  46.7% *TR*   1.4% *Tr*

  49.5% *tr*   2.4% *tR*

  n. 46.7% *TR*

  49.5% *tr*

  o. 1.4% *Tr*

  2.4% *tR*

  p. *Tr* and *tR*

  q. *T* and *R* are linked, as are *t* and *r*.

  r. Two genes on separate chromosomes can become linked through a translocation.

  s. The parents of the hybrid plant must both have contained a translocation for the same genetic regions.

t.   A corn cob is a structure that holds on its surface the progeny of the next generation.

u.

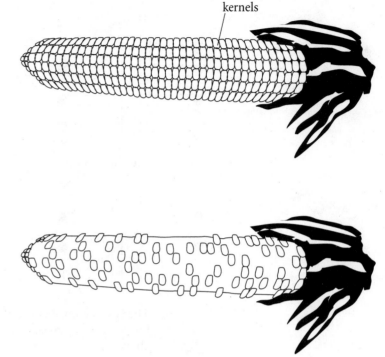

kernels

v.

w.   A kernel is one progeny on a corn cob.

x.   Absence of half the kernels can be caused by a meiotic process that leads to 50% aborted progeny.

y.   Approximately 50% of the progeny died. Meiotic processes in the female were the reason for the progeny death.

Solution to the Problem

a.   The progeny are not in a 1:1:1:1 ratio, indicating independent assortment; instead, the data indicate close linkage.

b.   The observations could have been caused by a translocation that brought the two loci close together.

c. Assume a translocation heterozygote in coupling. If pairing is as diagrammed below, then you would observe the following:

No Crossover

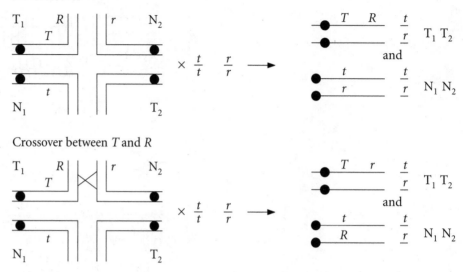

Crossover between *T* and *R*

d. The two recombinant classes result from a recombination event followed by proper segregation of chromosomes, as is diagrammed above.

e. The two minority classes are the result of recombination.

30. a. The cross was *nic-2 leu-3* × *nic-2⁺ leu-3⁺*. The expected results were

1/4 *nic-2 leu-3*   1/4 *nic-2⁺ leu-3*

1/4 *nic-2⁺ leu-3⁺*   1/4 *nic-2 leu-3⁺*

b. Only parentals were obtained. This could be observed if a translocation located both genes on the same chromosome. Alignment at metaphase I could be as follows, which would lead to abortion in the case of duplication and deletion progeny:

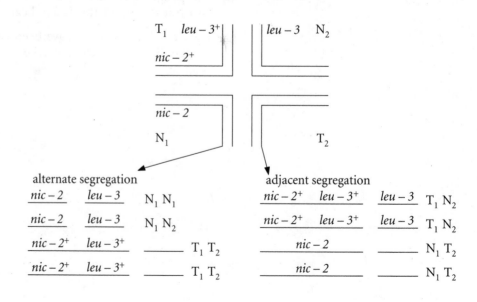

31. a. The homozygous wild-type parental could have been homozygous for a translocation. The translocation could result in the two genes' being on two chromosomes or on one. The latter is assumed below:

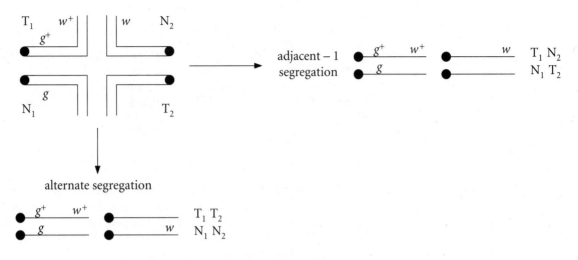

b. In order to solve this problem, list the gametes and their frequencies, and put this information into a Punnett square. Assume that unbalanced gametes are not viable, that adjacent-2 segregation does not occur, and that alternate and adjacent-1 segregation occur at equal frequencies. The gamete composition, but not the phenotypic outcomes, differ with the choice of whether the two genes are on two or one chromosomes.

	1/4 g^+w^+ T_1T_2	1/4 gw N_1N_2	1/4 g^+w^+w T_1N_2	1/4 g N_1T_2
1/4 g^+w^+ T_1T_2	1/16 ++	1/16 ++	abort	abort
1/4 gw N_1N_2	1/16 ++	1/16 gw	abort	abort
1/4 g^+w^+w T_1N_2	abort	abort	abort	1/16 ++
1/4 g N_1T_2	abort	abort	1/16 ++	abort

The viable progeny are

5/16 ++

1/16 gw

or 5 wild type to 1 green, waxy.

Of the wild type, four are semisterile and would have 50% abortion, and one is homozygous for the translocation and would produce 100% normal progeny. The green, waxy would produce 100% normal progeny. This assumes that abnormal gametes are capable of compensating for each other.

32. The cross was

P X^+Y (irradiated) $\times X^e X^e$

F_1 most $X^e Y$ yellow males

  two ?Y   gray males

a. The gray male 1 was crossed with a yellow female, yielding yellow females and gray males, which is reversed sex linkage. If the e^+ allele was translocated to the Y chromosome, the gray male would be $X^e Y\text{–}e^+$, or gray. When crossed with yellow females, the results would be

$X^e Y\text{–}e^+$ gray males

$X^e X^e$ yellow females

b. The gray male 2 was crossed with a yellow female, yielding gray and yellow males and females in equal proportions. If the e^+ allele was translocated to an autosome, the progeny would be as below, where "A" indicates an autosome involved in the translocation and the "/" separates male and female contributions:

P $A^{e^+} A\ XY \times AA\ X^e X^e$

F_1 $A^{e^+} X/A\ X^e$ gray female

  $A^{e^+} Y/A\ X^e$ gray male

  $A\ X/A\ X^e$ yellow female

  $A\ Y/A\ X^e$ yellow male

33. *Cross 1*: Independent assortment of 2 genes occurred.

Cross 2: Two genes are linked at 1 m.u. distance. Therefore, a reciprocal translocation took place and both genes were very close to the breakpoint. The black spores resulted from alternate segregation, the white from adjacent segregation.

Cross 3: Half the spores were normal, and nontranslocated, and half contained both translocated chromosomes.

34. The F_1 is heterozygous for both the translocation and *Pp*. Therefore, it is semisterile. A crossover that occurs within the region from the breakpoint to the centromere will be viable. Among the viable crossovers, two-thirds will maintain the link between the *P* allele and semisterility and one-third will disrupt that linkage. Therefore, the frequency of crossing-over that needs to be considered is 30 m.u. – 20 m.u. = 10 m.u. When the F_1 is backcrossed to the *pp* parent, which does not contain the translocation, the following progeny are obtained, where $\star$ denotes the translocation chromosome.

Parentals

45%	$P^\star/p$	purple, semisterile
45%	p/p	green, fully fertile

Recombinants

5%	P/p	purple, fully sterile
5%	$p^\star/p$	green, semisterile

a. green, semisterile = 5%

b. green, fully fertile = 45%

c. purple, semisterile = 45%

d. purple, fully fertile = 5%

e. To solve this problem, recognize that the gametes from the F_1 will occur in the same frequency as the progeny in the backcross. Green, fully fertile plants can arise from the union of two nontranslocated gametes, p, or from the union of two gametes bearing the translocation, $p^\star$. Therefore, the probability of obtaining green, fully fertile plants from an F_1 selfing is

$$\mathrm{p}(pp) + \mathrm{p}(pp^{\star\star}) = (0.45)(0.45) + (0.05)(0.05) = 0.2050$$

A third possibility does exist, although it is rather unlikely. If there is fusion of two unbalanced gametes in which the unbalanced genetic components are complementary, then a balanced embryo would result.

35. The original plant had two reciprocal translocation chromosomes that brought genes P and S very close together. Because of the close linkage, a ratio suggesting a monohybrid cross, instead of a dihybrid cross, was observed, both with selfing and with a testcross. All gametes are fertile because of homozygosity.

original plant: $P\ S/p\ s$

tester: $p\ s/p\ s$

F_1 progeny: heterozygous for the translocation:

The easiest way to test this is to look at the chromosomes of heterozygotes during meiosis I.

36. The breakpoint can be treated as a gene with two "alleles," one for normal fertility and one for semisterility. The problem thus becomes a two-point cross.

parentals	764	semisterile *Pr*
	727	normal *pr*
recombinants	145	semisterile *pr*
	186	normal *Pr*
	1822	

$100\%(145 + 186)/1822 = 18.17$ m.u.

37.

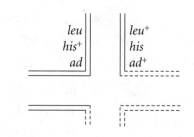

Because the short arm carries no essential genes, adjacent-1 segregation will yield progeny that are viable. Select for *leu⁺*, *his⁺*, and *ad⁺* by omitting those components from a minimal medium.

38. The percent degeneration seen in the progeny of the exceptional rat is 51% larger than that seen in the progeny of the normal. One explanation could be a chromosomal inversion. This could be verified by cytological observation of meiotic cells in the rat. A possibility would be a translocation, which could also be checked cytologically.

39. a. Heterozygous reciprocal translocations lead to duplications and deletions. Therefore, in asci in which crossing-over occurs within the translocated region, each crossing-over event would lead to two white ascospores that abort and two viable dark ascospores. In asci in which no crossing-over occurs within the translocated region, the ascospores would be normal color. Alternate segregation is assumed.

No CO	CO
●	●
●	○
●	○
●	●

b. Heterozygous pericentric inversions lead to duplications and deletions. Therefore, in asci in which crossing-over occurs within the pericentric inversion, each crossing-over event would lead to two white ascospores that abort and two viable dark ascospores. In asci in which no crossing-over occurs within the pericentric inversion, the ascospores would be normal color. Alternate segregation is assumed.

No CO	CO
●	●
●	○
●	○
●	●

c. Heterozygous paracentric inversions result in an acentric fragment that has lost some genetic material (deletion) and a dicentric chromosome that has gained some genetic material (duplication) if crossing-over occurs within the paracentric inversion. Therefore, in asci in which crossing-over occurs within the paracentric inversion, each crossing-over event would lead to two white ascospores that abort and two viable dark ascospores. In asci in which no crossing-over occurs within the paracentric inversion, the ascospores would be normal color.

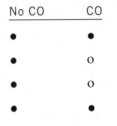

40. Species B is probably the "parent" species. A paracentric inversion in this species would give rise to species D. Species E could then occur by a translocation of *z x y* to *k l m*. Next, species A could result from a translocation of *a b c* to *d e f*. Finally, species C could result from a pericentric inversion of *b c d e*.

41. The assumptions are that half the gametes from a single heterozygous translocation are nonviable and that the two parents have the same chromosomes involved in translocations.

The progeny of crosses between parents with translocations is referring to the F_1 progeny, in which the parental generation was heterozygous. Designate the parents as follows:

A: (T1T2 N1N2)

B: (T1T2 N1N2)

The gametes will be

A: 1/4 (T1T2)   B: 1/4 (T1T2)
    1/4 (N1N2)   1/4 (N1N2)
  ★1/4 (T1N2)   ★1/4 (T1N2)
  ★1/4 (T2N1)   ★1/4 (T2N1)

where ★ equals an unbalanced gamete.

Fertilization between balanced gametes will occur 4(1/4)(1/4), or 4/16 of the time. An additional 2/16 of the fertilizations will lead to a balanced fetus even though both gametes were unbalanced, for example, (T2N1) × (T1N2). Therefore, 6/16 of the progeny are viable and 10/16 are nonviable.

9

Chromosome Mutation II: Changes in Number

▶ IMPORTANT TERMS
AND CONCEPTS

The **monoploid number** is the number of chromosomes in the basic set of chromosomes. It usually equals the number of chromosomes in a gamete. **Euploid** organisms have integer multiples of the monoploid number. A **diploid** organism has twice the monoploid number of chromosomes. A **haploid** organism has one monoploid number of chromosomes. A **polyploid** organism has more than twice the monoploid number of chromosomes. Examples of polyploids are **triploids, tetraploids, pentaploids,** and **hexaploids.**

An **aneuploid** has more or fewer chromosomes than an integer multiple of the monoploid number. Addition of one or more chromosomes produces a **hyperdiploid.** Loss of one or more chromosomes produces a **hypodiploid.** A **monosomic** lacks one chromosome. A **nullisomic** lacks both homologous chromosomes. A **trisomic** has an extra chromosome in a diploid. A **disomic** has an extra chromosome in a haploid.

Colchicine and its less toxic derivative colcemid disrupt the mitotic spindle and block mitosis. They produce polyploids.

Polyploids that have an even-integer multiple of the monoploid number are fertile because each chromosome can pair during meiosis. Those that have an odd-integer multiple of the monoploid number are infertile because one chromosome from each of the multiple homologous chromosomes cannot pair during meiosis.

An **autopolyploid** is composed of multiple sets of chromosomes from one species. The diploid derivative is fertile. An **allopolyploid** is composed of multiple sets of chromosomes from different species. An **amphidiploid** is an allopolyploid from two species. The diploid intermediate is not fertile.

Plants tolerate both polyploidy and aneuploidy much more easily than do higher animals.

Questions for Concept Map 9-1

1. Distinguish between "*x*" and "*n*" in humans. In an autopolyploid. In an allopolyploid.

2. What is the meaning of *genetic imbalance?*

3. Why are aneuploids considered to be genetically imbalanced while polyploids are not?

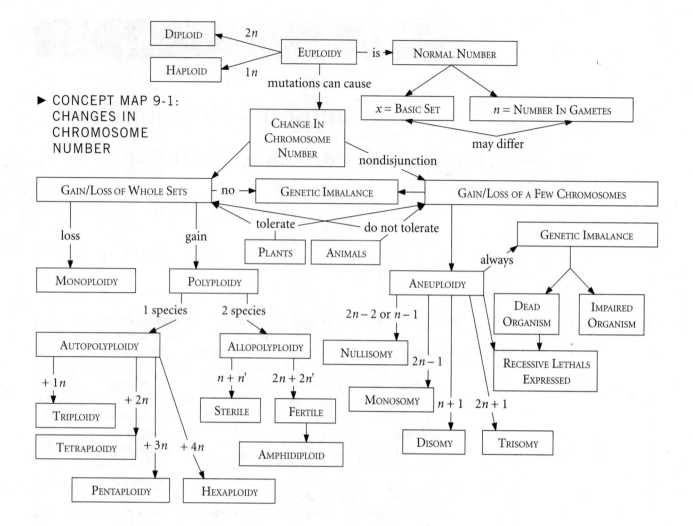

► CONCEPT MAP 9-1: CHANGES IN CHROMOSOME NUMBER

4. What is a recessive lethal?

5. Which chromosomes in humans are compatible with life in the trisomic state?

6. Which chromosomes in humans are compatible with life in the monosomic state?

7. Discuss the difference between the X chromosome and all the autosomes with regard to trisomy and monosomy. What is the basis of the difference?

8. Do you think that a nullisomic can ever be viable? Why?

9. Polyploids can be sterile or fertile, depending on the number of sets of chromosomes that they have. What is the basis for fertility and sterility?

10. If hybridization of two species results in sterility, how are fertile allopolyploids generated?

11. Discuss the "x" and "n" numbers of a hexaploid.

12. What is the mechanism that leads to aneuploidy?

13. What are three mechanisms that could cause polyploidy?

Answers to Concept Map 9-1 Questions

1. In humans, x and n are both 23. In an autopolyploid, x is half of n. In an allopolyploid, x is $n + n'$, where n and n' represent the contribution of two different species.

2. *Genetic imbalance* means that an organism does not have all the alleles, in the right number, that are seen in the wild type.

3. Aneuploids are genetically imbalanced because they have gained or lost one or more chromosomes. Polyploids have a consistent number of each chromosome.

4. A recessive lethal is an allele that kills an organism if it is the only allele present for that gene.

5. The trisomic state of X, Y, and chromosome number 21 is compatible with life in humans.

6. Monosomy of X is the only chromosome compatible with life in humans, and the vast number of these die in utero.

7. Humans can tolerate a missing or an extra X chromosome, while only an extra 21 is compatible with life amongst the autosomes. This is because X aneuploidy is buffered by the normal phenomenon of X-inactivation. With regard to chromosome 21, it is, in fact, the smallest autosome, despite its number, and it can be assumed that the smaller the chromosome the fewer the genes on it.

8. A nullisomic is lacking all the genes of one chromosome. Unless there are no essential genes on the chromosome that is completely missing, nullisomy would not be compatible with life.

9. The basis in polyploids for sterility or fertility is meiotic pairing. An even number of sets of chromosomes allows complete chromosome pairing, or fertility, while an odd number of sets of chromosomes leaves one complete set unpaired. This results in sterility because the unpaired chromosomes are randomly distributed during meiosis I to daughter cells. It is highly unlikely that a balanced gamete will result from this situation.

10. By chance alone an unreduced egg (one experiencing complete nondisjunction) is fertilized by an unreduced sperm, which is extremely rare.

11. In a hexaploid that is an autopolyploid, n is $3x$.

12. Aneuploidy results from nondisjunction.

13. Polyploidy could result from complete nondisjunction of all chromosomes, chemical treatment that leads to chromosome duplication without cell division, or fertilization of an egg by more than one sperm.

▶ SOLUTIONS TO PROBLEMS

Be sure that you have thoroughly read the entire chapter before you attempt any of the problems.

1. Klinefelter syndrome   XXY male

   Down syndrome   trisomy 21

   Turner syndrome   XO female

2. (1) Do somatic cell fusion between *AA* and *A'A'* plants.

(2) Cross *AA* with *A'A'*, and then double the chromosomes with colchicine treatment.

3. **a.** 3, 3, 3, 3, 3 3, 3 3, 0, 0

b. 7, 7, 7, 7, 8, 8, 6, 6

4. **a.** If a 6*x* were crossed with a 4*x*, the result would be 5*x*.

b. Cross *AA* with *aaaa* to obtain *Aaa*.

c. The easiest way is to expose the *Aa*★ plant cells to colchicine for one cell division. This will result in a doubling of chromosomes to yield *AAa*★*a*★.

d. Cross 6*x* (*aaaaaa*) with 2*x* (*AA*) to obtain *Aaaa*.

e. Obtain haploid cells from a plant and obtain resistant colonies by exposing them to the herbicide. Then expose the resistant colonies to colchicine to obtain diploids.

5. **b.**

6. To solve this problem, first recognize that *B* can pair with *B* one-third of the time (leaving *b* to pair with *b*) and that *B* can pair with *b* two-thirds of the time. If *B* pairs with *B*, the result is *Bb*, *Bb*, *Bb*, *Bb*, occurring one-third of the time. If *B* pairs with *b*, the resulting tetrad can be of two equally frequent types, depending on how the pairs align with respect to each other. One type is *BB*, *BB*, *bb*, *bb*; the second is *Bb*, *Bb*, *Bb*, *Bb*.

7. If there is no pairing between chromosomes from the same parent, then all pairs are *Aa*. Because the pairs can align with respect to each other, as diagrammed below,

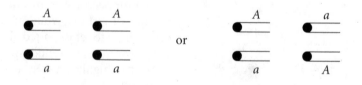

the gametes are 1*AA* : 2*Aa* : 1*aa*. With selfing, the progeny are

	1/4 *AA*	1/2 *Aa*	1/4 *aa*
1/4 *AA*	1/16 *AAAA*	2/16 *AAAa*	1/16 *AAaa*
1/2 *Aa*	2/16 *AAAa*	4/16 *AAaa*	2/16 *Aaaa*
1/4 *aa*	1/16 *AAaa*	2/16 *Aaaa*	1/16 *aaaa*

or   1/16 *AAAA*

4/16 *AAAa*

6/16 *AAaa*

4/16 *Aaaa*

1/16 *aaaa*

If there is no pairing between chromosomes from different parents, then all pairs are *AA* and *aa* and all gametes are *Aa*. Thus, 100% of the progeny are *AAaa*.

8. **a.** Let *T. turgidum* be designated by X, *T. monococcum* by Y, and *T. aestivum* by Z. Then note that bivalents contain two chromosomes while univalents contain one. You can write the following equations:

$$X + Y = 21$$

$$Z + Y = 28$$

$$Z + X = 35$$

Next, solve two of the three equations in terms of X:

$$X + Y = 21 \text{ becomes } Y = 21 - X$$

and

$$Z + X = 35 \text{ becomes } Z = 35 - X$$

Now, substitute these values for Y and Z into the third equation and solve it for X:

$$Z + Y = 28 \text{ becomes } (35 - X) + (21 - X) = 28$$

$$56 - 2X = 28$$

$$56 - 28 = 2X$$

$$28 = 2X$$

$$14 = X$$

Finally, substitute the value of X into the other two equations:

$$X + Y = 21 \text{ becomes } 14 + Y = 21$$

$$Y = 21 - 14 = 7$$

and

$$Z + X = 35 \text{ becomes } Z + 14 = 35$$

$$Z = 35 - 14 = 21$$

Because of the original definitions of X, Y, and Z, the following holds: *T. turgidum* gamete has 14 chromosomes, *T. monococcum* gamete has 7 chromosomes, and *T. aestivum* gamete has 21 chromosomes that were contributed to the hybrid. Therefore, the $2n$ value for *T. turgidum* is 28 chromosomes, for *T. monococcum* is 14 chromosomes, and for *T. aestivum* is 42 chromosomes.

b. and c. Let the 14 chromosomes in *T. monococcum* be designated by AA, which means it is diploid. Because 7 bivalents occurred in its hybridization with *T. turgidum*, the hybrid can be designated as AA + B. Because *T. turgidum* contributed half its chromosomes, it must be AA + BB, which indicates an allopolyploid.

When *T. turgidum* (AA + BB) was crossed with *T. aestivum*, the hybrid had 14 bivalents and 7 univalents. *T. turgidum* had to contribute A + B. Therefore, *T. aestivum* also had to contribute A + B to achieve 14 bivalents. It also contributed 7 univalents. In order to identify these 7 chromosomes, look at the cross between *T. aestivum* (AA + BB + ??) and *T. monococcum* (AA). The hybrids produced 7 bivalents, which have to be AA and 14 univalents, which have to be B + something else, which can be designated as C. Therefore, *T. aestivum* can be designated AA + BB + CC, which is an allopolyploid.

To check these conclusions, substitute them back into the original crosses:

$$AA + BB \times AA \rightarrow AA + B$$

$$AA + BB + CC \times AA \rightarrow AA + B + C$$

$$AA + BB + CC \times AA + BB \rightarrow AA + BB + C$$

9. Consider the following table, in which "L" and "S" stand for 13 large and 13 small chromosomes, respectively:

HYBRID	CHROMOSOMES
G. hirsutum × G. thurberi	S, S, L
G. hirsutum × G. herbaceum	S, L, L
G. thurberi × G. herbaceum	S, L

Each parent in the cross must contribute half its chromosomes to the hybrid offspring. It is known that *G. hirsutum* has twice as many chromosomes as the other two species. Furthermore, its chromosomes are composed of chromosomes donated by the other two species. Therefore, the genome of *G. hirsutum* must consist of one large and one small set of chromosomes. Once this is realized, the rest of the problem essentially solves itself. In the first hybrid, the genome of *G. thurberi* must consist of one set of small chromosomes. In the second hybrid, the genome of *G. herbaceum* must consist of one set of large chromosomes. The third hybrid confirms the conclusions reached from the first two hybrids.

The original parents must have had the following chromosome constitution:

G. hirsutum	26 large, 26 small
G. thurberi	26 small
G. herbaceum	26 large

G. hirsutum is a polyploid derivative of a cross between the two Old World species. This could easily be checked by looking at the chromosomes.

10. a. To do this problem you must first recognize that each allele can pair with any other allele within a gene. For a moment, pretend that you can distinguish all four alleles and, for simplicity's sake, number them 1 (*F*), 2 (*F*), 3 (*f*), and 4 (*f*). The combinations now become 1-2, 1-3, 1-4, 2-3, 2-4, and 3-4. In other words, for each gene there are six combinations. Changing the numbers into letters, the gametes for *F* would be 1/6 *FF*, 4/6 *Ff*, and 1/6 *ff*.

When two genes are considered, the gametes are

$$1/6 \ FF \ \begin{cases} 1/6 \ GG = 1/36 & FF\,GG \\ 4/6 \ Gg = 4/36 & FF\,Gg \\ 1/6 \ gg = 1/36 & FF\,gg \end{cases}$$

$$4/6 \ Ff \ \begin{cases} 1/6 \ GG = 4/36 & Ff\,GG \\ 4/6 \ Gg = 16/36 & Ff\,Gg \\ 1/6 \ gg = 4/36 & Ff\,gg \end{cases}$$

$$1/6\,ff \begin{cases} 1/6\,GG = 1/36 \quad ffGG \\ 4/6\,Gg = 4/36 \quad ffGg \\ 1/6\,gg = 1/36 \quad ffgg \end{cases}$$

b. The cross is *FFff GGgg* × *FFff GGgg*. For *FFFf GGgg*, consider each gene separately. The combination *FFFf* can be achieved in two ways:

$$p(FFFf) = [p(FF) \times p(Ff)] + [p(Ff) \times p(FF)]$$

$$= (1/6 \times 4/6) + (4/6 \times 1/6)$$

$$= 4/36 + 4/36 = 8/36 = 2/9$$

The combination *GGgg* can be achieved in three ways:

$$p(GGgg) = [p(GG) \times p(gg)] + [p(gg) \times p(GG)] + [p(Gg) \times p(Gg)]$$

$$= (1/6 \times 1/6) + (1/6 \times 1/6) + (4/6 \times 4/6)$$

$$= 1/36 + 1/36 + 16/36 = 1/2$$

Therefore

$$p(FFFf\,GGgg) = 2/9 \times 1/2 = 1/9$$

and

$$p(ffff\,gggg) = p(ffff) \times p(gggg)$$

$$= 1/6 \times 1/6 \times 1/6 \times 1/6 = 1/1296$$

11. **b, f,** and **h,** and sometimes **c**

12. One of the parents of the woman with Turner syndrome (XO) must have been a carrier for colorblindness, an X-linked recessive disorder. Because her father has normal vision, she could not have obtained her sole X from him. Therefore, nondisjunction occurred in her father. The sperm lacking an X chromosome fertilized the egg carrying the color-blindness allele. The nondisjunction event could have occurred during either meiotic division.

   If the colorblind patient had Klinefelter syndrome (XXY), then both X's must carry the allele for colorblindness. Therefore, nondisjunction had to occur in the mother. Remember that during meiosis I, given no crossover between the gene and the centromere, allelic alternatives separate from each other. During meiosis II, identical alleles on sister chromatids separate. Therefore, the nondisjunctive event had to occur during meiosis II because both alleles are identical.

13. a. If most individuals were female, this suggests that the normal allele has been lost (by nondisjunction or deletion) or is nonfunctional (through X-inactivation) in the colorblind eye.

    b. If most of the individuals were male, this suggests that the male might have two or more cell lines (he is a mosaic, $X^{normal}Y/X^{cb}Y$) or that he has two X chromosomes (he has Klinefelter syndrome) and that the same processes as in females could be occurring.

14. If the fluorescent spot indicates a Y chromosome, then two spots are indicative of nondisjunction. Presumably, exposure to dibromochloropropane increases the rate of nondisjunction. This could be tested in several ways. The most straightforward would be to expose male animals to the chemical, look for an increase in double-spotted sperm over

those not exposed, and also examine testicular cells to observe the rate of nondisjunction with and without exposure. Alternatively, specific crosses could be set up that would reveal nondisjunction upon exposure to the chemical. If X linkage in fruit flies is used for the assay, white-eyed females exposed to the chemical should have a higher rate of nondisjunction than do unexposed females. When crosses to red-eyed males are done, nondisjunction of the X chromosome would result in white-eyed females and red-eyed males.

15. **a. and b.** Nystagmus appears to be an X-linked recessive disorder. That means that the individual with Turner's syndrome had to have obtained her sole X from her mother. She did not obtain a sex chromosome from her father, which indicates that nondisjunction occurred in him. The nondisjunction could have occurred at either M_I or M_{II}.

16. One possibility is that the mean age of mothers at birth dropped significantly. Because the older mother is at higher risk for nondisjunction, this would result in the observation. Hospital records could be used to check the age of mothers at birth between 1952 and 1972, as compared with a 20-year period before 1952. Another possibility is an increase of amniocentesis amongst older mothers followed by induced abortion of trisomy-21 fetuses. Because pregnant women 35 and older routinely undergo amniocentesis, the rate for this population may have fallen while the rate for the younger population remained unchanged. This also could be checked through hospital records.

17. **a.** loss of one X in the developing fetus after the two-celled stage

b. nondisjunction leading to Klinefelter syndrome (XXY), followed by a nondisjunctive event in one cell for the Y chromosome after the two-celled stage, leading to XX and XXYY

c. nondisjunction for X at the one-celled stage

d. either fused XX and XY zygotes or fertilization of an egg and polar body by one sperm bearing an X and another bearing a Y, followed by fusion

e. nondisjunction of X at the two-celled stage or later

18. Remember that nondisjunction at meiosis I leads to the retention of both chromosomes in one cell, while at meiosis II it leads to the retention of both sister chromatids in one cell.

(1) trisomy 21; nondisjunction at meiosis II in the female

(2) trisomy 21; nondisjunction at meiosis I in the female

(3) normal; normal meiosis in both parents

(4) trisomy 21; nondisjunction at meiosis II in the male

(5) normal; nondisjunction in the female (meiosis I) and the male (either meiotic division)

(6) XYY nondisjunction for the sex chromosomes in the male meiosis II

(7) monosomy 21–trisomy 21 in a male zygote; occurrence of mitotic nondisjunction for the 21^c chromosome fairly early in development

(8) sexual mosaic; fused XX and XY zygotes or, as in Problem 17d, fused fertilized egg and fertilized polar body

19. *Type a*: The extra chromosome must be from the mother. Because the chromosomes are identical, nondisjunction had to have occurred at M_{II}.

Type b: The extra chromosome must be from the mother. Because the chromosomes are not identical, nondisjunction had to have occurred at M_I.

Type c: The mother correctly contributed one chromosome, but the father did not contribute any chromosome 4. Therefore, nondisjunction occurred in the male during either meiotic division.

Type d: One cell line lacks a maternal contribution while the other has a double maternal contribution. Because the two lines are complementary, the best explanation is that nondisjunction occurred in the developing embryo during mitosis.

Type e: Each cell line is normal, indicating that nondisjunction did not occur. The best explanation is that the second polar body was fertilized and was subsequently fused with the developing embryo. Alternatively, a pair of twins fused.

20. *Unpacking the Problem*

   a. **Homozygous** means that an organism has two identical alleles.

   A **mutation** is any deviation from wild type.

   An **allele** is one particular form of a gene.

   Closely linked means that two genes are physically very close to each other on the same chromosome.

   Recessive refers to a type of allele that is expressed only when it is the sole type of allele for that gene found in an individual.

   Wild type is the most frequent type found in a laboratory population or in a population in the "wild."

   Crossing-over refers to the physical exchange of alleles between homologous chromosomes.

   Nondisjunction is the failure of separation of either homologous chromosomes or sister chromatids in the two meiotic divisions.

   A **testcross** is a cross to a homozygous recessive organism for the trait or traits being studied.

   Phenotype is the appearance of an organism.

   Genotype is the genetic constitution of an organism.

   b. No, the genes in question are on an autosome, specifically, number 4.

   c. The chromosome number in *Drosophila* varies with the species. However, the species being used here has at least four autosomal pairs plus two sex chromosomes, so it must have at least 10 chromosomes.

d. P $e^+b/e^+b \times eb^+/eb^+$ (cross 1)

F$_1$ $e^+b/eb^+ \times eb/eb$ (cross 2)

F$_2$ $\dfrac{e^+b}{eb}$ expected parental

$\dfrac{eb^+}{eb}$ expected parental

$\dfrac{eb}{eb}$ unexpected recombinant, not observed

either $\dfrac{e^+b^+}{eb}$ unexpected recombinant, observed phenotype

or $\dfrac{e^+b}{eb^+} \quad eb$ unexpected nondisjunction, observed phenotype

Cross 3

Note that crossing-over is assumed not to occur.

either   or

P $\dfrac{e^+b^+}{eb} \times \dfrac{eb}{eb}$ P $\dfrac{e^+b}{eb^+} \quad eb \times \dfrac{eb}{eb}$

F$_1$ $\dfrac{e^+b^+}{eb}$ and $\dfrac{eb}{eb}$ F$_1$ 1 e^+b/eb bent
                                              1 $eb^+/eb/eb$ eyeless
not observed   1 $e^+b/eb^+/eb$ wild type
                                              1 eb/eb eyeless, bent
                                              1 eb^+/eb eyeless
                                              1 $e^+b/eb/eb$ bent
                                                       all observed

e.

f.

g. It is not at all surprising that the F$_1$ are wild type. This means that both mutations are recessive and complement.

h. tester $\dfrac{eb}{eb}$

gametes $\underline{eb}$

i. The two common gametes are b^+e and be^+. The two rare gametes are b^+e^+ and be.

j.

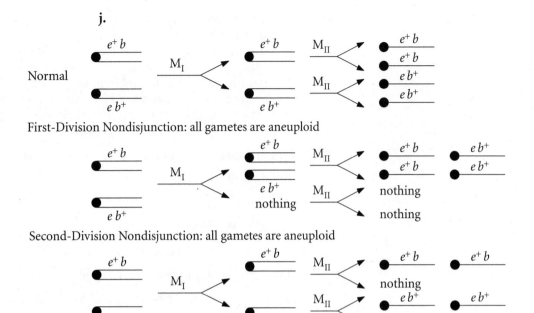

Normal

First-Division Nondisjunction: all gametes are aneuploid

Second-Division Nondisjunction: all gametes are aneuploid

k. This is answered in j, above.

l. Viable progeny may be able to arise from aneuploidic gametes because chromosome 4 is very small and may have very little genetic material on it. The progeny would be monosomic and trisomic.

m. Listed below are the gametes from i and j above, the contribution of the male parent, and the phenotype of the progeny.

FEMALE GAMETE	MALE GAMETE	PHENOTYPE
eb^+	eb	eyeless
e^+b	eb	bent
e^+b^+	eb	wild type
eb	eb	eyeless, bent
e^+b/eb^+	eb	wild type
e^+b/eb^+	eb	wild type
—	eb	eyeless, bent
—	eb	eyeless, bent
e^+b/e^+b	eb	bent
—	eb	eyeless, bent
eb^+/eb^+	eb	eyeless
—		

n. The ratio points to meiosis of a trisomic.

o. Research with artificial chromosomes has indicated that extremely small chromosomes segregate improperly at higher rates than longer chromosomes. It is suspected that the chromatids from homologous chromosomes need to intertwine in order to remain together until the onset of anaphase. Very short chromosomes are thought to have some difficulty in doing this and therefore have a higher rate of nondisjunction. In this instance, which deals with natural chromosomes as opposed to artificial chromosomes, very small chromosomes would be expected to have very little genetic material in them, and therefore their loss or gain may not be of too much importance during development.

p. See the last part of the drawing in answer d, above.

Solution to the Problem

The crosses are as follows:

Cross 1:   P $be^+/be^+ \times b^+e/b^+e$

   F_1 b^+e/be^+

Cross 2:   P   XX $b^+e/be^+ \times$ XY be/be

   F_1 expect 1 be^+/be :1 b^+e/be , XX and XY

   observed XX $b^+\!- e^+\!-$

a.   The common progeny are be^+/be and b^+e/be .

b.   The rare female could have come from crossing-over, which would have resulted in a gamete that was b^+e^+. The rare female also could have come from nondisjunction that gave a gamete that was be^+/b^+e. Such a gamete might give rise to viable progeny.

c.   If the female had been wild-type (b^+e^+/be) as a result of crossing over, her progeny would have been as follows:

Parental b^+e^+/be

        be/be

Recombinant be^+/be

        b^+e/be

These expected results are very far from what was observed, so the rare female was not the result of recombination.

   If the female had been the product of nondisjunction (be^+/b^+e /be), her progeny would be as follows:

1/6	b^+e/be	eyeless
1/6	$b^+e/be/be$	eyeless
1/6	$b^+/be/be$	bent
1/6	be^+/be	bent
1/6	be/be	bent eyeless
1/6	$b^+e/be^+/be$	wild type

These results are in accord with the observed results, indicating that the female was a product of nondisjunction.

21. a.   The cross is $PPp \times pp.$
   The gametes from the trisomic parent will occur in the following proportions:

1/6   *p*

2/6   *P*

1/6   *PP*

2/6   *Pp*

Only gametes that are *p* can give rise to potato leaves, because potato is recessive. Therefore, the ratio of normal to potato will be 5:1.

b. The cross is *Pp* × *pp*. The ratio of normal to potato will be 1:1.

22. The generalized cross is *AAA* × *aa*, from which *AAa* progeny were selected. These progeny were crossed with *aa* individuals, yielding the results in the table in the textbook. Assume for a moment that each allele can be distinguished from the other, and let 1 = *A*, 2 = *A* and 3 = *a*. The gametic combinations possible are

1-2 (*AA*)   and   3 (*a*)

1-3 (*Aa*)   and   2 (*A*)

2-3 (*Aa*)   and   1 (*A*)

Because diploid progeny were examined in the cross with *aa*, the haploid gametic ratio would be 2*A*:1*a*, and the diploid ratio would also be 2 wild type:1 mutant. The table indicates that *y* is on chromosome 1, *cot* is on chromosome 7, and *h* is on chromosome 10.

23. Radiation could have caused point mutations or induced recombination, but nondisjunction is the more likely explanation.

24. P   *a b⁺c d⁺ e* × *a⁺b c⁺d e⁺*

Selection for *a⁺ b⁺ c⁺ d⁺ e⁺*

Because this rare colony gave rise to both parental types among asexual (haploid) spores, the best explanation is that the rare colony initially contained both marked chromosomes due to nondisjunction. That is, it was disomic. Subsequent mitotic nondisjunction yielded the two parental types, possibly because the disomic was unstable.

25. Before attempting these problems, draw the chromosomes. The cross is

a. The aborted spores arise from nondisjunction. Nondisjunction at meiosis I would produce 4 *n+1* (*pan1⁺ leu2/pan1 leu2⁺*, viable):4 *n–1* (nonviable).

b. If, at meiosis II, the chromosome carrying *pan1* experienced nondisjunction, the progeny would be 2 *pan1 leu2⁺* (*n+1*): 2 *n–1*:4 *pan1⁺ leu2* (*n*).

c. Consider the following crossover (2-4):

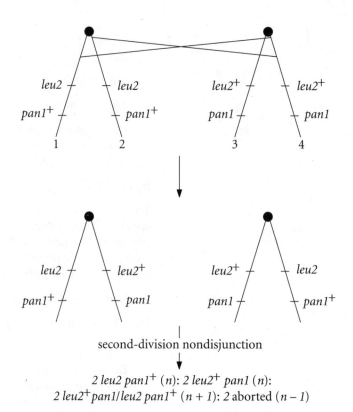

second-division nondisjunction

2 leu2 pan1⁺ (n): 2 leu2⁺ pan1 (n):
2 leu2⁺ pan1/leu2 pan1⁺ (n + 1): 2 aborted (n − 1)

The four chromatids would be (*leu2 pan1⁺*, *leu2⁺ pan1*) and (*leu2⁺ pan1*, *leu2 pan1⁺*). If the second chromosome experienced nondisjunction, it would give rise to 2 fully prototrophic:2 aborted spores. The first chromosome would give rise to 2 leu-requiring:2 pan-requiring spores.

26. The two chromosomes are

If one of the centromeres becomes functionally duplex before meiosis I, the homologous chromosomes will separate randomly. This will result in one daughter cell having one chromosome with one chromatid, and the second daughter cell having one chromosome with one chromatid and a second chromosome with two chromatids. Meiosis II will lead to a nullisomic (white) and a monosomic (buff) from the first daughter cell and a disomic (black) and a monosomic (buff) from the second daughter cell.

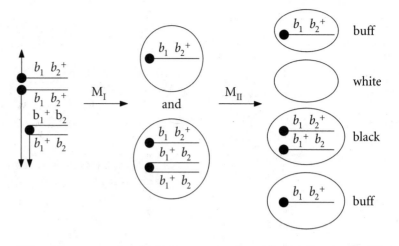

If both centromeres divide prematurely, each daughter cell will get two chromosomes, each with one chromatid. All the ascospores will be black.

If nondisjunction occurred at meiosis I, the result would be half black (disomic) and half white (nullisomic) spores. Nondisjunction at meiosis II for one of the cells would result in two buff spores (normal meiosis II) and one white (nullisomic) and one buff (two copies of the same gene).

Normal meiosis will yield all buff spores because no crossing-over occurs between the two genes.

27. Aneuploidy is the result of nondisjunction. Therefore, any system that will detect nondisjunction will work. One of the easiest follows. Cross white-eyed females with red-eyed males and look for the reverse of X linkage in the progeny. Compare populations unexposed to any environmental pollutants with those exposed to different suspect agents.

28. a. Before considering function, simply list the gametes and their frequency in both parents. In this cross, the parents have the same chromosome constitution, so they have the same outcomes:

FEMALE AND MALE

2/6	P
2/6	Pp
1/6	PP
1/6	p

Next, consider whether the gametes are functional in each sex:

FEMALE	FUNCTIONAL	RELATIVE PROPORTION	FINAL FREQUENCY
2/6 P	100%	4	4/9
2/6 Pp	50%	2	2/9
1/6 PP	50%	1	1/9
1/6 p	100%	2	2/9

MALE	FUNCTIONAL	RELATIVE PROPORTION	FINAL FREQUENCY
2/6 P	100%	2	2/3
2/6 Pp	0%	0	0
1/6 PP	0%	0	0
1/6 p	100%	1	1/3

Now, a table can be constructed for the frequency of functional gametes from each sex:

		FEMALE			
		4/9 P	2/9 Pp	1/9 PP	2/9 p
MALE	2/3 P	8/27 PP	4/27 PPp	2/27 PPP	4/27 Pp
	1/3 p	4/27 Pp	2/27 Ppp	1/27 PPp	2/27 pp

The ratio of purple to white is 25:2.

The same method was used to solve the remaining parts of this problem. The results are given below.

b. 17 purple:10 white

c. 4 purple:5 white

29. a. *B. campestris* was crossed with *B. napus*, and the hybrid had 29 chromosomes consisting of 10 bivalents and 9 univalents. *B. napus* had to have contributed a total of 19 chromosomes to the hybrid. Therefore, *B. campestris* had to have contributed 10 chromosomes. The $2n$ number in *B. campestris* is 20.

When *B. nigra* was crossed with *B. napus*, *B. nigra* had to have contributed 8 chromosomes to the hybrid. The $2n$ number in *B. nigra* is 16.

B. oleracea had to have contributed 9 chromosomes to the hybrid formed with *B. juncea*. The $2n$ number in *B. oleracea* is 18.

b. First list the haploid and diploid number for each species:

SPECIES	HAPLOID	DIPLOID
B. nigra	8	16
B. oleracea	9	18
B. campestris	10	20
B. carinata	17	34
B. juncea	18	36
B. napus	19	38

Now, recall that a bivalent in a hybrid indicates that the chromosomes are essentially identical. Therefore, the more bivalents formed in a hybrid, the closer the two parent species.

Three crosses result in no bivalents, suggesting that the parents of each set of hybrids are not closely related:

CROSS	HAPLOID NUMBER
B. juncea × *B. oleracea*	18 vs. 9
B. carinata × *B. campestris*	17 vs. 10
B. napus × *B. nigra*	19 vs. 8

Three additional crosses resulted in bivalents, suggesting a closer relationship among the parents:

CROSS	HAPLOID NUMBER	BIVALENTS	UNIVALENTS
B. juncea × *B. nigra*	18 vs. 8	8	10
B. napus × *B. campestris*	19 vs. 10	10	9
B. carinata × *B. oleracea*	17 vs. 9	9	8

Note that in each cross the number of bivalents is equal to the haploid number of one species. This suggests that the species with the larger haploid number is a hybrid composed of the second species and some other species. In each case, the haploid number of the unknown species is the number of univalents. Therefore, the following relationships can be deduced:

B. juncea is an amphidiploid formed in the cross of *B. nigra* and *B. campestris*.

B. napus is an amphidiploid formed in the cross of *B. campestris* and *B. oleracea*.

B. carinata is an amphidiploid formed in the cross of *B. nigra* and *B. oleracea*.

These conclusions are in accord with the three crosses that did not yield bivalents:

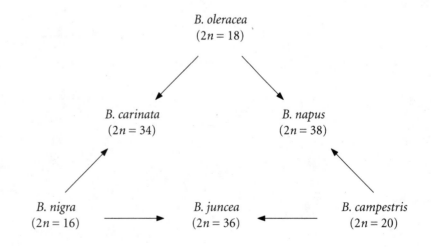

30. Recall that ascospores are haploid. The normal genotype associated with the phenotype of each spore is given below.

1	2	3
$b^+ f^+$	$b f^+$	$b^+ f$
$b^+ f^+$	$b f^+$	$b^+ f$
$b^+ f^+$	abort	$b^+ f^+$
$b^+ f^+$	abort	$b^+ f^+$
abort	$b^+ f$	$b f$
abort	$b^+ f$	$b f$
abort	$b^+ f$	$b f^+$
abort	$b^+ f$	$b f^+$

a. For the first ascus, the most reasonable explanation is that nondisjunction occurred at the first meiotic division. Second-division nondisjunction and second-division segregation are two explanations of the second ascus. Crossing-over best explains the third ascus.

b.

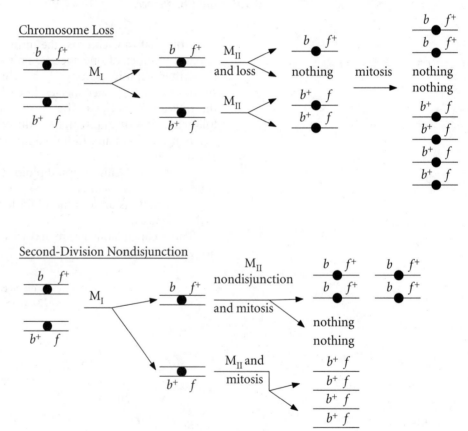

Chromosome Loss

Second-Division Nondisjunction

10

Recombination in Bacteria and Their Viruses

► IMPORTANT TERMS AND CONCEPTS

The prokaryotes are composed of **blue-green algae (cyanobacteria)** and **bacteria. Bacteriophages,** or **phages,** are viruses that reproduce in bacteria.

Conjugation is the one-way transfer of DNA from one bacterium to another. The ability to transfer DNA by conjugation is dependent on the presence of the **F,** or **fertility, factor.**

The F factor is an **episome,** a genetic particle that can exist either free in the cytoplasm or integrated into the host chromosome.

Cells carrying the F factor are F^+ (free in cytoplasm), **Hfr** (integrated into the bacterial chromosome), or **F'** (F factor in cytoplasm, with chromosomal genes inserted into it). Cells lacking the F factor are F^-.

Cells with the F factor produce **pili,** which are proteinaceous structures that attach to the F^- cell. Cells with the F factor also produce a **conjugation tube,** through which genes are transferred.

F^+ cells transfer genes located on the F factor. Hfr cells transfer chromosomal genes and, rarely, the F factor. F' cells transfer both chromosomal and F factor genes in a process called **sexduction.**

After the transfer of genes, recombination can occur between the transferred genes (**exogenote**) and the F^- **endogenote.** There is a **gradient of transfer** of chromosomal genes.

Transformation is the process by which cells take up naked DNA from their environment. The DNA subsequently recombines into the bacterial recipient.

Phage infection of a bacterial cell can result in cell destruction, **lysis,** or a change in the characteristics of the bacterial cell, **lysogeny.** Lysis results in a **plaque,** or clear area, in the bacterial lawn.

A **virulent** phage cannot integrate into the bacterial chromosome and therefore always causes lysis. A **temperate** phage can integrate into the bacterial chromosome and therefore causes lysogeny. When integrated, the virus is called a **prophage.**

Phage characteristics include plaque morphology, host range, and burst size.

Transduction is the movement of genes from one bacterial cell to another with the phage as the vector. **Generalized** transduction results from lysis, frequently without a preceding lysogeny; each bacterial gene has an equal chance of being transduced. **Specialized** transduction results from lysogeny; genes close to the site of prophage insertion are transduced.

► CONCEPT MAP 10-1:
  RECOMBINATION IN
  BACTERIA

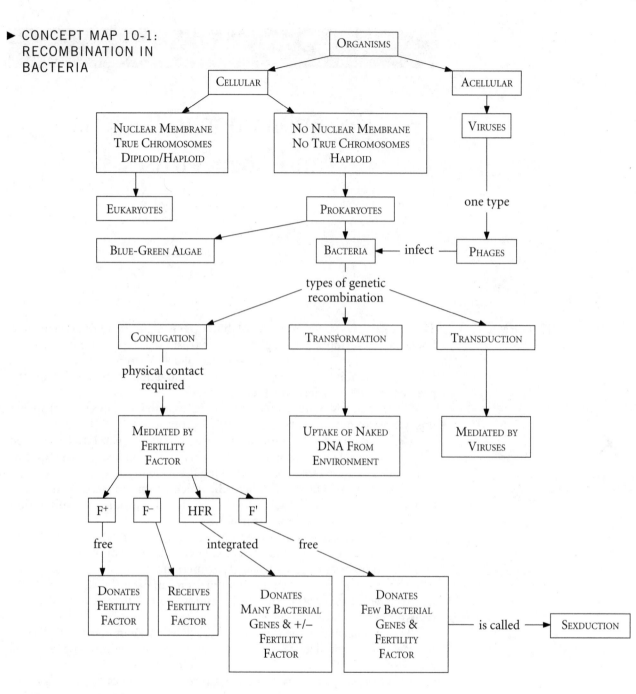

Questions for Concept Map 10-1

1. Do you consider viruses to be alive? Why?

2. What distinguishes eukaryotes from prokaryotes?

3. When and where does recombination occur in eukaryotes?

4. Can recombination occur in a haploid organism?

5. If recombination can occur in a haploid organism, can the genes involved be haploid?

6. Of the three types of recombination that do occur, which mechanism, if any, could be used to introduce genes into eukaryotes?

7. Are the three types of recombination seen in all prokaryotes?

8. What structure acts as a connector between two organisms during conjugation? What is responsible for the making of that structure?

9. What is a major difference between recombination in eukaryotes and prokaryotes?

10. Distinguish between F^+ and F^-. Between Hfr and F'. Between F^+ and F'.

11. Why is the fertility factor seldom transferred by an Hfr?

12. What are the similarities between transduction and sexduction? The differences?

13. What are the two types of transduction?

14. What types of viruses carry out transduction in bacteria?

15. Can three strands of DNA actually undergo pairing at the same time?

16. During the three types of recombination seen in bacteria, are genes that are closely linked more or less likely to be recombined together?

Answers to Concept Map 10-1 Questions

1. It can be argued that viruses are alive or not alive. The important part of your answer is the reasons that you give to support it.

2. Eukaryotes have a nuclear membrane, membrane-bound vesicles, and true chromosomes (DNA + histones in a nucleosome structure). Prokaryotes do not have a nuclear membrane, membrane-bound vesicles, or true chromosomes (they have naked DNA or RNA).

3. Recombination occurs in the primary oocytes or primary spermatocytes in the gonads.

4. Recombination can occur in a haploid organism.

5. When recombination occurs in a haploid organism, the genes involved are diploid.

6. Both transformation and generalized transduction have been used to introduce genes into eukaryotes.

7. The three types of recombination are not seen in all prokaryotes.

8. The pilus acts as a connector between two organisms during conjugation. It exists only when the F factor is in one bacterial cell.

9. Eukaryotic recombination is part of a process in which two individuals donate genetic material to make a third individual. Prokaryotic recombination involves the transfer of genetic material from one individual to another, which then can become a recombinant.

10. In F^+ cells, a fertility factor is free in the cytoplasm, while F^- cells do not have a fertility factor. The fertility factor in Hfr cells is integrated within the bacterial DNA, while the cells that are F' have an improperly excised fertility factor free in the cytoplasm. Cells that are F^+ contain only fertility factor genes in the circular plasmid free in the cytoplasm, while cells that are F' have both fertility factor and bacterial genes in the circular plasmid free in the cytoplasm.

11. The fertility factor is seldom transferred by an HFR cell because it is the last genetic material to be transferred, and usually the connection between cells is lost before that occurs.

12. Transduction is the transfer of bacterial genes through a viral particle, while sexduction is the transfer of bacterial genes through a fertility factor during conjugation.

13. The two types of transduction are generalized and specialized.

14. The viruses that carry out transduction in bacteria are virulent for generalized transduction and lysogenic for specialized transduction.

15. Three strands of DNA cannot actually undergo pairing at the same time for the same region. The bacterial double helix must unwind and unpair in a small region to allow the introduced DNA to pair with one strand.

16. During recombination in bacteria, genes that are closely linked are more likely to be recombined together.

► CONCEPT MAP 10-2: EPISOMES

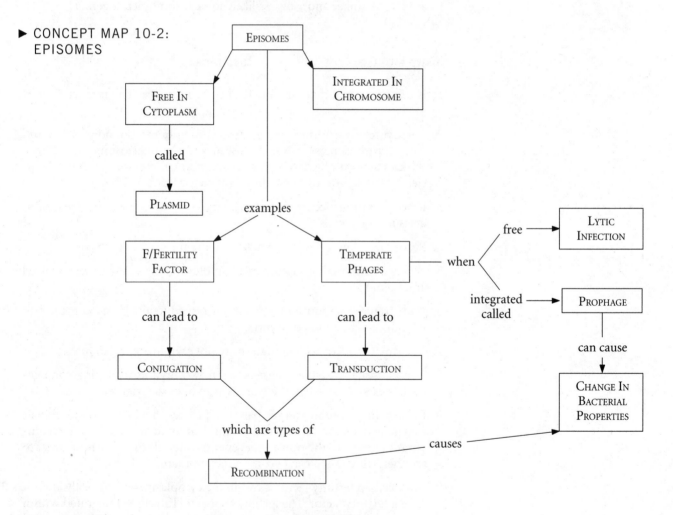

Questions for Concept Map 10-2

1. What is the definition of an episome?

2. How do temperate phages meet the definition of episomes?

3. What other type of phages exists besides temperate phages?

4. If temperate phages and the fertility factor are both examples of episomes, would you expect that they would lead to recombination by essentially the same process? Why?

5. Assume for a moment that a plasmid is a defective temperate phage. What two main characteristics has the plasmid lost in comparison to temperate phages?

6. Assume that virulent phages are defective temperate phages. What main characteristic has the virulent phage lost in comparison with temperate phages?

7. If prophages change bacterial properties, would you expect that the fertility factor changes bacterial properties? If yes, can you name one?

8. What is the name of the process that converts a prophage into a particle free in the cytoplasm?

9. What type of recombination can be caused by temperate phages?

10. Would you expect that eukaryotic cells could have the equivalent of plasmids and temperate phages?

11. If eukaryotic cells have the equivalent of plasmids and temperate phages, would you expect them to be capable of changing the cellular properties in some respect?

12. Do eukaryotic cells have virulent viruses?

13. What is the meaning of *lysis?*

14. Why are temperate phages described as *lysogenic?*

Answers to Concept Map 10-2 Questions

1. An episome is DNA that can be free in the cytoplasm or integrated within the host cell chromosome.

2. Temperate phages normally integrate into bacterial chromosomes, but they can also excise and begin a lytic infection.

3. Besides temperate phages there are virulent phages.

4. Temperate phages and the fertility factor do not lead to recombination in the same way. Temperate phages can introduce *double-stranded DNA* through either generalized or specialized transduction, while the fertility factor introduces *single-stranded DNA* through conjugation, which is then converted to double-stranded DNA.

5. If a plasmid is a defective temperate phage, it has lost the ability to synthesize a protein coat and cause lysis of the host cell.

6. If virulent phages are defective temperate phages, they have lost the ability to lysogenize, or integrate within the bacterial chromosome, and to repress their virulent functions.

7. Both prophages and the fertility factor change bacterial properties. The bacterial cell forms a pilus if it has the fertility factor.

8. Excision converts a prophage into a particle free in the cytoplasm.

9. Both generalized and specialized transduction can be caused by temperate phages.

10. Eukaryotic cells *do* have the equivalent of plasmids and temperate phages.

11. Eukaryotic plasmids and temperate phages have the capability of changing the cellular properties in some respect.

12. Eukaryotic cells have virulent viruses.

13. *Lysis* is cell rupture followed by the release of many viral particles.

14. Temperate phages are described as *lysogenic* because they can give rise to cell lysis following excision.

▶ CONCEPT MAP 10-3:
   PHAGES

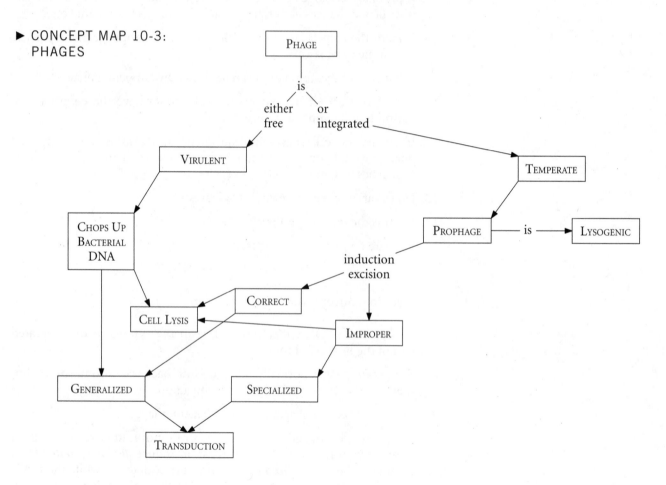

Questions for Concept Map 10-3

1. Which type of phage is called *lytic?*

2. Which type of phage is called *nonlysogenic?*

3. Which type of phage is called *lysogenic?*

4. Does a virulent phage always cause lysis?

5. What is cell lysis?

6. After cell lysis, can a cell continue to exist?

7. Does a temperate phage always undergo integration?

8. Does a temperate phage always eventually cause lysis?

9. When a virulent phage enters a cell, what does it do to the host chromosome?

10. When a virulent phage enters a cell, what does it make?

11. When a virulent phage causes lysis, what is released from the cell?

12. How does a virulent phage have the potential to cause transduction?

13. Why is the transduction caused by a virulent phage called *generalized?*

14. What is the source of bacterial genes with generalized transduction?

15. What is induction?

16. What is excision?

17. Does correct excision lead to specialized transduction? Why?

18. What is being excised during induction and excision?

19. After correct excision, what happens to the bacterial chromosome?

20. What does a temperate phage make and release after correct excision?

21. What characterizes improper excision?

22. After improper excision, what happens to the bacterial chromosome?

23. What does a temperate phage make and release after improper excision?

24. Which is more likely to occur: correct or improper excision?

25. When transduction occurs, is a complete phage genome, either virulent or temperate, present?

26. Transduction is one form of recombination seen in bacteria. What is the source of the new genes that appear in a bacterial cell?

27. What is the minimum number of crossing-over events that must occur for transduction, transformation, or conjugation? Would three crossing-over events work for any of those processes?

Answers to Concept Map 10-3 Questions

1. Virulent phages are called *lytic.*

2. Virulent phages are called *nonlysogenic.*

3. Temperate phages are called *lysogenic.*

4. A virulent phage always causes lysis, unless it is defective.

5. Cell lysis is the rupture of a cell, followed by the release of many viral particles.

6. After cell lysis, a cell cannot continue to exist.

7. A temperate phage does not always undergo integration.

8. A temperate phage eventually causes lysis, but that may occur many generations after the initial infection.

9. When a virulent phage enters a cell, it chops up the host chromosome into small pieces of DNA through the use of specialized enzymes.

10. When a virulent phage enters a cell, it needs to make the enzymes that destroy the host cell DNA. It may also need to make its own polymerases.

11. When a virulent phage causes lysis, many viral particles are released from the cell.

12. A virulent phage can cause transduction through partial digestion of the host chromosome followed by packing of the host DNA into the viral protein by accident.

13. The transduction caused by a virulent phage is called *generalized* because any host gene may be included within the viral particle.

14. The source of bacterial genes with generalized transduction is the host DNA that has been partially digested.

15. Induction is the process whereby a prophage is activated to go through excision followed by lysis.

16. Excision is the leaving of the host chromosome by either a fertility factor or a phage.

17. Correct excision does not lead to specialized transduction because no bacterial genes leave the host chromosome during correct excision.

18. The integrated particle is being excised during induction and excision.

19. After correct excision, the bacterial chromosome is partially digested in the lytic phase of a temperate phage.

20. A temperate phage makes and releases mature viral particles after correct excision.

21. Improper excision is characterized by the removal of some bacterial genes with the excised particle.

22. After improper excision, the bacterial chromosome is partially digested during the lytic phase of temperate phages.

23. A temperate phage makes and releases viral particles that contain a subset of viral genes and a subset of bacterial genes.

24. Correct excision is more likely to occur than improper excision.

25. When transduction occurs, an incomplete phage genome, either virulent or temperate, is present.

26. The source of the new genes that appear in a bacterial cell during transduction is the previous bacterial host.

27. The minimum number of crossing-over events that must occur for transduction, transformation, or conjugation is two. Three crossing-over events would not work.

▶ SOLUTIONS TO PROBLEMS

Be sure that you have thoroughly read the entire chapter before you attempt any of the problems.

1. An Hfr strain has the fertility factor, F, integrated into the chromosome. An F$^+$ strain has the fertility factor free in the cytoplasm. An F$^-$ strain lacks the fertility factor.

2. All cultures of F⁺ strains have a small proportion of cells in which the F factor is integrated into the bacterial chromosome. These cells transfer markers from the host chromosome to a recipient during conjugation.

3. **a.** Hfr cells involved in conjugation transfer host genes in a linear fashion. The genes transferred depend on both the Hfr strain and the length of time during which the transfer occurred. Therefore, a population containing several different Hfr strains will appear to have an almost random transfer of host genes. This is similar to generalized transduction, in which the viral protein coat forms around a specific amount of DNA rather than specific genes. In generalized transduction, any gene can be transferred.

   b. F′ factors arise from improper excision of an Hfr from the bacterial chromosome. They can have only specific bacterial genes on them because the integration site is fixed for each strain. Specialized transduction resembles this in that the viral particle integrates into a specific region of the bacterial chromosome and then, upon improper excision, can take with it only specific bacterial genes. In both cases, the transferred gene exists as a second copy.

4. Generalized transduction occurs with lytic phages that enter a bacterial cell, fragment the bacterial chromosome, and then, while new viral particles are being assembled, improperly incorporate some bacterial DNA within the viral protein coat. Because the amount of DNA, not the information content of the DNA, is what governs viral particle formation, any bacterial gene can be included within the newly formed virus. In contrast, specialized transduction occurs with improper excision of viral DNA from the host chromosome in lysogenic phages. Because the integration site is fixed, only those bacterial genes very close to the integration site will be included in a newly formed virus.

5. While the interrupted-mating experiments will yield the gene order, it will be relative only to fairly distant markers. Thus, the precise location cannot be pinpointed with this technique. Generalized transduction will yield information with regard to very close markers, which makes it a poor choice for the initial experiments because of the massive amount of screening that would have to be done. Together, the two techniques allow, first, for a localization of the mutant (interrupted-mating) and, second, for precise determination of the location of the mutant (generalized transduction) within the general region.

6. This problem is analogous to forming long gene maps with a series of three-point testcrosses. Write the four sequences so that they overlap:

    M Z X W C

          W C N A L

                A L B R U

                      B R U M Z

   The regions with the bars above or below are identical in sequence. Therefore, the order of markers on the circular map is M-Z-X-W-C-N-A-L-B-R-U-M-Z.

7. An F⁻ strain will respond differently to an F⁺ (L) or an Hfr (M) strain while Hfr × Hfr, Hfr × F⁺, F⁺ × F⁺, and F⁻ × F⁻ will give O. Thus strains

2, 3, and 7 are F⁻. Strains 1 and 8 are F⁺, and strains 4, 5, and 6 are Hfr.

8. **a.**

AGAR TYPE	SELECTED GENES
1	c^+
2	a^+
3	b^+

b. The order of genes is revealed in the sequence of colony appearance. Because colonies first appear on agar type 1, which selects for c^+, c must be first. Colonies next appear on agar type 3, which selects for b^+, indicating that b follows c. Allele a^+ appears last. The gene order is $c\ b\ a$. The three genes are roughly equally spaced.

c. In this problem you are looking for cotransfer, which results in no growth because the Hfr strain is d^-. Therefore, the farther a gene is from d, the more growth that will occur. From the data, d is closest to b. It is also closer to a than it is to c. The gene order is $c\ b\ d\ a$.

d. With no A or B in the agar, the medium selects for A⁺ B⁺, and the first colonies should appear at about 17.5 minutes.

9. First, carry out a series of crosses in which you select in a long mating each of the auxotrophic markers. Thus, select for $arg^+ T^r$. In each case score for penicillin resistance. Although not too informative, these crosses will give the marker which is closest to pen^r by showing which marker has the highest linkage. Then do a second cross concentrating on the two markers on either side of the pen^r locus. Suppose that the markers are ala and glu. You can first verify the order by taking the cross in which you selected for ala^+, the first entering marker, and scoring the percentage of both pen^r and glu^+. Because of the gradient of transfer, the percentage of pen^r should be higher than the percentage of glu^+ among the selected ala^+ recombinants.

Then, take the mating in which glu^+ was the selected marker. Since this marker enters last, one can use the cross data to determine the map units by determining the percentage of colonies that are $ala^+ pen^r$, and by the number of $ala^- pen^r$ colonies, as shown in Figure 10-13.

10. **a.** Determine the gene order by comparing $arg^+ bio^+ leu^-$ with $arg^+ bio^- leu^+$. If the order were $arg\ leu\ bio$, four crossovers would be required to get $arg^+ bio^+ leu^-$, while only two would be required to get $arg^+ bio^- leu^+$. If the order is $arg\ bio\ leu$, four crossovers would be required to get $arg^+ bio^- leu^+$, and only two would be required to get $arg^+ bio^+ leu^-$. The gene order is $arg\ bio\ leu$.

b. The arg–bio distance is estimated by the $arg^+ bio^- leu^-$ colony type. RF = 100%(48)/376 = 12.76 m.u.

The bio–leu distance is estimated by the $arg^+ bio^+ leu^-$ colony type. RF = 100%(8)/376 = 2.12 m.u.

11. To solve this problem, draw the Hfr and recipient chromosomes in both crosses and note the number of crossovers needed to get $Z_1^+ Z_2^+$ for the two possible gene orders.

Order 1:

Hfr	Z_1^-	Z_2^+	ade^+	str^s
recipient	Z_1^+	Z_2^-	ade^-	str^r

Order 2:

Hfr	Z_2^-	Z_1^+	ade^+	str^s
recipient	Z_2^+	Z_1^-	ade^-	str^r

From the number of crossovers required to get Z^+ ade^+ str^r, the order must be *ade* Z_2 Z_1.

12. In crosses A and B, the only types that will grow are pro^+ (lac-x^+ lac-y^+) ade^+. Both crosses require a crossover between the *pro* and the *lac* genes, and between the *lac* genes and the *ade* gene. In cross A, if *x* is to the left of *y*, one crossover is required; and if *y* is to the left of *x*, two crossovers are required. The opposite is true for cross B. Single crossovers are more frequent than double crossovers.

X	Y	CROSS A	CROSS B	CONCLUSION
1	2	173	27	1 is to the left of 2
1	3	156	34	1 is to the left of 3
1	4	46	218	4 is to the left of 1
1	5	30	197	5 is to the left of 1
1	6	168	32	1 is to the left of 6
1	7	37	215	7 is to the left of 1
1	8	226	40	1 is to the left of 8
2	3	24	187	3 is to the left of 2
2	8	153	17	2 is to the left of 8
3	6	20	175	6 is to the left of 3
4	5	205	17	4 is to the left of 5
5	7	199	34	5 is to the left of 7

The sequence is *pro*-4-5-7-1-6-3-2-8-*ade*.

13. The most straightforward way would be to put an Hfr at both ends of the same sequence and measure the time of transfer between two specific genes. For example,

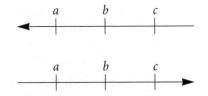

14. **a.** To survive on the selective medium all cultures must be *ery r*. Keep in mind that 300 of these cells were tested under four separate conditions.

    If 263 colonies can grow when only *ery* is added, they must be arg^+ aro^+ ery^r. The remaining 37 cultures are mutant for one or both genes. One additional colony can grow if arg is added to the medium ($264 - 263 = 1$). It must be arg^- aro^+. A total of 290 colonies are arg^+ because they can grow when ery and aro are added to the medium. Of these, 27 are aro^- ($290 - 263 = 27$). The genotypes and their frequencies are summarized below:

$$263 \quad ery^r \, arg^+ \, aro^+$$

$$27 \quad ery^r \, arg^+ \, aro^-$$

$$1 \quad ery^r \, arg^- \, aro^+$$

$$\underline{9} \quad ery^r \, arg^- \, aro^-$$

$$300$$

b. Recombination in the *aro–arg* region is represented by two genotypes: $aro^+ \, arg^-$ and $aro^- \, arg^+$. The frequency of recombination is

$$100\%(1 + 27)/300 = 9.3 \text{ m.u.}$$

Recombination in the *ery–arg* region is represented by two genotypes: $aro^+ \, arg^-$ and $aro^- \, arg^-$. The frequency of recombination is

$$100\%(1 + 9)/300 = 3.3 \text{ m.u.}$$

Recombination in the *ery–aro* region is represented by three genotypes: $arg^+ \, aro^-$, $arg^- \, aro^-$, and $arg^- \, aro^+$. Recall that the DCO must be counted twice. The frequency of recombination is

$$100\%(27 + 9 + 2)/300 = 12.6 \text{ m.u.}$$

c. The ratio is 28:10, or 2.8:1.0.

15. The best explanation is that the integrated pro^+ was incorporated onto an F′ factor that was transferred into recipients early in the mating process. These cells now carry the F factor and are able to transmit F^+ in the second cross as part of the F′ factor, which still carries pro^+.

16. The high rate of integration and the preference for the same site originally occupied by the sex factor suggest that the F′ contains some homology with the original site. The source of homology could be a fragment of the sex factor, or it could be the chromosomal copy of the bacterial gene (more likely).

17. Here, we first need to carry out a cross with the Hfr and F⁻, in which we select for $ala^+ str^r$. If the Hfr donates the *ala* region late, then we should do a short, interrupted mating. If it donates this region early, then we should use a Rec⁻ strain that cannot incorporate a fragment of the donor chromosome by recombination. The colonies from the cross should then be used in a second mating to another ala^- strain to see whether we can donate easily the *ala* gene, which would indicate that we have received an F′ *ala*. To do this we need another marker. If we had an F⁻ that was also ala^- and T^r, then we could do the second cross selecting for $ala^+ \, T^r$.

18. a. and b.

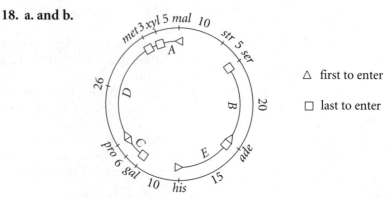

△ first to enter

□ last to enter

c. A: select for *mal*⁺

B: select for *ade*⁺

C: select for *pro*⁺

D: select for *pro*⁺

E: select for *his*⁺

19. a. If the two genes are far enough apart to be located on separate DNA fragments, then the frequency of double transformants should be the product of the frequency of the two single transformants, or $(4.3\%) \times (0.40\%) = 0.017\%$. The observed double transformant frequency is 0.17%, a factor of 10 greater than expected. Therefore, the two genes are located close together and are cotransformed at a rate of 0.17%.

b. Here, when the two genes must be contained on separate pieces of DNA, the rate of cotransformation is much lower, confirming the conclusion in part a.

20. a. Notice that each gene was transferred into about one-tenth of the cells (single drugs tested). Also notice that pairwise testing gives low values whenever B is involved but fairly high rates when any drug but B is involved. This suggests that the gene for B resistance is not close to the other three genes and that the low rates come from double crossovers.

b. To determine the relative order of genes for resistance to A, C, and D, notice that the frequency of resistance to AC is approximately the frequency of resistance to ACD. Also notice that AD resistance is roughly 50% higher and CD resistance is about 25% higher. This suggests that the gene for D resistance is between the other two genes.

21. The expected number of double recombinants is $(0.01)(0.002)(100,000) = 2$. Interference $= 1 - (\text{observed DCO/expected DCO}) = 1 - 5/2 = -1.5$. By definition, the interference is negative.

22. a. m–r: The two parentals are $+ + +$ and *m r tu*. The crossovers between *m* and *r* are

m + *tu*   162

m + +   520

+ *r tu*   474

+ *r* +   172

———

1328

Therefore the distance is $100\%(1328)/10,342 = 12.8$ m.u.

r–tu: use the same approach as above to show that the distance is $100\%(2152)/10,342 = 20.8$ m.u.

m–tu: by the approach above, the distance is $100\%(2812)/10,342 = 27.2$ m.u.

b. Because *m* and *tu* are farthest apart, the sequence is *m r tu*. At this point, the distance between *m* and *tu* can be corrected for double crossovers (classes + *r* + and *m* + *tu*). The final *m–tu* distance is the sum of the two smaller distances, or $12.8 + 20.8 = 33.6$.

c. Recall that I = 1 − c.c. = 1 − (observed DCO/expected DCO). The observed DCO is 162 + 172 = 334. The expected DCO would be (0.128)(0.208)(10,342) = 275. c.c. = 1.2. I = 1 − 1.2 = −.2. A negative value for I indicates that the occurrence of one crossover makes a second crossover more likely to occur than it would have been without that first crossover. That is, more double crossovers occur than are expected.

23. a.   I: minimal plus proline and histidine

II: minimal plus purines and histidine

III: minimal plus purines and proline

b. The order can be deduced from cotransfer rates. It is *pur-his-pro*.

c. The closer the two genes, the higher the rate of cotransfer. *His* and *pro* are closest.

d. *Pro*⁺ transduction requires a crossover on both sides of the *pro* gene. Because *his* is closer to *pro* than *pur*, you get the following:

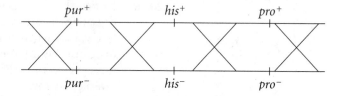

A *pur*⁺ *his*⁻ *pro*⁺ genotype requires four crossovers. If you select for *pro*⁺ *pur*⁺, then the probability of *his*⁺ is very much larger than the probability of *his*⁻.

24. In several percent of the cases, *gal*⁺ transductants arise from recombination between the λdgal transducing phage and the chromosome, without the transducing phage remaining integrated in the chromosome.

25. a. Specialized transduction is at work here. It is characterized by the transduction of one to a few markers.

b. The prophage is located in the *cys–leu* region, which is the only region that gave rise to colonies when tested against the six nutrient markers.

26. a.

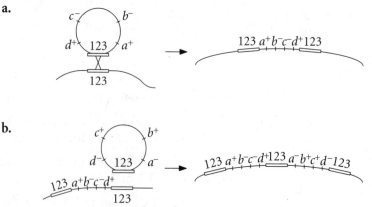

b.

pur his pro

c.

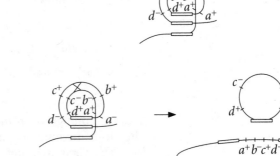

d.

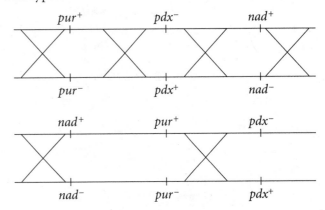

27. **a.** If *trp1* and *trp2* are alleles, then a cross between strains A and B will never result in *trp⁺*, unless recombination occurs within the *trp* gene. If they are not allelic, there will be *trp⁺* colonies. Check these colonies for the possibility of recombination.

   b. Infect strain C with the Z phage and use the progeny to infect strain B cells (which are immune because they are lysogenic for Z). The strain B cells should be plated onto minimal medium and minimal medium plus cysteine.

   If the order is *cys trp2 trp1*, two crossovers will result in *cys⁺ trp2⁺ trp1⁺*, and the number of colonies on the two media should be approximately the same.

   If the order is *cys trp1 trp2*, four crossovers are required for *cys⁺ trp1⁺ trp2⁺*. Therefore, the number of colonies on medium containing cysteine would be greater than the number of colonies on minimal medium.

28. Recognize that if a compound is not added and growth occurs, the *E. coli* has received the genes for it by transduction. Thus, the BCE culture must have received *a⁺* and *d⁺*. The BCD culture received *a⁺* and *e⁺*. The ABD culture received *c⁺* and *e⁺*. The order is thus *d a e c*. Notice that *b* is never cotransduced and is therefore distant from this group of genes.

29. **a.** $100\%(3 + 10)/50 = 26\%$

   b. $100\%(10 + 13)/50 = 46\%$

   c. *pdx* is closer as determined by cotransduction rates.

   d. If the order is *pur pdx nad*, four crossovers are required to get *pur⁺ pdx⁺ nad⁺*. If the order is *nad pur pdx*, two crossovers are required to get wild type:

Because completely wild type occurred less frequently than the other gene combinations, the order is *pur pdx nad*.

30. a. The colonies are all *cys*+ and either + or − for the other two genes.

  b. (1) *cys+ leu+ thr+* and *cys+ leu+ thr−*

  (2) cys+ *leu+ thr+* and *cys+ leu− thr+*

  (3) *cys+ leu+ thr+*

  c. Because none grew on minimal medium, no colony was *leu+ thr+*. Therefore, medium (1) had *cys+ leu+ thr−*, and medium (2) had *cys+ leu− thr+*. The remaining cultures were *cys+ leu− thr−*, and this genotype occurred in 100% − 56% − 5% = 39% of the colonies.

  d.

  leu   cys   thr

31. To isolate the specialized transducing particles of phage φ80 that carried *lac+*, the researchers would have had to lysogenize the strain with φ80, induce the phage with UV, and then use these lysates to transduce a Lac− strain to Lac+. Lac+ colonies would then be used to make a new lysate, which should be highly enriched for the *lac+* transducing phage.

The Structure of DNA

▶ IMPORTANT TERMS
AND CONCEPTS

The **transforming principle** is DNA.

DNA is made of four **nucleotides.** A nucleotide contains a phosphate group, a deoxyribose sugar, and one of four **bases.** A **nucleoside** contains a deoxyribose sugar and one of four bases. The bases are **adenine, guanine, cytosine,** and **thymine.** Adenine and guanine are **purines**; thymine and cytosine are **pyrimidines.** The number of thymine bases plus cytosine bases equals the number of adenine bases plus guanine bases.

DNA is a right-handed **double helix.** The nucleotides are connected by **phosphodiester bonds**, and the chains of the helix are held together by **hydrogen bonds.** The two chains are **antiparallel.** The orientation of deoxyribose within a chain provides the polarity of the chain.

Replication of the DNA is **semiconservative.** Each chain acts as a **template** during replication. Replication proceeds from the origin **bidirectionally.** A **primer**, composed of RNA, begins the replication process. The primer is made by an **RNA polymerase**, often called **primase.** **DNA polymerase** catalyzes the reactions that lead to **polymerization** of nucleotides. **DNA ligase** ligates two DNA molecules together. **DNA helicases** disrupt hydrogen bonds between the two strands, leading to kinks and twists in the DNA. **DNA topoisomerase** converts DNA from one topological form to another, removing kinks and twists. **DNA gyrase** can induce twisting and coiling of DNA, leading to **supercoiling.**

Each chromatid is composed of a single DNA double helix.

Questions for Concept Map 11-1

1. What does *DNA* mean?

2. What is a nucleotide?

3. What is the source of the phosphate group?

4. Name four triphosphates.

5. How do a nucleotide and a nucleoside differ?

6. What are the three components of a nucleotide?

7. In what general class of chemicals is deoxyribose?

8. What does it mean when it is stated that deoxyribose establishes the polarity of a DNA strand?

► CONCEPT MAP 11-1:
THE STRUCTURE AND
REPLICATION OF DNA

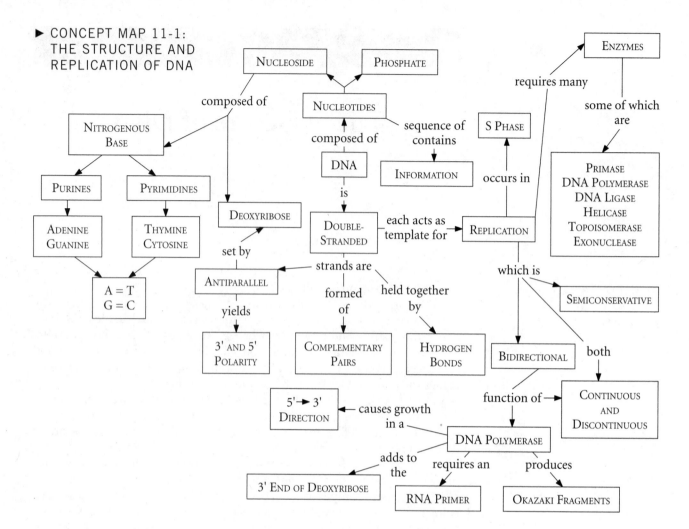

9. What is a nitrogenous base?

10. What two general classes exist for nitrogenous bases?

11. Which are bigger, purines or pyrimidines?

12. What are the two purines? The two pyrimidines?

13. Whose rule is it that A never pairs with C?

14. If the percentage of A in a species is 30%, what is the percentage of G?

15. Is the AT/GC ratio the same for all species?

16. Is the AT/GC ratio the same for all members of a species?

17. The sequence of nitrogenous bases along the length of DNA is information that is used by the cell. How?

18. What is the process whereby the sequence of nitrogenous bases is changed?

19. Do all organisms have DNA as their genetic material?

20. Is DNA always double-stranded?

21. What does *antiparallel* mean?

22. What is complementary pairing?

23. How does complementary pairing provide a potential mechanism for DNA replication?

24. What is a template?

25. What holds the two DNA strands together?

26. What holds the three components of a single DNA strand together?

27. Are hydrogen bonds strong?

28. Are covalent bonds strong?

29. For DNA replication to occur, must the two strands separate?

30. If so, which types of bonds are breaking?

31. What is the point of separation called?

32. When does DNA replication occur in eukaryotes?

33. List the enzymes involved in DNA replication, and be sure you know the function of each.

34. Why is DNA replication called *semiconservative?* What were two other hypothetical modes of replication that were once viewed as possible?

35. Why is DNA replication called *bidirectional?*

36. Where along the DNA does DNA replication begin?

37. What is a replication "bubble"?

38. Why is DNA replication described as both *continuous* and *discontinuous?*

39. What is DNA polymerase?

40. What are Okazaki fragments?

41. Why is an RNA primer required for DNA replication?

42. What happens to the RNA primer after replication?

43. To which end of the new DNA strand is an incoming triphosphate added?

44. In what direction does growth of the new strand occur?

45. If an organism lives in a hot springs, would you expect the AT/GC ratio to be high or low?

46. Which pair of nucleotides is more strongly bound together, A-T or G-C?

47. Which pair of nucleotides, A-T or G-C, is heavier?

48. From what you know about nucleosomes, do they also replicate during DNA replication?

49. Are nucleosomes bound to DNA during DNA replication?

50. How many DNA helices are there in a chromosome with one chromatid?

51. How many DNA helices are there in a chromosome with two chromatids?

52. Which of the following two sequences would you be more likely to find in an organism that grows in a hot spring? Why?

    a. 3'ATATCTGATTTAT-5'

    b. 3'GGGCGTGGGCGGA-5'

53. a. By most genetic measures, chimpanzees and gorillas are closer to each other than humans are to either. Several years ago (1990), DNA hybridization was used as a measure of genetic relatedness, and the results indicated that humans and chimpanzees are the closest pair. The results were confirmed independently. This conclusion has been hotly contested by a number of researchers. What are their possible arguments?

    b. Recently (1992), a new analysis of the old fossil evidence supported the molecular conclusion that chimpanzees are closer to humans genetically than they are to gorillas. Does this new analysis change your answer to the above question?

54. If a DNA sequence is 3'-ATTGC-5', what is its complementary sequence?

55. Assume that a single break occurred in a DNA chain. Which enzyme is capable of connecting the two pieces?

56. Buoyant density is a measure of molecular density during centrifugation. The more dense a molecule, the faster it will pellet in a given solution. The buoyant density (ρ) of DNA molecules in 6 M CsCl increases with the molar percentage of G+C nucleotides according to the following formula:

$$\rho = 1.660 + 0.00098 \ (G+C)$$

Find the molar percentage of G+C in DNA from *Streptococcus pneumoniae*, in which $\rho = 1.700$. What is the A+T molar content?

57. Which enzyme initiates transcription in bacteria?

58. Which enzyme fills gaps during DNA replication in bacteria?

59. Which enzyme is the primary DNA replication enzyme in bacteria?

60. What class of enzymes is responsible for changing DNA topology?

61. Which enzyme unwinds DNA during replication?

62. Graph DNA content within chromosomes as a cell goes through mitosis and then through meiosis.

Answers to Concept Map 11-1 Questions

1. *DNA* means "deoxyribonucleic acid."

2. A nucleotide is the basic building block of DNA. It consists of a nitrogenous base, a deoxyribose, and a phosphate group.

3. The phosphate group comes from a triphosphate (dATP, dGTP, dCTP, dTTP). The energy release from splitting the triphosphate is used to form the covalent bond between the nucleotide and the growing strand of DNA.

4. The four triphosphates are deoxyadenosine 5'-triphosphate (dATP), deoxyguanosine 5'-triphosphate (dGTP), deoxycytidine 5'-triphosphate (dCTP), and deoxythymidine 5'-triphosphate (dTTP).

5. A nucleotide and a nucleoside differ by the presence (nucleotide) or absence (nucleoside) of a phosphate group.

6. The three components of a nucleotide are a nitrogenous base, a deoxyribose, and a phosphate group.

7. Deoxyribose is a carbohydrate, more specifically, a sugar.

8. Deoxyribose is an asymmetric ring molecule. The carbon groups are traditionally labeled from 1′ through 5′. The 3′ carbon is the point where incoming nucleotides attach to the growing DNA strand, and the 5′ carbon is the point on the incoming nucleotide that attaches to the strand. This sets up the polarity of the entire DNA strand.

9. A nitrogenous base is the portion of DNA that hangs off the sugar-phosphate backbone of the molecule. It is the sequence of nitrogenous bases in DNA that contains information for proteins and other molecules.

10. The two general classes of nitrogenous bases are purines and pyrimidines.

11. Purines consist of two rings and are therefore bigger than pyrimidines, which consist of only one ring.

12. The two purines are adenine (A) and guanine (G). The two pyrimidines are thymine (T) and cytosine (C).

13. It is Chargaff's rule that A never pairs with C.

14. If the percentage of A in a species is 30%, the percentage of G is 20%.

15. The AT/GC ratio varies with the species.

16. The AT/GC ratio is the same for all members of a species.

17. The sequence of nitrogenous bases along the length of DNA contains information that is converted ultimately into either a sequence of RNA (tRNA and rRNA) or a sequence of amino acids.

18. The process whereby the sequence of nitrogenous bases is changed is called *mutation*.

19. Not all organisms have DNA as their genetic material. Some (viruses) have RNA, instead.

20. DNA is not always double-stranded. It is temporarily single-stranded during replication and transcription, and in some organisms (viruses) it is single-stranded in the virus particles.

21. Each of the two strands of DNA in the double helix has a polarity determined by the deoxyribose. However, the two strands undergo complementary pairing, with one strand upside down in reference to the other. If you have a DNA helix stretching from the top to the bottom of this page, then one strand is 3′ at the top and the other is 5′ at the top of the page. The two backbones of the two strands can be thought of as oppositely pointed arrows, and this gives rise to the term *antiparallel*.

22. Complementary pairing is the pairing that occurs between A and T, and G and C. There is always a purine paired with a pyrimidine, which keeps the diameter of the double-stranded DNA constant.

23. Complementary pairing provides a potential mechanism for DNA

replication because a purine always pairs with a pyrimidine. Furthermore, two of the nitrogenous bases can form two hydrogen bonds (A and T), and two of the bases can form three hydrogen bonds (G and C). By maximizing hydrogen bonds and observing the purine-pyrimidine pairing, it is easy to see that two separated old strands of DNA can serve as templates for two new strands during replication.

24. In DNA replication, a template is a pattern used to make new DNA.

25. The two DNA strands are held together by hydrogen bonds.

26. The three components of a single DNA strand are held together by covalent bonds.

27. Hydrogen bonds are very weak singly but are very strong in the aggregate.

28. Covalent bonds are quite strong.

29. For DNA replication to occur, the two strands must separate in the region actively engaged in replication.

30. During this separation, hydrogen bonds are breaking.

31. The point of separation is called the *replication fork*.

32. DNA replication occurs during S (synthesis) in eukaryotes.

33. Primase initiates DNA synthesis by laying down an RNA primer (about 10 bases long). DNA polymerase synthesizes the vast majority of DNA. Exonuclease removes the RNA primer and any errors that have been made in complementary pairing (through its proofreading function). DNA ligase links shorter pieces of DNA together. Helicase and topoisomerase help unwind the double-stranded DNA molecule and "straighten" any entanglements that occur in the DNA.

34. DNA replication is called *semiconservative* because each helix consists of one old strand of DNA that acted as a template for the new strand. Two other modes of replication that were thought possible initially are dispersive and conservative replication.

35. DNA replication is called *bidirectional* because it proceeds in both directions from each origin of replication.

36. DNA replication begins at multiple fixed points along the molecule called *origins*.

37. When an origin opens and replication begins bidirectionally, the superficial appearance is that of a bubble appearing in the helix.

38. DNA replication is described as both continuous and discontinuous because the leading strand simply continues replicating, while the lagging strand must be synthesized in short pieces (discontinuously) as the bubble continues to expand.

39. DNA polymerase is the major enzyme that synthesizes DNA.

40. Okazaki fragments are the short pieces of DNA along the lagging strand.

41. RNA primer is required for DNA replication because DNA polymerase cannot initiate DNA synthesis; it can only add onto a growing chain.

42. After replication, the RNA primer is removed by an exonuclease, and the gap is filled by DNA polymerase.

43. An incoming triphosphate is added to the 3′ end of a growing DNA strand.

44. Growth of the new strand is said to occur in a 5′ → 3′ direction.

45. If an organism lives in a hot springs, it has a problem keeping its two strands hydrogen-bonded because energy in the form of heat tends to break hydrogen bonds. The AT/GC ratio would be expected to be low.

46. G-C pairs are more strongly hydrogen-bonded (by three bonds) than A-T pairs (with two bonds).

47. A G-C pair is 1 mole heavier than an A-T pair. It is this weight difference that allows for the separation in cesium chloride centrifugation of various DNA molecules that differ in their relative content of these pairs.

48. Because nucleosomes are tightly associated with DNA in eukaryotes, they must also undergo replication during DNA replication. However, it is not proper to speak of any protein replication; rather, simply more nucleosomes are formed during DNA replication.

49. Nucleosomes are thought to be loosely bound to DNA during DNA replication.

50. There is one DNA helix in a chromosome with one chromatid.

51. There are two DNA helices in a chromosome with two chromatids.

52. b. The sequence with a high G-C content, because of the three hydrogen bonds per G-C base pair, is more resistant to melting than is the sequence with a high A-T content, since there are only two hydrogen bonds per A-T base pair.

53. a. The objections have ranged from the highly technical to the highly slanderous. One scientific objection is that DNA hybridization is a much more complex process than has been recognized. Another scientific objection revolves around the statistical analysis that was conducted. A third is that because the results contradict the results obtained by several other approaches, DNA hybridization simply is not a good measure of genetic relatedness.

    b. While the scientific objections that were raised have yet to be answered, the furor over the molecular analysis and the specific findings may have biased the reanalysis of the fossil evidence. Alternatively, the molecular analysis may have tipped the scales for some who were bothered by the old analysis of the fossil data and yet could not point to specific pieces of fossil evidence that indicated definitively that chimpanzees were closer to humans than to gorillas.

    The student needs to remember that scientific interpretation is conducted by humans. All humans have a set of biases; in some those biases are much stronger than in others. In the past, bias has led to the dismissal of elegant experiments (the work of Avery, MacLeod, and McCarty on DNA as the transforming principle, as an example), and the complete acceptance of unproven facts (the initial acceptance of semiconservative replication although no proof existed when it was proposed, as an example). There is every reason to think that these two "nonscientific" processes are still occurring within science today.

54. 3′-GCAAT-5′.

55. DNA ligase.

56. The equation is $1.700 = 1.660 + 0.00098(G + C)$. $G + C = 40.8$ percent. Because $(A + T) + (G + C) = 1.0$, the A + T molar percentage is 59.2.

57. Primase.

58. DNA polymerase I.

59. DNA polymerase III.

60. Topoisomerases.

61. Helicase.

62.

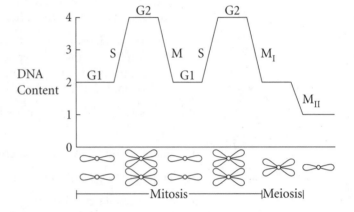

▶ SOLUTIONS TO PROBLEMS

Be sure that you have thoroughly read the entire chapter before you attempt any of the problems.

1. The DNA double helix has two types of bonds, covalent and hydrogen. Covalent bonds occur within each linear strand and strongly bond nucleotides, sugars, and phosphate groups (both within each component and between components). Hydrogen bonds occur between the two strands and involve a nucleotide from one strand with a nucleotide from the second in complementary pairing. These hydrogen bonds are individually weak but collectively quite strong.

2. Conservative replication is a form of DNA synthesis in which the two template strands remain together but dictate the synthesis of two new DNA strands, which then form a second DNA helix. The end point is two double helices, one containing only old DNA and one containing only new DNA. This hypothesis was found not to be correct. Semiconservative replication is a form of DNA synthesis in which the two template strands separate and each dictates the synthesis of a new strand. The end point is two double helices, both containing one new and one old strand of DNA. This hypothesis was found to be correct.

3. A primer is a short segment of RNA that is synthesized by RNA polymerase using DNA as a template during DNA replication. Once the primer is synthesized, DNA polymerase then adds DNA to the 3′ end of the RNA. Primers are required because the major DNA polymerase involved with DNA replication is unable to initiate DNA synthesis and, rather, requires a 3′ end. The RNA is subsequently removed and replaced with DNA so that no gaps exist in the final product.

4. Helicases are enzymes that disrupt the hydrogen bonds which hold the

two DNA strands together in a double helix. This breakage is required for both RNA and DNA synthesis. Topoisomerases are enzymes that create and relax supercoiling in the DNA double helix. The supercoiling itself is a result of the twisting of the DNA helix that occurs when the two strands separate.

5. Because the DNA polymerase is capable of adding new nucleotides only at the 3′ end of a DNA strand, and because the two strands are antiparallel, at least two molecules of DNA polymerase must be involved in the replication of any specific region of DNA. When a region becomes single-stranded, the two strands have an opposite orientation. Imagine a single-stranded region that runs from right to left. The 5′ end is at the right, with the 3′ end pointing to the left; synthesis can initiate and continue uninterrupted toward the right end of this strand. Remember: new nucleotides are added in a 5′→3′ direction, so the template must be copied from its 3′ end. The other strand has a 5′ end at the left with the 3′ end pointing right. Thus, the two strands are oriented in opposite directions (antiparallel), and synthesis (which is 5′→3′) must proceed in opposte directions. For the leading strand (say, the top strand) replication is to the right, following the replication fork. It is continuous and may be thought of as moving "downstream." Replication on the bottom strand cannot move in the direction of the fork (to the right), since, for this strand, that would mean copying the template from its 5′ end. Therefore, this strand must replicate discontinuously: as the fork creates a new single-stranded stretch of DNA, this is replicated *to the left* (away from the direction of fork movement). For this lagging strand, the replication fork is always opening new single-stranded DNA for replication *upstream* of the previously replicated stretch, and a new fragment of DNA is replicated back to the previously created fragment. Thus, one (Okazaki) fragment follows the other in the direction of the replication fork, but each fragment is created in the opposite direction.

6. A = T, G = C, and A + T + G + C = 100%. If T = 15% and A = 15%, A + T = 30%. G + C must = 70%, and G = C = 35%.

7. Because the percent G equals the percent C, the percent of each in the molecule is 1/2(48%) = 24%. Because A + T + G + C = 100%

A + T = 100% − (G + C) = 52%.

Therefore, the frequency of both A and T is 1/2(52%) = 26%.

8.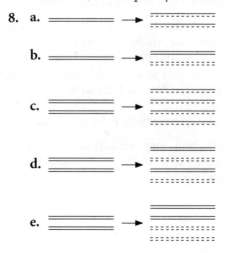

f.   Models b and e are ruled out by the experiment. The results were compatible with semiconservative replication, but the exact structure could not be predicted from the results. In this experiment, it was proved that DNA replication does *not* occur conservatively at either the DNA or chromosomal level.

9. The results suggest that the DNA is replicated in short segments that are subsequently joined by enzymatic action (DNA ligase). Because DNA replication is bidirectional, because there are multiple points along the DNA where replication is initiated, and because DNA polymerases work only in a $5' \rightarrow 3'$ direction, one strand of the DNA is always in the wrong orientation for the enzyme. This requires synthesis in fragments.

10. Replication requires that the enzymes and initiation factors of replication have access to the DNA. One possible answer is that the heterochromatic regions, being more condensed than euchromatin, require a longer period to decondense to the point where replication can be initiated. A second possibility is that the heterochromatic regions, which are generally located at centromeres and telomeres in many organisms, "anchor" the chromosome to the nuclear membrane. Embedded in part in the nuclear membrane, these regions could require extra time to disentangle so that replication may proceed. A third possibility involves the mechanism of heterochromatization. Genes that are inactive in a given cell type are generally heterochromatic. This mechanism may also specifically delay replication. Many other possibilities can be proposed.

11. a.   A very plausible model is of a triple helix, which would look like a braid, with each strand interacting by hydrogen bonding to the other two.

    b.   Replication would have to be terti-conservative. The three strands would separate, and each strand would dictate the synthesis of the other two strands.

    c.   The reductional division would have to result in three daughter cells, and the equational would have to result in two daughter cells, in either order. Thus, meiosis would yield six gametes.

12. Chargaff's rules are that A = T and G = C. Because this is not observed, the most likely interpretation is that the DNA is single-stranded. The phage would first have to synthesize a complementary strand before it could begin to make multiple copies of itself.

13. Remember that there are two hydrogen bonds between A and T, while there are three hydrogen bonds between G and C. Denaturation involves the breaking of these bonds, which requires energy. The more bonds that need to be broken, the more energy that must be supplied. Thus the temperature at which a given DNA molecule denatures is a function of its base composition. The higher the temperature of denaturation, the higher the percentage of G-C pairs.

14. Deletions and duplications at the DNA level would look exactly like the same events during homologous pairing of chromosomes at the duplex DNA level:

Inversions would have two possibilities, depending on the relative length of the inverted and noninverted regions:

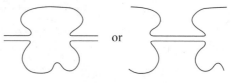

15. **a.** The first shoulder appears before strand interaction takes place, suggesting that the complementary regions are in the same molecule. This is called a palindrome:

A T G C A T G G C C A ——— T G G C C A T G C A T

T A C G T A C C G G T ——— A C C G G T A C G T A

When the strands separate, each strand can base-pair with itself:

A T G C A T G G C C A

T A C G T A C C G G T

b. The second shoulder represents sequences that are present in many copies in the genome (repeated sequences). Because they are at a higher concentration than are sequences present in only one copy per haploid genome (unique sequences), they have a higher probability of encountering one another during a given time period.

16. First realize that the repeated and unique sequences could be completely interspersed, partially interspersed, or completely segregated from each other in the chromosomal DNA. The distribution of these two types of sequences will affect the reannealing curves observed.

   Extract the DNA from the cells. Do separate renaturation curves for unsheared DNA and DNA sheared to fragments of just a few kilobases. If there is generalized interspersion, then the part of the curve corresponding to the annealing of unique sequences will be farther to the left for the unsheared sample compared with the sheared sample. This is because the interspersed repeated sequences in the unsheared sample will facilitate pairing.

   You might also examine the rapidly annealing DNA from the unsheared sample by electron microscopy. Interspersion would result in partial duplexes, with unmatched single-stranded regions (unique-sequence DNA) interspersed.

17. This observation suggests that the mouse cancer cells have copies of the virus genome integrated into them. Thus, it may be that the viral genes somehow alter cell function, triggering malignancy. Alternatively, the viral genome may carry one or more genes that directly result in malignancy. In either case, viral infection may also be the mechanism by which human malignancy is triggered in some instances.

18. The data suggest that each chromosome is composed of one long, continuous molecule of DNA and that chromosomes are fragmented during translocation.

19. Let the broken line indicate DNA that has incorporated bromodeoxyuridine and the unbroken line indicate normal DNA.

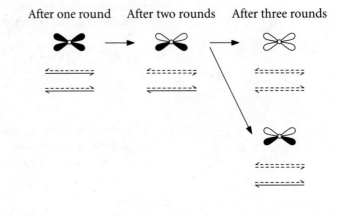

The Nature of the Gene

▶ IMPORTANT TERMS AND CONCEPTS

Genes control biochemical reactions by controlling the production of enzymes. **Enzymes** are proteins, which consist of one or more chains of **amino acids** linked together by **peptide bonds. A polypeptide** is a chain of amino acids. One or more polypeptides exist in each enzyme. A protein-encoding gene specifies the sequence of amino acids in a polypeptide.

A change in base sequence can lead to a change in amino acid sequence, which can result in an alteration of function of the enzyme.

Both **recombination** and **mutation** can occur within a gene. The smallest unit of both recombination and mutation is a single nucleotide pair.

A **cistron** is the equivalent of a gene. It is a genetic region within which normally no complementation occurs between mutations.

Questions for Concept Map 12-1

1. What does the sequence of nitrogenous bases contain?

2. What is the name of the process that leads to a change in the sequence?

3. If there is a change in nitrogenous base sequence, is there always a change in amino acid sequence?

4. If there is a change in amino acid sequence, how is the new sequence related to the old sequence?

5. What is the term that describes a change in nitrogenous base sequence resulting in a change in amino acid sequence in the same relative position?

6. If there is a change in nitrogenous base and amino acid sequence, is there always a change in function associated with it?

7. What is the smallest unit of change in base sequence?

8. What are the two major types of proteins?

9. Which type of protein is responsible for metabolic pathways?

10. Which type of metabolism leads to the synthesis of steroids?

11. Which type of metabolism leads to the degradation of steroids?

12. Are enzymes always composed of just one polypeptide?

▶ CONCEPT MAP 12-1:
  GENES

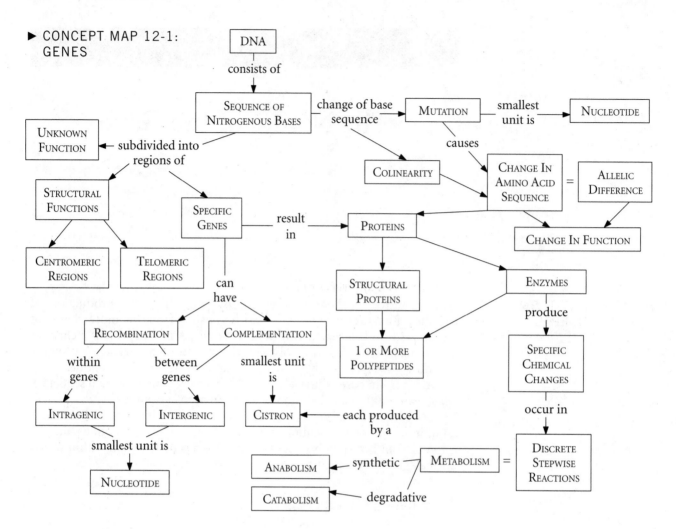

13. Name three structural proteins.

14. What produces a single polypeptide?

15. Can recombination occur between genes? Complementation?

16. Can recombination occur within genes? Complementation?

17. Distinguish between *intragenic* and *intergenic*.

18. What is the smallest unit of recombination?

19. Define cistron.

20. Two deaf parents gave rise to a normal-hearing child. What is the minimum number of cistrons involved?

21. Two mutants known to be in the same cistron were able to complement each other. How can you explain this?

Answers to Concept Map 12-1 Questions

1. The sequence of nitrogenous bases contains information.

2. *Mutation* is the name of the process that leads to a change in the sequence.

3. If there is a change in nitrogenous base sequence, there is not always a change in amino acid sequence. The reason for this will be clear in Chapter 13.

4. If there is a change in amino acid sequence, the new sequence is an allelic alternative to the old sequence. The two sequences are alleles of each other.

5. *Colinearity* is the term that describes a change in nitrogenous base sequence resulting in a change in amino acid sequence in the same relative position.

6. If there is a change in nitrogenous base and amino acid sequence, there is not always a change in function associated with it. Again, this will become clear in Chapter 13.

7. The smallest unit of change in base sequence is the nucleotide.

8. The two major types of proteins are enzymes and structural proteins.

9. Enzymes are responsible for metabolic pathways.

10. Anabolism leads to the synthesis of steroids.

11. Catabolism leads to the degradation of steroids.

12. Enzymes are not always composed of just one polypeptide.

13. Three structural proteins are globin, actin, and keratin.

14. A cistron produces a single polypeptide.

15. Recombination and complementation can occur between genes.

16. Recombination can occur within genes, but complementation cannot.

17. *Intragenic* means "within a gene" and *intergenic* means "between genes."

18. The smallest unit of recombination is a nucleotide. Benzer found that recombination occurred between two nucleotides at a rate of 0.01%.

19. A cistron is the smallest unit of function. It produces a polypeptide.

20. Two.

21. The end product of the wild-type gene contained at least two identical polypeptides. The mutant product of one allele interacted with the mutant product of the second allele in such a way that normal or nearly normal function resulted.

▶ SOLUTIONS TO PROBLEMS

Be sure that you have thoroughly read the entire chapter before you attempt any of the problems.

1. The primary structure of a protein is the sequence of amino acids along its length. It is held together by covalent bonds. The secondary structure of a protein is caused by hydrogen bonding between CO and NH groups on different amino acids. Frequently, α helices and β pleated sheets result. The tertiary structure of a protein is caused by electrostatic, hydrogen, and van der Waals bonds between the R groups of amino acids. If the final, functional protein is composed of more than one polypeptide, then the protein has a quaternary structure.

2. The one gene–one enzyme hypothesis is that each gene is responsible for the synthesis of a specific protein.

3. An auxotroph is a strain that requires at least one nutrient for growth beyond that normally required for the organism.

4. Sickle-cell anemia results from the replacement of a valine for a glutamic acid at position 6 in the β chain of hemoglobin.

5. Yanofsky analyzed mutations in the *trpA* gene and ordered them using transduction. He also determined the amino acid sequence of the altered gene products. By this, he demonstrated an exact correlation between the sequence of mutation sites and the sequence of altered amino acids.

6. Complementation is the cooperation of the products of two or more genes to produce the final phenotype, while recombination is the exchange of DNA segments between chromosomes. Recombination can occur both within and between genes, while complementation occurs between gene products.

7.

	B	K
rII	large plaques	no plaques
rII⁺	small plaques	plaques

8. The defective enzyme that results in albinism may not be able to detoxify a chemical component of Saint-John's-wort that the wild-type enzyme can detoxify. In fact, the plant contains a chemical that is made toxic by light (a phototoxin).

9. Lactose is composed of one molecule of galactose and one molecule of glucose. A secondary cure would result if all galactose and lactose were removed from the diet. The disorder would be expected not to be dominant, because one good copy of the gene should allow for at least some, if not all, breakdown of galactose. In fact, the disorder is recessive.

10. Amniocentesis can be used to detect chromosomal abnormalities and any single-gene abnormalities for which there is a test. Thus, Down syndrome, Turner syndrome, and other chromosomal abnormalities can be detected. Furthermore, biochemical disorders such as galactosemia, Tay-Sachs disease, sickle-cell anemia, and phenylketonuria can be detected. Amniocentesis would thus be useful whenever there is a family history of chromosomal or biochemical disorders and whenever there is a history of parental exposure to mutagens (X rays or chemicals). It would also be useful when the maternal age is over 35, because the rate of nondisjunction is elevated in this population.

11. a. The main use is in detecting carrier parents and in diagnosing a disorder in the fetus.

    b. Because the values for normal individuals and carriers overlap for galactosemia, there is ambiguity if a person has 25 to 30 units as a result. That person could be either a carrier or normal.

    c. These wild-type genes are phenotypically dominant but are incompletely dominant at the molecular level. A minimal level of enzyme activity apparently is enough to ensure normal function and phenotype.

12. One less likely possibility is a germ-line mutation. More likely is that each parent was blocked at a different point in a metabolic pathway. If one were *AA bb* and the other were *aa BB*, then the child would be *Aa Bb* and would have sufficient levels of both resulting enzymes to produce pigment.

13. Assuming homozygosity for the normal gene, the mating is *AA bb* × *aa BB*. The children would be normal, *Aa Bb* (see Problem 12).

14. **a.** Complementation refers to genes within a cell, which is not what is happening here. Most likely, what is known as cross-feeding is occurring, whereby a product made by one strain diffuses to another strain and allows growth of the second strain. This is equivalent to supplementing the medium. Because cross-feeding seems to be taking place, the suggestion is that the strains are blocked at different points in the metabolic pathway.

    b. For cross-feeding to occur, the growing strain must have a block that occurs earlier in the metabolic pathway than does the block in the strain from which it is obtaining the product for growth.

    c. *E-D-B*

    d. Without some tryptophan, no growth at all would occur, and the cells would not have lived long enough to produce a product that could diffuse.

15.

EXPERIMENT	RESULT	INTERPRETATION
v into *v* hosts	scarlet	defects in same gene
cn into *cn* hosts	scarlet	defects in same gene
v into wild-type hosts	wild type	wild type provides *v* product
cn into wild-type hosts	wild type	wild type provides *cn* product
cn into *v* hosts	scarlet	*v* cannot provide *cn* product; *cn* later than *v* in metabolic pathway
v into *cn* hosts	wild type	*cn* provides *v* product; *v* defect earlier than *cn*

A simple test would be to grind up *cn* animals, inject *v* larvae with the material, and look for wild-type development.

16.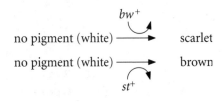

Scarlet plus brown results in red.

17. **a.** The later a compound is in a pathway, the more mutants for which it can support growth. Therefore,

    b. A block is indicated by the fact that no growth takes place if a compound is provided that is already being made. Growth occurs if a compound that cannot be made by the organism is supplied.

Mutant 1: grows on D and G; the block is at the conversion of B to D.

Mutant 2: grows on B, D and G; the block is at the conversion of C to B.

Mutant 3: grows on G; the block is at the conversion of D to G.

Mutant 4: grows on B, C, D and G; the block is at the conversion of A to C.

Mutant 5: grows on all but E; the block is at the conversion of E to A.

c. 1,3 + 2,4: 1,3 would accumulate B and require G; 2,4 would require B, D, or G but could then make G for 1,3 growth. Therefore, growth would occur.

1,3 + 3,4: 1,3 would require G and accumulate B; 3,4 would require G and accumulate A. Neither can help the other to grow.

1,2 + 2,4 + 1,4: 1,2 would require D or G and would accumulate B; 2,4 would require B, D, or G and accumulate C; 1,4 would require D or G and would accumulate B. Therefore growth would occur.

18. a. When $m = 0.5$, $e^{-m} = 0.60$. Therefore, 60 percent of the meioses will not have a crossover.

b. Because $RF = 1/2(1 - e^{-m})$, $RF = 1/2(1 - 0.6) = 0.20$. That is, there are 20 m.u. between the two genes.

c. The recombinants will be $1/2(val\text{-}1\ val\text{-}2)$ and $1/2(val\text{-}1^+\ val\text{-}2^+)$. Therefore, $1/2RF = (val\text{-}1^+\ val\text{-}2^+) = 10$ percent.

d. Remember that accumulation of substance x means the gene responsible for converting substance x to the next metabolic substance is defective. Also, if substance y permits growth, it is beyond a block in a metabolic pathway.

$$\longrightarrow B \xrightarrow[val\text{-}1]{} A \xrightarrow[val\text{-}2]{} valine$$

19. a. A defective enzyme A (from $m_2 m_2$) would yield red petals.

b. Purple, because it has a wild-type allele for each gene.

c.

9	M_1-	M_2-	purple
3	$m_1 m_1$	M_2-	blue
3	M_1-	$m_2 m_2$	red
1	$m_1 m_1$	$m_2 m_2$	white

d. Because they do not produce a functional enzyme, and the wild-type does.

20. a. white

b. blue

c. purple

d. P $bb\ DD \times BB dd$

F$_1$ $Bb\ Dd \times Bb Dd$

F$_2$ 9 $B- D-$ purple

3 *bb D–*   white

3 *B– dd*   blue

1 *bb dd*   white

The ratio of purple:blue:white would be 9:3:4.

21. The *cis* and *trans* burst size should be the same if the mutants are in different cistrons, and if they are in the same cistron, the *trans* burst size should be zero. Therefore, assuming *rV* is in A, *rW* also is in A, and *rU*, *rX*, *rY*, and *rZ* are in B.

22. The cross is *pan2x pan2y⁺ × pan2x⁺ pan2y*.

   a. If one centromere precociously divides, that will put three chromatids in one daughter cell and one in the other.

   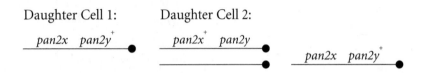

   After meiosis II and mitosis, the first daughter cell would give rise to two pale (*pan2x + pan2y⁺*) and two white, aborted (nullisomic) ascospores, while the second daughter cell would give rise to two black (*pan2x pan2y⁺ / pan2x⁺ pan2y*) and two pale (*pan2x⁺ pan2y*) ascospores. The same result (4 pale:2 colorless:2 black) would occur if only the other centromere divided early.

   b. If both centromeres divided precociously, each daughter cell would be *pan2x pan2y⁺/pan2x⁺ pan2y*. This would lead to 4 colorless (nullisomic) and 4 black (disomic) ascospores.

23. **a.** The mutant does not complement any other mutant. The best interpretation is that it is a deletion.

   b. Complementation groups do not complement within a group (−) but do complement between groups (+). Notice that mutants 1, 5, 8, and 9 complement all others but do not complement within the group. The same holds for mutants 2, 3, 4, and 12 as a group and mutants 6, 7, 10, 11, and 13 as a group. These are the three complementation groups.

   c. Mutants 1 and 2 are in different cistrons, so the cross can be written 1 + × + 2. Assuming independent assortment, the progeny would be

1/4	1 +	*eye⁻*
1/4	1 2	*eye⁻*
1/4	+ 2	*eye⁻*
1/4	+ +	*eye⁺*

   or 3 *eye⁻*:1 *eye⁺*

   Mutants 2 × 6 also complement each other. If independent assortment existed, a 3:1 ratio would be observed. Because the ratio is 113:5, there is no independent assortment and the cistrons are linked. Only one of the two recombinant classes can be distinguished: 5 *eye⁺*.

Because the recombinants should be of equal frequency, the total number of recombinants is 10 out of 118, which leads to 100% (10/118) = 8.47 m.u. distance between the cistrons.

Mutant 14 includes the same cistron (no complementation) as mutant 1.

d. There are three complementation groups; therefore there are three loci, plus mutant 14. Because two of the groups are independently assorting, mutant 14 is a very large deletion spanning the three loci (and they are therefore on the same chromosome), or it is a separate fourth locus that in some fashion controls the expression of the other three loci, or it is a double mutant with a point mutation within one complementation group (1, 5, 8, 9) and a deletion spanning the two linked complementation groups.

e. Two groups are linked, with 8.5 m.u. between them: (2, 3, 4, 12 and 6, 7, 10, 11, 13), and the third group is either on a separate chromosome or more than 50 m.u. from the two other groups.

24. The best interpretation is that *r*W is a deletion spanning *r*Z and *r*D. Alternatively, it could be a double point mutation in *r*Z and *r*D. If it is a double point mutation, then a cross of *r*W with a mutant between *r*Z and *r*D (*r*Z–D) should yield wild type at a low frequency (double crossover). If it is a deletion, no wild type will be observed.

P + *r*Z–D + × *r*Z + *r*D (*r*W)

Double recombinants: + + + and *r*Z *r*Z–D *r*D

25. a. A point mutation within a deletion yields no growth (–), while a point mutation external to a deletion yields growth (+). Only mutant *c* fails to complement deletion 1, so it is within the deletion and the other mutants are not. Mutant *d* is in the overlapping region of deletions 2 and 3, while mutant *e* is in the small region of deletion 2 that is not overlapped by other deletions. The partial order is therefore *c e d*. Mutant *a* is within the nonoverlapped deletion 3, and mutant *b* is within deletions 3 and 4. The final map is

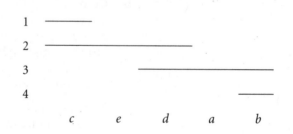

b. The suggestion is that deletion 4 spans both cistrons. This does not affect the conclusions from part a.

26. a. Here, + indicates nonoverlapping and – indicates overlapping. Therefore, 1 overlaps 3 and 5, 2 overlaps 5 only, 3 overlaps all but 2, 4 overlaps 3 only, and 5 overlaps all but 4. Putting all these pieces together yields the deletion map

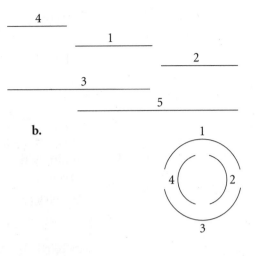

b.

27. **a.** The values equal half the recombinants. Because the problem asks for relative frequency, however, it is not necessary to convert to real map units. Therefore, the distances are

1–2: 14   2–3: 12

1–3: 2   2–4: 6

1–4: 20   3–4: 18

The only map possible is

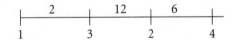

b. No.

28. This is like any other mapping problem when the reciprocal cannot be detected: the detected class is multiplied by 2. Rewrite the crosses and results, using parentheses to indicate unknown order, so that it is clear what the results mean:

Cross 1: his-2⁺ (a, b⁺) nic-2⁺ × his-2 (a⁺, b) nic-2

Cross 2: his-2⁺ (a, c⁺) nic-2 × his-2 (a⁺, c) nic-2⁺

Cross 3: his-2 (b, c⁺) nic-2 × his-2⁺ (b⁺, c) nic-2⁺

Results:

GENOTYPE	CROSS 1	CROSS 2	CROSS 3
his-2 nic-2	DCO between *a* and *b*, and *b* and *nic*	1 CO between *a* and *c*	DCO between *b* and *c*, and *b* and *nic*
his-2⁺ nic-2⁺	DCO between *a* and *b*, and *his* and *a*	as above	DCO between *b* and *c*, and *his* and *c*
his-2 nic-2⁺	1 CO between *a* and *b*	DCO between *a* and *c*, and *nic* and *c*	1 CO between *b* and *c*

GENOTYPE	CROSS 1	CROSS 2	CROSS 3
his-2⁺ nic-2	3 CO between *a* and *b, his* and *a*, and *b* and *nic*	DCO between *a* and *c, his* and *a*	3 CO between *b* and *c, his* and *c,* and *b* and *nic*

Putting all this information together, the only gene sequence possible is *his-2 a c b nic-2*. To calculate the genetic distances, it is necessary to multiply the frequency of recombinants by 2 because reciprocals are not seen:

a–b: 100%(2)(15)/41,236 = 0.072 m.u.

a–c: 100%(2)(6)/38,421 = 0.031 m.u.

b–c: 100%(2)(5)/43,600 = 0.023 m.u.

The final map is

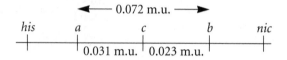

29. **a.** The allele s^n will show dominance over s^f because there will be only 40 units of square factor in the heterozygote.

   b. Here, the functional allele is recessive.

   c. The allele s^f may become dominant over time in two ways: (1) it could mutate slightly, so that it produces more than 50 units, or (2) other modifying genes may mutate to increase the production of s^f.

30. Benzer used the Poisson distribution to make his calculations. While he knew the number of 1 and 2 occurrence sites, he did not know the total number of occurrence sites. Therefore, he had to assume that the Poisson distribution was applicable.

   The equation is

$f(i) = e^{-m}m^i/i!$

   The terms:

$f(0) = e^{-m}m^0/0! = e^{-m}$

$f(1) = e^{-m}m^1/1! = e^{-m}m^1 = 117$ (from Figure 12-35)

$f(2) = e^{-m}m^2/2! = e^{-m}m^2/2 = 53$ (from Figure 12-35)

   These equations can be used to determine the value of *m*:

$f(2)/f(1) = (e^{-m}m^2/2)/(e^{-m}m^1) = 53/117$

$m/2 = 53/117$

$m = 0.9059$

   The number of 0 occurrences is

$f(0)/f(1) = e^{-m}/e^{-m}m$

$f(0) = f(1)/m = 117/0.9059 = 129.15$

31. a. *Cross 1 × 2*: All purple F_1 indicates that two genes are involved. Call the defect in 1 *aa* and the defect in 2 *bb*. The cross is

P   *aa BB × AA bb*

F_1 *Aa Bb*

If the two genes assort independently, a 9:7 ratio of purple : white would be seen. A 1:1 ratio indicates tight linkage. The cross above now needs to be rewritten

P   *a B/a B × A b/A b*

F_1 *a B/A b*

F_2 1 *a B/a B* (white):2 *a B/A b* (purple):1 *A b/A b* (white)

Cross 1 × 3: Again, an F_1 of all purple indicates two genes. The 9:7 F_2 indicates independent assortment. Therefore, let the cross 3 defect be symbolized by *d*:

P   *a B/a B   DD × A B/A B   dd*

F_1 *a B/A B   Dd × a B/A B   Dd*

F_2 9 *A– BB D–*   (purple)

       3 *aa BB D–*   (white)

       3 *A– BB dd*   (white)

       1 *aa BB dd*   (white)

Cross 1 × 4: All white F_1 and F_2 indicates that the two mutations are in the same gene. The cross is

P   *a B/a B   DD × a B/a B   DD*

F_1 same as parents

F_2 same as parents

b. *Cross 2 × 3*:

P   *A b/A b   DD × A B/A B   dd*

F_1 *A b/A B Dd*   (purple)

F_2 9 *A– B– D–*   (purple)

       3 *A b/A b D–*   (white)

       3 *A– B– dd*   (white)

       1 *A b/A b dd*   (white)

Cross 2 × 4: same as cross 1 × 2

32. Mutants *a* and *e* have point mutations within the same cistron. The other point mutations are all in different cistrons. There are at least four cistrons involved with leucine synthesis.

With the exception of two crosses (*a × e*, *b × d*), the frequency of prototrophic progeny is approximately 25 percent. This indicates independent assortment of *a* plus *e* with *b*, *d*, and *c*; and *c* with *b* and *d*. Cistrons *b* and *d* are linked: RF = 100%(4)/500 = 0.8 m.u.

33. a. There are three cistrons:

Cistron 1: mutants 1, 3, and 4

Cistron 2: mutants 2 and 5

Cistron 3: mutant 6

b. Diagram the heterozygotes that yield *star*⁺ gametes, using parentheses to indicate unknown order:

1–6: $A (1^+, 6) B/a (1, 6^+) b \longrightarrow a (1^+, 6^+) B$

2–4: $A (2^+, 4) B/a (2, 4^+) b \longrightarrow a (2^+, 4^+) B$

To determine cistron order within the 1–6 cross, note that the order *A 1 6 B* would require three crossovers, while the order *A 6 1 B* would require one crossover. The order is more likely *A/a* 6 (1, 3, 4) *B/b*.

To determine cistron order from the 2–4 cross, note that the order *A 2 4 B* would require three crossovers, while the order *A 4 2 B* would require one crossover. The order is more likely *A/a* (4, 1, 3) (2, 5) *B/b*. The final order is *A/a* 6 (1, 3, 4) (2, 5) *B/b*.

34. *Unpacking the Problem*

a.

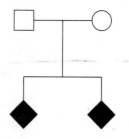

b. By age 10, most of the brain damage caused by an excess level of phenylalanine has already occurred.

c. The bacteria were needed to convert phenylalanine, for which no test existed, to phenylpyruvic acid, for which a test existed.

d. The level of phenylalanine in the urine is only an *estimate* of the level in the blood. It is the blood level that affects brain development, not the urine level.

e. The phenylpyruvic acid in urine was formed from phenylalanine in the blood.

f. Blood concentrations would be expected to be much higher than urine concentrations.

g. The green substance in the ferric chloride test was not in itself important. However, the presence of the colored substance was a definitive indicator of the presence of phenylpyruvic acid.

h. Both phenylalanine and phenylpyruvic acid would be expected to be very low in the blood and urine of unaffected children because both substances are intermediates in a biochemical pathway.

i. The odor in the urine was due to the presence of phenylpyruvic acid.

j. Both parents were heterozygotes: *Aa*.

k. Because the allele is rare, the genotype is also rare. It would be very rare to have both parents of a family heterozygous.

l.   The progeny of heterozygous parents would be expected to occur as follows: 1*AA*: 2 *Aa*: 1 *aa*.

m.   Most families in the population would have parents who were *AA*.

n.   Inheritance was inferred because of the observation that in families with retardation associated with green-staining urine, the affected:unaffected ratio was 1:3. This is the expected ratio in a heterozygous-by-heterozygous cross.

o.   The disorder was inferred to be a Mendelian recessive because both parents were free of the disorder but had children with the disorder.

p.   Most of the nervous system develops before birth.

q.   Adults with PKU do not need to remain on the diet because their nervous systems are completely developed.

r.   The maternal blood circulatory system contains much more fluid than the fetal circulatory system.

s.   Macromolecules and small molecules can pass through the placental barrier.

t.   The urine of the fetus passes from the child to the mother. Maternal phenylalanine and phenylpyruvic acid pass from the mother to the child.

u.   An essential amino acid is an amino acid that must be derived from a dietary source because an organism is incapable of making it, or it is an amino acid that is made by the organism in insufficient amounts for its stage of development.

v.   Phenylalanine is not an essential amino acid.

w.   Phenylalanine blood concentration increases in the absence of the enzyme that converts it to the next product of the metabolic pathway (tyrosine).

x.   PKU is recessive because one normal copy of the gene produces enough enzyme to convert all excess phenylalanine to tyrosine.

y.   PKU is the result of a blocked biochemical pathway. The blockage occurs because of two copies of an allele that does not make a functional enzyme.

z.   See Figure 12-20.

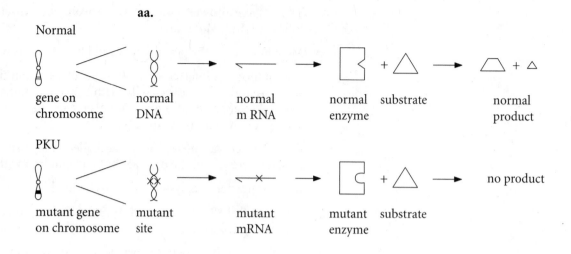

aa.

Normal

gene on chromosome · normal DNA → → normal m RNA → → normal enzyme + substrate → normal product

PKU

mutant gene on chromosome · mutant site → → mutant mRNA → → mutant enzyme + substrate → no product

Solution to the Problem

a.   The mothers still have an excess of phenylalanine in their blood, and that excess is passed through the placenta into the fetal circulatory system, where it causes brain damage prior to birth.

b.   A fetus with two mutant copies of the allele that causes PKU makes no functional enzyme. However, the mother of such a child is heterozygous and makes enough enzyme to block any brain damage; the excess phenylalanine in the fetal circulatory system enters the maternal circulatory system and is processed by the maternal gene product. After birth, which is when PKU damage occurs in a PKU child, dietary restrictions block a buildup of phenylalanine in the circulatory system until brain development is completed.

The fetus of a PKU mother is exposed to the very high level of PKU in her circulatory system during the time of major brain development. Therefore, brain damage occurs before birth, and no dietary restrictions after birth can repair that damage.

c.   The obvious solution to the brain damage seen in the babies of PKU mothers is to return the mother to a restricted diet during pregnancy in order to block high levels of exposure to her child.

d.   PKU is characterized as a rare recessive disorder. A child with PKU has two parents that carry a mutant allele for the metabolism of phenylalanine. When two individuals who are heterozygous for PKU have a child, the risk that the child will have PKU is 25%. A PKU child is unable to make a functional enzyme that converts phenylalanine to tyrosine. As a result, an excess level of phenylalanine is found in the blood, and the excess is detected as an increase in phenylpyruvic acid in both the blood and the urine. The excess phenylpyruvic acid blocks normal development of the brain, resulting in retardation.

DNA Function

RNA is a single-stranded nucleic acid composed of a phosphate group, a ribose sugar, and one of four bases: adenine, guanine, cytosine, or uracil. The ribose sugar provides the polarity to the molecule. Some RNA viruses are double-stranded.

Transcription results in an RNA strand that is complementary to a DNA strand. Only one DNA strand is usually transcribed off a given region of a DNA helix. Along the helix, however, transcription can switch between one strand and the other numerous times, depending on the location of **promoters.** A promoter is the site for the initiation of transcription. For protein-encoding genes, the ultimate product of transcription is **mRNA.** Both **tRNA** and **rRNA** are also transcribed, but not translated.

Translation is the ribosome-mediated production of a polypeptide, using the information encoded in mRNA. A **codon** is a sequence of three DNA bases that specifies an amino acid. The **genetic code** is the set of correspondences between nucleotide triplet codons and amino acids. The code is **degenerate.**

Wobble is the sloppy pairing between the anticodon of a tRNA molecule and the codon in the mRNA.

RNA processing consists of the addition or removal of one or more bases in an RNA molecule plus, in eukaryotes, the splicing out of **intron** transcript sequences. An intron is a sequence within the amino acid–encoding region of a gene that is transcribed but not translated. An **exon** is the coding region in eukaryotic genes that is transcribed and translated into an amino acid sequence.

► CONCEPT MAP 13-1:
INFORMATION
TRANSFER (THE
CENTRAL DOGMA,
REVISED)

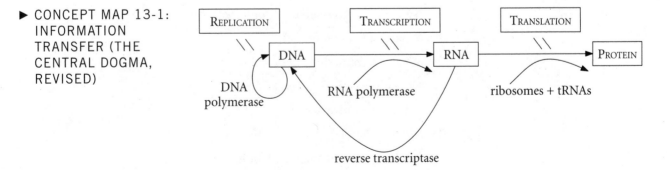

The central dogma was proposed by Crick to describe the one-way transfer of information from DNA to protein. It was later revised to take into account the RNA viruses that are capable of making a DNA copy of their genome with the use of the enzyme reverse transcriptase.

Questions for Concept Map 13-1

1. What are the two major processes involved in making proteins within replicating cells?

2. Describe the structure of DNA.

3. Which enzymes are involved in DNA replication?

4. What are the subunits of which DNA is composed?

5. During replication, if an error is made, what is the result called?

6. If an error is made during the replication of a gene, what does that do to the resulting protein structure?

7. What is the term that is used to describe an organism with alternative forms for the same gene?

8. What is the term used to describe the two alternative forms of a gene?

9. Which enzymes are involved in RNA transcription?

10. What are the subunits of which RNA is composed?

11. What are the major differences between DNA and RNA?

12. What event occurs between transcription and translation in eukaryotes?

13. Which enzymes are involved in translation?

14. What is the product of translation?

15. What are the subunits of proteins?

16. What are the two major subdivisions of proteins?

17. What is a ribosome?

18. Where are ribosomes assembled?

19. What are the three major components of translation?

20. Where does translation occur in prokaryotes? In eukaryotes?

21. What is a codon?

22. What is an anticodon?

23. Name two places where codons are found.

24. What is a reading frame?

25. What is a frameshift?

26. How is the code degenerate?

27. Is the code universal?

28. What is wobble?

29. What is made during translation?

30. Of what is the translation product composed?

31. What are the subunits of the translation product?

32. What is the "oldest" end of the translation product?

33. What is a nonsense codon?

34. What is the function of a ribosome during translation?

35. What is the A site on a ribosome?

36. What is the P site on a ribosome?

37. What type of bond is formed during translation?

38. What is the role of tRNA synthetases?

39. Is there more than one tRNA synthetase?

40. What is the role of peptidyl transferase?

41. What is the first amino acid incorporated in all polypeptides?

42. The sequence on a nontemplate strand of DNA is 5'-ATGC-3'. Write the sequence of the template strand, the codon, the mRNA, the tRNA, and the anticodon.

43. If the anticodon sequence is 3'-AUG-5', what is the sequence of the nontemplate DNA?

44. A mutation in the β-globin gene, a component of hemoglobin, results in a shortened polypeptide even though the immature (unspliced) mRNA transcript was the correct length. Propose two different causes for the short polypeptide.

45. What would be the effect of an inversion within a gene?

46. What would be the effect of an inversion of an entire gene, including all its control sequences?

47. Are tRNA and rRNA transcribed? Translated?

48. Distinguish between missense and nonsense mutations.

Answers to Concept Map 13-1 Questions

1. The two major processes involved in making proteins within replicating cells are transcription and translation.

2. DNA is a double-stranded helix.

3. Enzymes involved in DNA replication include helicase, topoisomerase, primase (RNA polymerase), DNA polymerase, exonuclease, and DNA ligase.

4. DNA is composed of subunits called *nucleotides.*

5. During replication, if an error is made, the result is called a *mutation.*

6. If an error is made during the replication of a gene, the resulting protein structure may have a change in amino acid sequence.

7. The term that is used to describe an organism with alternative forms for the same gene is *heterozygous.*

8. The term used to describe the two alternative forms of a gene is *alleles.*

9. The major enzyme involved in RNA transcription is RNA polymerase, of which there are three forms in eukaryotes but only one in prokaryotes.

10. The subunits of which RNA is composed are nucleotides.

11. The major differences between DNA and RNA are the number of strands involved (two for DNA, one for RNA), the nucleotides of which they are composed (thymidine for DNA, uracil for RNA), and the sugar moiety used (deoxyribose for DNA, ribose for RNA).

12. The event that occurs between transcription and translation in eukaryotes is processing.

13. Translation involves tRNA molecules and ribosomes. The major enzymes are synthetases and peptidyl transferase.

14. The product of translation is a polypeptide.

15. The subunits of proteins are amino acids.

16. The two major subdivisions of proteins are structural proteins and enzymatic proteins.

17. A ribosome is a subcellular organelle that is composed of proteins and rRNA.

18. Ribosomes are assembled in the nucleolus.

19. The three major components of translation are ribosomes, tRNAs, and mRNA.

20. Translation occurs in the general cytoplasm of prokaryotes, which is not separated by a membrane from the location of the DNA. In eukaryotes, translation occurs after the mRNA passes through the nuclear membrane into the cytoplasm.

21. A codon is a three-base sequence that specifies an amino acid.

22. An anticodon is a three-base sequence located on the tRNA molecule that is complementary to a codon.

23. Codons are located in mRNA and in the nontemplate strand of DNA.

24. A reading frame is a sequence of codons.

25. A frameshift is a shift out of the reading frame so that every amino acid is incorrect. It is caused by addition and deletion mutations that do not involve three bases or an integer multiple of three bases.

26. The code is degenerate because more than one codon can specify a particular amino acid.

27. The code is universal.

28. Wobble is the relaxing of complementary pairing requirements between the 5′ base of the anticodon and the 3′ base of the codon so that, for example, G of the anticodon can pair with either U or C of the codon.

29. Polypeptides are made during translation.

30. The polypeptides are composed of amino acids.

31. The subunits of the translation product are amino acids.

32. The "oldest" end of a polypeptide is the amino end.

33. A nonsense codon is a stop codon.

34. A ribosome is the site of translation. The rRNA and protein are thought to have a catalytic function.

35. The A site on a ribosome is the acceptor site, the place where charged tRNAs enter and pair with the codon of a message.

36. The P site on a ribosome is the peptidyl site, the place where a growing amino acid chain is covalently bound to a tRNA.

37. Peptide bonds are formed during translation.

38. The tRNA synthetases "charge" tRNAs with amino acids.

39. There are many more than one tRNA synthetase because each synthetase recognizes only one codon and tRNA.

40. Peptidyl transferase is an enzyme that breaks a bond between the last amino acid added to the chain and its tRNA, and transfers that amino acid to the next incoming amino acid, forming a peptide bond between them.

41. The first amino acid incorporated in all polypeptides is methionine.

42. 5′-ATGC-3′   nontemplate

    3′-TACG-5′   template

    5′-AUGC-3′   codon and mRNA

    3′-UACG-5′   tRNA and anticodon

43. 5′-TAC-3′.

44. One cause would be the introduction of a stop codon. A second cause could be a change of the splicing site so that splicing eliminates more bases in the mRNA than it should.

45. At the very least, the amino acid sequence would be changed for the inverted region. Stop codons might also be generated.

46. No effect should be seen.

47. Both are transcribed; neither is translated.

48. A missense mutation changes a codon so that it stands for a wrong amino acid. A nonsense mutation changes a codon so that it means "stop" to the translation system.

▶ CONCEPT MAP 13-2:
THE STRUCTURE AND
SYNTHESIS OF RNA

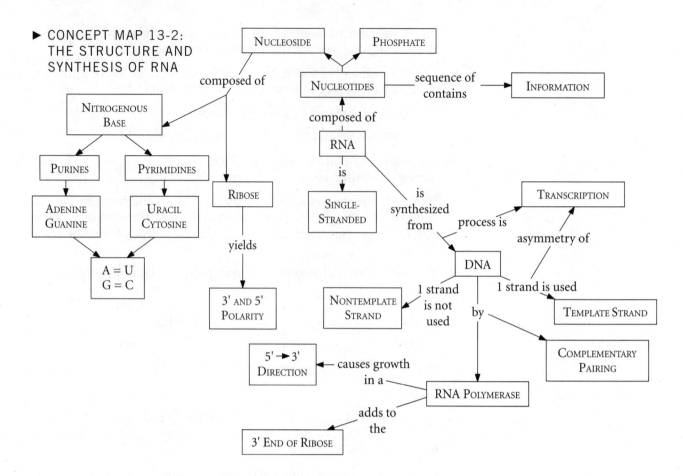

Questions for Concept Map 13-2

1. What does *RNA* mean?

2. Of what is RNA composed?

3. Which nitrogenous bases are in RNA?

4. How does RNA differ from DNA?

5. RNA is single-stranded. However, there are regions within the single-stranded RNA that are complementary. Do you think RNA is capable of pairing within a strand? Why?

6. If the DNA sequence is 3′-ATCCGA-5′, what is the RNA sequence?

7. Is a primer required for RNA synthesis?

8. In an RNA strand, what is the "oldest" end of the molecule?

9. How is complementary pairing involved with RNA synthesis?

10. Distinguish between template and nontemplate strand.

11. Diagram the switching of the template strand along the length of a chromosome and the subsequent RNA synthesis.

12. What molecule provides polarity to an RNA strand?

Answers to Concept Map 13-2 Questions

1. *RNA* means "ribonucleic acid."

2. RNA is composed of the sugar ribose, a nitrogenous base, and a phosphate group.

3. The nitrogenous bases in RNA are adenine, guanine, cytosine, and uracil.

4. RNA differs from DNA in that it is single-stranded rather than double, it has ribose instead of deoxyribose, and it has uracil in place of thymidine.

5. RNA is very capable of self-pairing. The tRNA structure is proof of that.

6. If the DNA sequence is 3'-ATCCGA-5', the RNA secondary sequence is 5'UAGGCU-3'.

7. A primer is not required for RNA synthesis. RNA polymerase is capable of initiating synthesis without one.

8. Recall that addition of bases is at the 3' hydroxy group in ribose (or deoxyribose for DNA). The "oldest" end of the RNA molecule is the 5' end.

9. The sequence of bases in the template strand of DNA is read by RNA polymerase, and complementary RNA bases are added.

10. The template strand of DNA is the strand that is copied in complementary RNA. The nontemplate strand of DNA is also complementary to the template strand.

11.

    ↑ = initiation

12. Ribose provides polarity to an RNA strand.

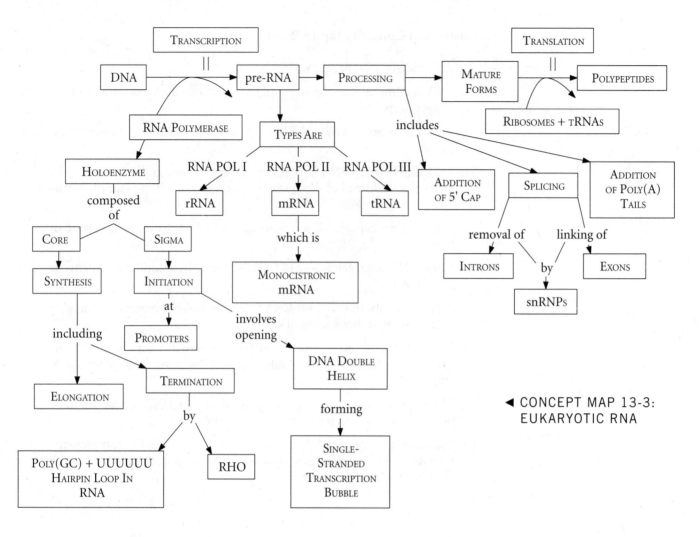

◀ CONCEPT MAP 13-3:
EUKARYOTIC RNA

Questions for Concept Map 13-3

1. Is there any difference in structure between eukaryotic and prokaryotic RNA?

2. Is there any difference in synthesis between eukaryotic and prokaryotic RNA?

3. Is there any difference in the enzymes for RNA synthesis between eukaryotic and prokaryotic RNA?

4. Is there any difference in processing between eukaryotic and prokaryotic RNA?

5. What are the three types of eukaryotic RNA? Do prokaryotes have these types also?

6. What is monocistronic RNA? Polycistronic RNA?

7. What are the components of RNA polymerase?

8. What are promoters?

9. Do prokaryotes have promoters?

10. What does initiation of RNA synthesis consist of in eukaryotes?

11. Would you expect the same events in prokaryotes?

12. What is termination?

13. What are the two mechanisms used for termination?

14. Is sigma involved in termination?

15. Define *intron* and *exon*.

Answers to Concept Map 13-3 Questions

1. The RNA in eukaryotes and prokaryotes is composed of the same nucleotide subunits.

2. Other than the enzymes used, one major difference in synthesis between eukaryotic and prokaryotic RNA is that prokaryotes have polycistronic mRNA while eukaryotes have monocistronic mRNA. Both have polycistronic rRNA. Another major difference is that in eukaryotes synthesis occurs in the nucleus, and the mRNA must reach the cytoplasm for translation to occur, while in prokaryotes there is no nuclear membrane and one end of an mRNA molecule can still be being synthesized while the other end is being translated.

3. The major difference in synthesis of RNA between eukaryotes and prokaryotes is that the former has three enzymes that synthesize the three types of RNA, while the latter has only one enzyme that synthesizes all three types.

4. There is no processing of prokaryotic RNA. In eukaryotic RNA a 5′ cap is added, a poly(A) tail may be added, and any introns are spliced out of the molecule.

5. The three types of eukaryotic RNA are mRNA (which is translated into a polypeptide sequence), tRNA (which participates in the translation of mRNA), and rRNA (which is part of the ribosome structure and also participates in the translation of mRNA). Prokaryotes have these types also.

6. Monocistronic RNA is RNA synthesized off one gene (cistron). Polycistronic RNA is a piece of RNA synthesized off two or more genes (cistrons).

7. RNA polymerase is composed of the sigma factor, which identifies the starting point of synthesis, and the core, a protein structure that conducts the actual synthesis.

8. Promoters are regions preceding (upstream to, 3′ to on the template strand) a gene that indicate to RNA polymerase that synthesis of the gene should begin a somewhat fixed distance from them.

9. Prokaryotes have promoters.

10. In eukaryotes, initiation of RNA synthesis consists of identifying where a promoter is located, binding to the promoter, and opening the DNA double helix locally so that complementary pairing with incoming RNA bases can occur.

11. The same events occur in prokaryotes.

12. Termination is the stopping of RNA synthesis.

13. The two mechanisms used for termination are the rho factor and a sequence in the DNA that results in formation of an RNA hairpin loop. The sequence in RNA is poly(GC) followed by a minimum of six uracil bases.

14. Sigma is not involved in termination, only in initiation.

15. An intron is a sequence within a gene that must be spliced out of the RNA prior to translation. An exon is an amino acid–coding region within a gene. Introns are interspersed in most, but not all, eukaryotic genes. They do not occur in prokaryotes.

▶ SOLUTIONS TO PROBLEMS

Be sure that you have thoroughly read the entire chapter before you attempt any of the problems.

1. Because RNA can hybridize to both strands, the RNA must be transcribed from both strands. This does not mean, however, that both strands are used as a template *within each gene*. The expectation is that only one strand is used within a gene but that different genes are transcribed in different directions along the DNA. The most direct test would be to purify a specific RNA coding for a specific protein and then hybridize it to the λ genome. Only one strand should hybridize to the purified RNA.

2. a. The data do not indicate whether one or both strands are used for transcription in either case.

   b. If the RNA is double-stranded, the percentage of purines (A + G) would equal the percentage of pyrimidines (U + C), and the AG/UC ratio would be 1.0. This is clearly not the case for *E. coli*, which has a ratio of 0.80. Therefore, *E. coli* RNA is single-stranded. The ratio for *B. subtilis* is 1.02. Either the RNA is double-stranded, or there are an equal number of purines and pyrimidines in the strand.

3. A single nucleotide change should result in three adjacent amino acid changes in a protein. One and two adjacent amino acid changes would be expected to be much rarer than the three changes. This is directly the opposite of what is observed in proteins.

4. It suggests very little evolutionary change between *E. coli* and humans with regard to the translational apparatus. The code is universal, the ribosomes are interchangeable, the tRNAs are interchangeable, and the enzymes involved are interchangeable.

5. There are three codons for isoleucine: 5'-AUU-3', 5'-AUC-3', and 5'-AUA-3'. An unacceptable anticodon is 3'-UAU-5' because it would also recognize the codon for *Met,* 5'-AUG-3'. Possible anticodons (using Figure 13-18) are 3'-UAA-5' (complementary), 3'-UAG-5' (complementary), and 3'-UAI-5' (wobble).

6. a. As shown in Figure 13-18, there are eight cases in which knowing the first two nucleotides does not tell you the specific amino acid.

   b. If you knew the amino acid, you would not know the first two nucleotides in the cases of Arg, Ser, and Leu (Figure 13-18).

7. The codon for amber is UAG. Listed below are the amino acids that

would have been needed to be inserted to continue the wild-type chain and their codons:

glutamine	CAA, CAG★
lysine	AAA, AAG★
glutamic acid	GAA, GAG★
tyrosine	UAU★, UAC★
tryptophan	UGG★
serine	AGU, AGC, UCU, UCC, UCA, UCG★

In each case, the codon that has an asterisk by it would require a single base change to become UAG.

8. a. The codons for phenylalanine are UUU and UUC. Only the UUU codon can exist with randomly positioned A and U. Therefore, the chance of UUU is (1/2)(1/2)(1/2) = 1/8.

    b. The codons for isoleucine are AUU, AUC, and AUA. AUC cannot exist. The probability of AUU is (1/2)(1/2)(1/2) = 1/8, and the probability of AUA is (1/2)(1/2)(1/2) = 1/8. The total probability is thus 1/4.

    c. The codons for leucine are UUA, UUG, CUU, CUC, CUA, and CUG, of which only UUA can exist. It has a probability of (1/2)(1/2)(1/2) = 1/8.

    d. The codons for tyrosine are UAU and UAC, of which only UAU can exist. It has a probability of (1/2)(1/2)(1/2) = 1/8.

9. a. 1U:5C: The probability of a U is 1/6, and the probability of a C is 5/6.

CODONS	AMINO ACID	PROBABILITY	SUM
UUU	Phe	(1/6)(1/6)(1/6) = 0.005	Phe = 0.028
UUC	Phe	(1/6)(1/6)(5/6) = 0.023	
CCC	Pro	(5/6)(5/6)(5/6) = 0.578	Pro = 0.694
CCU	Pro	(5/6)(5/6)(1/6) = 0.116	
UCC	Ser	(1/6)(5/6)(5/6) = 0.116	Ser = 0.139
UCU	Ser	(1/6)(5/6)(1/6) = 0.023	
CUC	Leu	(5/6)(1/6)(5/6) = 0.116	Leu = 0.139
CUU	Leu	(5/6)(1/6)(1/6) = 0.023	

1 Phe:25 Pro:5 Ser:5 Leu

    b. Using the same method as above, the final answer is 4 stop:80 Phe:40 Leu:24 Ile:24 Ser:20 Tyr:6 Pro:6 Thr:5 Asn:5 His:1 Lys:1 Gln.

    c. All amino acids are found in the proportions seen in Figure 13-18.

10. a. $(GAU)_n$ codes for Asp (GAU), Met (AUG), and stop (UGA). $(GUA)_n$ codes for Val (GUA), Ser (AGU), and stop (UAG). One reading frame contains a stop codon.

b.   Each of the three reading frames contains a stop codon.

c.   The way to approach this problem is to focus initially on one amino acid at a time. For instance, line 4 of Table 13-10 indicates that the codon for Arg might be AGA or GAG. Line 7 indicates it might be AAG, AGA, or GAA. Therefore, Arg is at least AGA. That also means that Glu is GAG (line 4). Lys and Glu can be AAG or GAA (line 7). Because no other combinations except the ones already mentioned result in either Lys or Glu, no further decision can be made with respect to them. However, taking wobble into consideration, Glu may also be GAA, which leaves Lys as AAG.

Next, focus on lines 1 and 5. Ser and Leu can be UCU and CUC. Ser, Leu, and Phe can be UUC, UCU, and CUU. Phe is not UCU, which is seen in both lines. From line 14, CUU is Leu. Therefore, UUC is Phe, and UCU is Ser.

The footnote for Table 13-10 says that line 13 is in the correct order. In line 13, if UCU is Ser (see above), then Ile is AUC, Tyr is UAU, and Leu is CUA.

Continued application of this approach will allow the assignment of an amino acid to each codon.

11. *Mutant 1*: A simple substitution of Arg for Ser exists, suggesting a nucleotide change. Two codons for Arg are AGA and AGG, and one codon for Ser is AGU. The final U for Ser could have been replaced by either an A or a G.

Mutant 2: The Trp codon (UGG) changed to a stop codon (UGA or UAG).

Mutant 3: Two frameshift mutations occurred:

5'-GCN CCN (–U)GGA GUG AAA AA(+U or C) UGU/C CAU/C-3'.

Mutant 4: An inversion occurred after Trp and before Cys. The DNA original sequence (with both strands shown for the area of inversion) was

3'-CGN GGN ACC TCA CTT TTT ACA/G GTA/G-5'

5'——————— AGU GAA AAA———————3'

Therefore, the complementary RNA sequence was

5'-GCN CCN UGG AGU GAA AAA UGU/C CAU/C-3'

The DNA inverted sequence became

3'-CGN GGN ACC AAA AAG TGA ACA/G GTA/G-5'
                              ^   ^

Therefore, the complementary RNA sequence was

5'-GCN CCN UGG UUU UUC ACU UGU/C CAU/C-3'
                  ^   ^

12. old: AAA/G A̅GU CCA UCA CUU AAU GCN GCN AAA/G

new: AAA/G GUC CAU CAC UUA AUG GCN GCN AAA/G
                                              +

Plus (+) and minus (–) are indicated in the appropriate strands.

13. If the anticodon on a tRNA molecule also was altered by mutation to be four bases long, with the fourth base on the 5' side of the anticodon, it

would suppress the insertion. Alterations in the ribosome can also induce frameshifting.

14. 3′ CGT   ACC   ACT   GCA   5′

     5′ GCA   TGG   TGA   CGT   3′

     5′ GCA   UGG   UGA   CGU   3′

     3′ CGU   ACC   ACU   GCA   5′

NH$_3$-Ala-Trp stop

15. e. With an insertion, the reading frame is disrupted. This will result in a drastically altered protein from the insertion to the end of the protein (which may be much shorter or longer than wild type because of altered stop signals).

16. f, d, j, e, c, i, b, h, a, g.

17. Cells in long-established culture lines usually are not fully diploid. For reasons that are currently unknown, adaptation to culture frequently results in both karyotypic and gene dosage changes. This can result in hemizygosity for some genes, which allows for the expression of previously hidden recessive alleles.

18. a. and b. The sequence of double-stranded DNA is as follows:

    TAC ATG ATC ATT TCA CGG AAT TTC TAG CAT GTA

    ATG TAC TAG TAA AGT GCC TTA AAG ATC GTA CAT

First look for stop signs. Next look for the initiating codon, AUG (3′-TAC-5′ in DNA). Only the upper strand contains a code exactly five amino acids long:

DNA   3′ TAC   GAT   CTT   TAA   GGC   ACT   5′

RNA   5′ AUG   CUA   GAA   AUU   CCG   UGA   3′

protein   Met   Leu   Glu   Ile   Pro   stop

The strand, obviously, is read from right to left as written in your text and is written above in reverse order from your text.

c. Remember that polarity must be taken into account. The inversion is

              stop   start

                                        ←

DNA 5′ TAC ATG CTA GAA ATT CCG TGA AAT GAT CAT GTA 3′

                ^   ^

                                                       start

RNA   stop 3′-CUU UAA GGC ACU UUA CUA GUA-5′

amino acids   HO-7   6   5   4   3   2   1-NH$_3$

d. DNA   3′ ATG TAC TAG TAA AGT GCC TTA AAG ATC GTA CAT 5′

     mRNA 5′ UAC AUG AUC AUU UCA CGG AAU UUC UAG   3′

                  ^ 1   2   3   4   5   6   7 ^

               start   stop

Codon 4 is 5′-UCA-3′, which codes for Ser. Anticodon 4 is 3′-AGU-5′.

Recombinant DNA Technology

Recombinant DNA molecules are made from nonhomologous DNA.

Vectors are small, well-characterized molecules that contain an origin of replication so that inserted fragments can be replicated. Prokaryotic vectors include plasmids, the λ phage, cosmids, and single-stranded phages. Expression vectors allow for the transcription and translation of inserted DNA because they contain the start signals for both processes. Shuttle vectors allow for transcription and translation in two or more host cells.

Restriction enzymes are naturally occurring bacterial enzymes that make sequence-specific cuts in DNA. The resulting fragments, which have **sticky (cohesive) ends**, can then be inserted into vectors. Terminal transferase is used to generate sticky ends in their absence through a process known as **tailing.**

Restriction maps can be generated through the use of restriction enzymes. The maps vary with the species.

Cloning is the production of a large number of specific segments of DNA through the insertion of the DNA into a vector that allows for replication.

Cloning strategies include **shotgunning** and the construction of a **gene bank** or **gene library.**

Southern blotting is the transfer of single-stranded DNA from an electrophoretic gel to a nitrocellulose filter. **Northern blotting** is the transfer of single-stranded RNA from an electrophoretic gel to a nitrocellulose filter. **Western blotting** is the transfer of proteins from an electrophoretic gel to a nitrocellulose filter.

Complementary DNA, or **cDNA,** is made from mature mRNA by using the bacterial enzyme **reverse transcriptase.**

Chromosome walking allows the analysis of very large segments of DNA.

Two sequence-determination methods exist: **base-destruction sequencing** and **dideoxy sequencing.**

The **polymerase chain reaction** (PCR) allows for the amplification of a specific region of DNA.

► CONCEPT MAP 14-1:
   RECOMBINANT DNA
   TECHNOLOGY

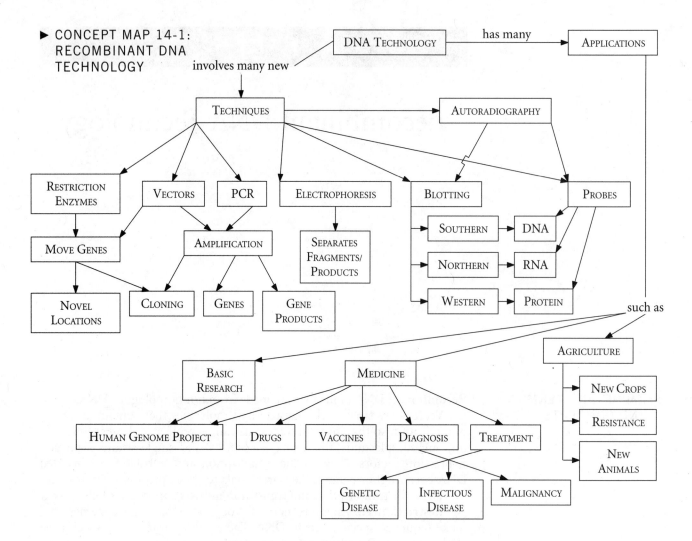

Questions for Concept Map 14-1

1. What is a restriction enzyme?

2. Name three restriction enzymes.

3. If a restriction enzyme cuts at a specific seven-base sequence, how frequently will that enzyme cut genomic DNA?

4. What is the source of the restriction enzymes?

5. Is DNase a restriction enzyme?

6. Distinguish between PCR amplification of a sequence and amplification that occurs through bacterial cloning.

7. What type of vectors result in an amplification of gene products?

8. What does *PCR* mean?

9. Would you expect a problem with contamination using bacterial gene cloning?

10. Would you expect a problem with contamination using PCR?

11. What is gel electrophoresis?

12. What is the "gel" in gel electrophoresis?

13. What is another name for a *vector?*

14. What was the necessity for developing blotting?

15. Distinguish between Southern, Northern, and Western blotting.

16. Could Eastern blotting exist one day?

17. How do both Southern and Northern blotting differ from Western blotting?

18. What types of crops have been developed using DNA technology?

19. What types of new animals have been developed using DNA technology?

20. What types of drugs have been developed using DNA technology?

21. What is the human genome project?

22. How many genomes does each human have?

23. Is there a difference between curing a genetic disease and eliminating a genetic disease within an individual?

Answers to Concept Map 14-1 Questions

1. A restriction enzyme is an enzyme produced by bacteria to fight viral infection that cuts DNA at a specific sequence within the DNA. It is an endonuclease.

2. Three commonly used restriction enzymes are *Hpa*II, *Hae*III, and *Bam*H1.

3. A restriction enzyme that cuts at a specific seven-base sequence will cut genomic DNA on average every $(1/4)^7$ bases.

4. Bacteria are the source of the restriction enzymes.

5. DNase is not a restriction enzyme.

6. Only rather short sequences can be amplified by PCR. It is conducted by the use of what is called a *thermocycler,* which controls the temperature over time, raising it to a precise value at one point in the cycle and lowering it to a precise value at another point in the cycle. Bacterial cloning is amplification that occurs within a bacterial cell. Much longer pieces of DNA can be amplified. Whereas an almost pure product is obtained with PCR, the product obtained by bacterial cloning needs to be purified a great deal.

7. Expression vectors result in an amplification of gene products These are vectors in which transcription and translation can occur.

8. *PCR* means "polymerase chain reaction."

9. There is a slight problem with contamination using bacterial gene cloning.

10. Contamination using PCR is a major problem because PCR is extremely sensitive. There is also the problem that not all sequences are amplified to the same extent. By chance alone, rather rare sequences may sometimes be amplified over those that are in a much higher concentration in the original mixture.

11. Gel electrophoresis is the movement of charged particles through a gel, which acts like a molecular sieve, in response to an electrical current. The particles move as a function of their charge, their shape, and their size. Smaller particles, in general, move faster than larger particles. The movement is logarithmic over time.

12. The "gel" in gel electrophoresis can be composed of a number of products, including starch, agarose, and polyacrylamide.

13. Another name for a vector is a *plasmid* or a phage chromosome.

14. The necessity for developing blotting lay in the fragility of the gel. Whereas a membrane can be probed numerous times, a gel will usually survive a single probing, but not always. Therefore, much less information can be obtained directly from a gel, compared with a membrane.

15. Southern blotting is used for DNA, Northern blotting is used for RNA, and Western blotting is used for proteins.

16. Eastern blotting probably will not exist someday because there are no other major classes of biochemicals that can be electrophoresed beyond DNA, RNA, and proteins.

17. Both Southern and Northern blotting use complementarity to detect a sequence, while Western blotting uses an antibody-antigen reaction.

18. Crop development using DNA technology includes insertion of resistance genes, genes that delay ripening, genes that help with survival against frost, and genes that are involved with nitrogen fixation.

19. Animals have had inserted into them genes that produce hormones in milk, genes that result in a much lower body fat, and genes that cause specific genetic disorders.

20. Drug development using DNA technology includes production of human insulin for diabetics, various other human hormones including human growth hormone, and interferon for use against viral infection.

21. The human genome project is a massive, multinational project designed to sequence the entire human genome.

22. Each human actually has two genomes.

23. *Curing*, in the broadest sense of the term, means to eliminate the effects of the disease within the individual. This means to replace an allele in the tissues in which it is normally expressed. To *eliminate* a genetic disease, the allele must be replaced in germ tissue.

▶ SOLUTIONS TO PROBLEMS

Be sure that you have thoroughly read the entire chapter before you attempt any of the problems.

1. *Unpacking the Problem*

  a. Plasmids are circular pieces of DNA with an origin of replication that exists free in the cytoplasm. Some are capable of integrating into the chromosome. In recombinant DNA technology, they are used to introduce new genes into a cell, to produce many copies of a gene, or to produce a large amount of the gene product.

b. If you wish to find a completely new plasmid, you would have the best chance if you were to examine bacterial species that have not been previously examined by others.

c.

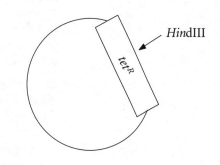

d. Those genes impart a selective advantage to the bacteria that have them.

e. A restriction enzyme makes a double-stranded cut within a sequence of DNA. The enzymes are used in recombinant DNA technology to insert and remove genes in plasmids.

f. The single *Hin*dIII site in pBP1 allows for a simple opening up of the plasmid so that a DNA fragment made with *Hin*dIII can be inserted.

g. The single *Hin*dIII site within the tet^R allows for selection of bacteria that are either resistant (having no inserted gene) or nonresistant (having an inserted gene) to tetracycline.

h. When pBP1 is cut with *Hin*dIII, the plasmid is linearized by a single cut within the tet^R gene.

i. Electrophoresis of pBP1 that has been cut with *Hin*dIII results in a single band.

j. Genomic DNA was cut with *Hin*dIII and mixed with pBP1 that was cut with the same enzyme. DNA ligase was added to recircularize all the molecules. The plasmids were then introduced into bacteria, and the bacteria were cloned. Clones were tested for resistance to tetracycline. Those clones that were not resistant presumably carried fragments of the genomic DNA.

k. To find the gene of interest, a probe composed of the sequence, a cDNA copy of it, or possibly an antibody to the gene product was used.

l. Clone 15 contains the sequence for the gene of interest.

m. *Drosophila* were ground up and DNA was precipitated from the solution, purified, and cut with *Hin*dIII.

n. The functional differences between restriction enzymes are two: (1) the sequence cut and (2) the type of cut made (staggered or blunt).

o. The ethidium bromide, which fluoresces when bound to DNA and exposed to ultraviolet light, is used to visualize the location of bands.

p. You might try beginning with the index.

q. The Southern blot was done by laying a membrane (a piece of nitrocellulose paper or nylon) over the electrophoretic gel. A solution bathing the gel was of the proper pH to make the DNA single-stranded. Paper towels were placed on top of the nitrocellulose membrane so that osmosis would draw the solution through the gel, through the membrane, and into the paper towels. The single-stranded DNA passed out of the gel but adhered to the membrane.

r. The autoradiogram was made by placing a piece of X-ray film (in the dark) on top of the membrane. As radioactive decay of the DNA took place, the energy released struck the X-ray film and was recorded by the film. The film was developed as is any piece of film. Each radioactive decay event leaves a very small dot on the film, indicating the location of the decay event.

s. The spots on an autoradiogram indicate a radioactive decay event.

t. The common use of *label* refers to a tag placed on some object that identifies it in some way. A label on a DNA fragment identifies where the DNA is located.

u. A kilobase is 1000 nucleotide pairs of DNA. A piece of DNA can also be measured by molecular weight, through how far it moves in an electrophoretic gel, or by length in angstroms.

v. The current passing through an electrophoretic gel will cause all electrically charged particles to move through the gel. DNA is negatively charged and moves toward the anode.

w. A nonhomologous vector is a vector with a completely different DNA sequence.

x. A population of DNA molecules of the same size is actually in the band on the gel. This can be demonstrated by cutting out the band and analyzing for DNA.

y. A restriction map shows the location of all cuts made by specific restriction enzymes along a specific piece of DNA.

z. The vector pBP1 has 5.0 kilobases of DNA and the genomic fragment has 2.5 kilobases of DNA. Any variation from this total of 7.5 kb would be suspect.

aa. A double digest results in more bands than a single digest because each enzyme cuts in a different place, so more cuts are made.

bb. If the donor DNA has additional restriction sites, then these sites will be detected in a restriction digest.

cc. Two bands of the same size in two different lanes may or may not be the same DNA, depending on the enzyme used. The *Hin*dIII cuts result in a 5-kb band in the control and in clone 15, representing the vector in each case. The *Eco*RV cut of clone 15 and the *Hin*dIII + *Eco*RV of the control both result in a 3-kb band, but the two bands contain different DNA.

dd. In a circular molecule, the number of cuts equals the number of bands.

Solution to the Problem

a.

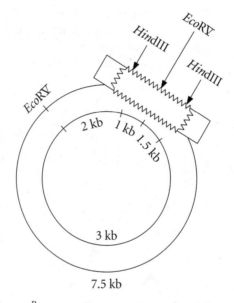

5 kb   +   2.5 kb

7.5 kb

b. The *tet^R* gene will detect the ends of the vector pBP1. For *Hin*dIII, the 5-kb band will be radioactive. For *Eco*RV, all bands will be radioactive. For *Hin*dIII + *Eco*RV, the 3-kb and 2-kb bands will be radioactive.

c. The entire *tet^R* gene is used as a probe. No bands will be radioactive in the control lanes. The clone 15 lanes will all have at least one radioactive band. For *Hin*dIII, 2.5-kb band is radioactive. For *Eco*RV, the 4.5 kb and 3.0-kb bands will be radioactive. For *Hin*dIII + *Eco*RV, the 1.5- and 1-kb bands are radioactive.

2. GTTAAC occurs, on average, every 4^6 bases, which is 4.096 kb. GGCC occurs, on average, every 4^4 bases, which is 0.256 kb.

3. Isolate the DNA and separate the two strands. Test each strand for the ability to hybridize with mRNA produced by the phage.

4.

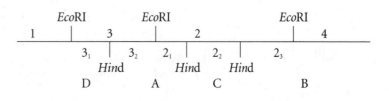

5. a. The easiest way to detect a clone coding for actin would be to screen with actin antibodies.

   b. Either Southern blotting or colony hybridization would identify a clone coding for a specific tRNA.

6. Reading from the top the sequence is

Hind-Hae-Hae-Hind-Hae-Hae-Hae-Hind-Hind-Hae-Hind-Hae-EcoRI

7. Reading from the bottom up,

   left column: GGTACAACTATATATCAATTATAAAC

   right column: GGATCTATTCTTATGATTATATAG

8. The XY male contains every sequence found in the XX female, plus the DNA sequence of the Y chromosome. Therefore, the DNA from a female can be used to purify DNA from the Y chromosome. Isolate, denature, and fix the DNA from a female to a filter. Pour denatured DNA from a male through the filter several times to remove all sequences complementary to the female. The remaining single-strand DNA will be Y chromosome DNA.

9. This problem assumes a random distribution of nucleotides.

   *Alu*I $(1/4)^4$ = every 256 nucleotide pairs

   *Eco*RI $(1/4)^6$ = every 4096 nucleotide pairs

   *Acy*I $(1/4)^4(1/2)^2$ = every 1024 nucleotide pairs

10. a.

2.5	*Hind*	3.0	*Sma*	2.0

   b.

2.5	*Hind*	3.0	*Sma*	2.0
		1.5 *EcoR* 1.5		

11. a. 1200 nucleotides.

   b. At 2 hours the viral transcript contains a nontranscribed intron sequence, which does not hybridize to the cDNA. The RNase removed the sequence, leaving behind 500- and 700-base fragments:

   By 10 hours, the transcript has been spliced out, and a perfect hybrid forms between the 1200-base viral mRNA and the cDNA:

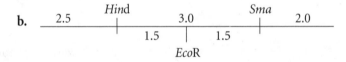

   c. It takes a minimum of 2 hours to transcribe and splice the mRNA and then translate it into protein.

12. Assuming that the same genes from different species have approximately the same base sequence, use the β-tubulin gene cloned from

Neurospora to isolate the β-tubulin gene from *Podospora*.

Isolate plasmids from *Neurospora* using ethidium bromide and cesium chloride.

Cleave the *Podospora* DNA with *Eco*RI. Next, use denatured DNA from the cloned *Neurospora* gene to isolate the sequence from single-stranded *Podospora* DNA. Finally, melt the double-stranded DNA, releasing the desired sequence.

Once the desired sequence has been isolated, cleave the *E. coli* plasmid with *Eco*RI. Mix this DNA with the isolated sequence and add ligase.

Some of the plasmids will contain the desired sequence; they will lack resistance to tetracycline. The rest of the plasmids will not have the desired sequence; they will be resistant to tetracycline.

Introduce the plasmids into *E. coli* cells that are Kan^S Tet^S using calcium chloride. This will make them leaky, and they will take up the plasmid. Select for Kan^R, killing off all *E. coli* that did not incorporate the plasmid.

Identify which resulting clones contain the desired sequence by isolating DNA from each clone, transferring it to a nitrocellulose filter, denaturing it, and probing with P^{32}-labeled DNA from *Neurospora*.

13 a.   There is one *Bgl*II site, and the plasmid is 14 kb.

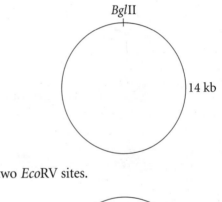

b.   There are two *Eco*RV sites.

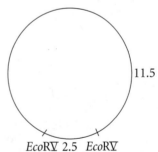

c.   The arrangement of the sites must be as indicated below.

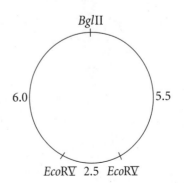

d. The *Bgl*II site must be within the *tet*^R site.

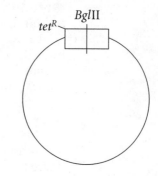

e. There was an insert at the *Bgl*II site in the clone.

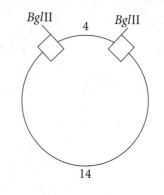

f. There was an *Eco*RV site within the insert.

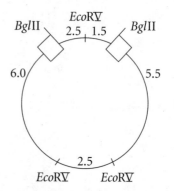

14. a. The restriction map of pBR322 with the mouse fragment inserted is shown below. The 2.5-kb and 3.5-kb fragments would hybridize to the pBr322 probe.

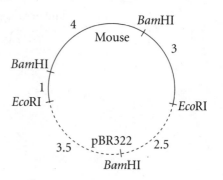

b. A protein 400 amino acids long requires a minimum of 1200 nucleotide bases. Only fragment 3 is long enough (3000 bp) to contain two or more copies of the gene.

15. a. To ensure that a colony is not, in fact, a prototrophic contaminant, the prototrophic line should be sensitive to a drug to which the recipient is resistant. A simple additional marker would achieve the same end but would not be quite so definitive as drug resistance.

b. To ensure that a colony is not in fact a revertant, the donor line should have a resistance gene that is closely linked to the gene of interest, and the recipient should be sensitive. A simple additional marker would achieve the same end but would not be quite so definitive as drug resistance. Alternatively, use a nonrevertible auxotroph as the recipient.

16. a. The transformed phenotype should map to the same locus, whether mapping is done by traditional methods or by restriction mapping, if the transforming gene replaced the original gene.

b. The transformed phenotype should map to a different locus, whether mapping is done by traditional methods or by restriction mapping, if the transforming gene did not replace the original gene.

17. If the size of a chromosome is altered, the size of the resulting band should also be altered. You could use known translocations to identify specific bands. You could also use known additions and deletions to identify specific chromosomes. If probes exist for specific mapped genes, these could be used to identify chromosomes.

18. a. and b.

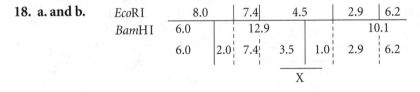

19. a. The total weight of the bands is 2.3 kilobases. The combined digest results in four bands, and *Taq*I results in two bands. This suggests that *Taq*I cuts once in the linearized fragment and that *Hae*III cuts once within each *Taq*I fragment. Note that 0.8 + 0.4 = 1.2, and that 0.6 + 0.5 = 1.2. The map of all the fragments is

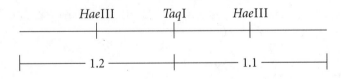

Only the 0.8 and 0.5 fragments contain the ends. Therefore, the complete map is as follows:

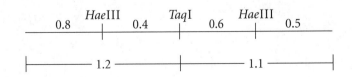

b. There are three bands, only two of which have labeled ends. The map is

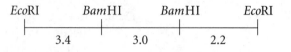

c. The 1.1 *Taq*I fragment would hybridize to the 2.2 fragment.

d. The 3.0 genomic fragment is very likely an intron that would therefore have no representative in the cDNA.

20. a. Homozygosity is represented by one band. Therefore, individuals 1, 3, 4, and 5 are homozygous.

b. Individual 3 makes no RNA. Therefore, the most likely conclusion is that the subject is homozygous for a mutation in the promoter that blocks transcription.

Individual 4 makes a protein smaller than the two alleles seen in normal people but the mRNA is of the same size. This suggests that a chain termination (stop) mutation occurred.

Individual 5 makes a normal-length protein. The best explanation is that a missense mutation occurred that eliminated an essential site of the protein.

21. a. Note that the genomic fragment is end-labeled before the *Bam* digest. This means that the 10-kb *Bam* fragment is labeled only on its external end (compared with the internal site that has a *Bam* end). Only fragments extending from the external end toward the *Bam* site are detectable in the partial digest with *Hpa*II. The bands, then, indicate intervals between *Hpa*II sites. The genomic map is

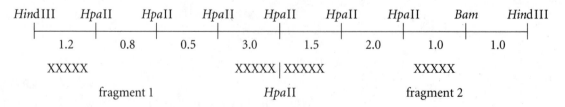

b. and c. The *Hpa*II cut within the cDNA is done before end labeling. Thus, as given in the question, both internal and external ends (i.e., both fragments) are labeled. In the following genomic DNA, the cDNA ends are marked with X's. (An alternative interpretation regarding the 1-kb segments is given later.)

The cDNA map is as follows:

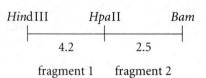

The genomic and cDNA maps differ dramatically because two introns have been removed from the genomic DNA to produce the cDNA. Additionally, the genomic map contains a sequence external to the gene, which is not included in the cDNA. This sequence may be irrelevant to the gene or may be a control sequence (a promoter or a terminator).

An alternative interpretation is shown below, in which the *Bam* site is missing in the cDNA and the fragment ends in a *Hin*dIII site. Either map is possible because you are not told how the cDNA sequence was removed from its vector.

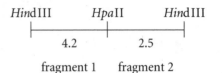

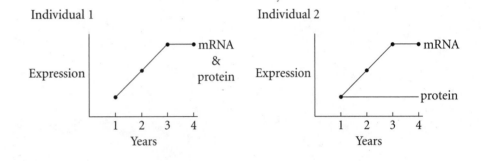

22. a.

b. and c. Very low levels of active enzyme are caused by the introduction of an *Xho* site within the *D* gene. Because of the sizes of the fragments involved and the lack of a frameshift (indicated by some enzyme expression), the most likely event was a point mutation.

This point mutation results in a normal level of RNA, which means that the transcription rate was not affected. Any splicing that occurs as part of processing does not alter the size of the band at the detectable level, which suggests that the point mutation probably does not affect splicing. However, other processing might be affected by the mutation. For example, poly(A) tails are associated with a long-lived RNA, and a failure to place a poly(A) tail on the mutant would explain the lower enzyme activity; there is less translation per mRNA copy. Another processing event that could be disrupted is the placement of a signal protein on the gene product, so that it moves to the correct site at a lower rate. Protein splicing may also be affected, so that less of the mature protein product is made. There may be normal translation and processing, but the active site may be altered by the point mutation, so that there is less enzyme activity. Finally, the degradation of the protein might be accelerated.

d. Individual 1 would be defined as homozygous normal, while individual 2 is homozygous mutant. If individual 2 were heterozygous, he would have an additional band at 12 kb in the Southern blot.

Applications of Recombinant DNA Technology

Two methods exist for making specific alterations of genes: **site-directed mutagenesis** and **gene synthesis.**

Once specific genes have been altered, they can be inserted into a cell through various techniques: transformation (exposing the cell to "naked" DNA), the use of vectors, injection, and transduction (transport into the cell by a virus), to name a few.

Transgenic organisms are organisms that develop from a cell into which new DNA has been introduced. Both **gene inactivation** and **regulation** have been studied in this way. **Gene therapy** has been conducted in animals.

Amniocentesis involves the removal and analysis of fetal amniotic fluid and cells. A number of genetic diseases can be diagnosed this way.

RFLP analysis can be used to screen for genetic disease.

Pulsed field gel electrophoresis is used to electrophorese chromosome-sized pieces of DNA.

Chromosome jumping establishes widely spaced markers along a chromosome that can be used for finely detailed mapping.

Questions for Concept Map 15-1

1. How is a gene cloned?

2. What does *in vitro* mean?

3. What does *mutagenesis* mean?

4. What is the name associated with probing for DNA? RNA?

5. If one probes chromosomes, what does one look at to see the results?

6. If one probes cells, what does one look at to see the results?

7. How does cloning allow the production of novel genotypes?

8. What is reverse genetics?

9. What are the main differences, and their importance, between the two types of viruses used for gene therapy?

10. What is transformation?

► CONCEPT MAP 15-1:
RECOMBINANT DNA
TECHNOLOGY

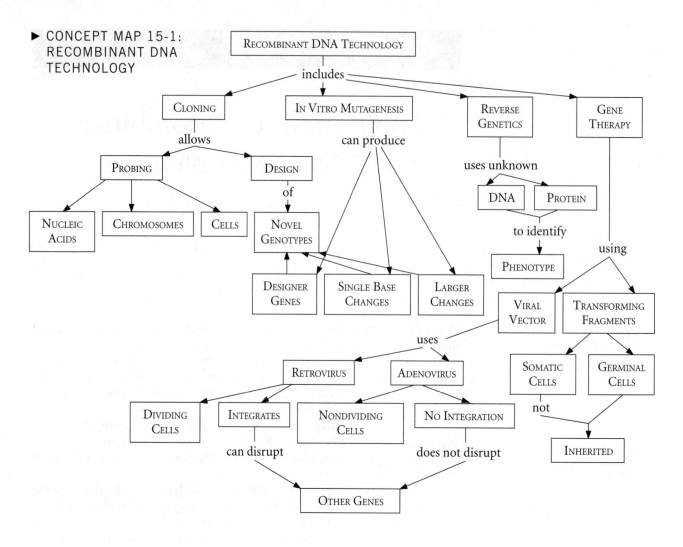

11. Why are therapeutic changes to somatic cells not inherited?

12. What techniques are used for diagnosis of human genetic disorders?

13. RNA from the vesicular stomatitis virus (VSV), which causes characteristic visible effects in cultured cells, was mixed with the protein coat of the rhabdovirus RD114, along with reverse transcriptase. RD114 is also an RNA virus. Virions were formed and used to infect cell lines generated by the fusion of human and mouse cells (recall that these cells will selectively lose human chromosomes). The cell lines were scored for their ability to support VSV infection. In the table below, a + indicates either chromosome present or infection and a − indicates either chromosome absent or no infection.

   a. Which human chromosome contains a gene that codes for a cell receptor for RD114?

   b. What is the role of the reverse transcriptase in this experiment?

   c. What would be the implication if no human chromosome could be linked with the presence of a cell receptor to the virus?

HUMAN CHROMOSOME

CELL LINE	1	2	4	6	7	10	12	14	17	18	19	21	22	X	VSV
1	+	+	+	−	−	+	+	−	−	−	+	−	−	−	+
2	−	+	−	+	−	+	−	+	+	−	+	−	+	−	+
3	+	−	+	−	+	−	+	−	+	+	+	+	−	+	+
4	+	−	+	−	+	+	+	+	−	−	−	+	−	−	−
5	+	+	−	−	−	+	−	−	+	+	−	−	+	+	−

14. *Herpes simplex* virus, type I, which causes cold sores, has a gene for thymidine kinase (*TK*). This virus was used to infect mouse cells that were previously grown under conditions that would kill any cell that expressed *TK* (they were *TK*⁻). The mouse genome was known to contain only one copy of the gene before beginning this experiment. After infection, the cells were grown under conditions in which *TK* expression was required (they became *TK*⁺). Using some of "the tricks of the trade," researchers isolated the chromosomes from those mouse cells that stably expressed *TK* and inserted them into Chinese hamster cells that were previously grown under conditions that would kill any cell that expressed *TK* (they were *TK*⁻). The Chinese hamster cells were then grown under conditions in which *TK* expression was required for survival (they became *TK*⁺). Several independently derived Chinese hamster cell lines that stably expressed *TK* were isolated. Chromosome preparations were made from each of these cell lines and stained with both Giemsa and Hoechst dye (a fluorescent dye that detects AT-rich segments of DNA). In each cell line, a single Hoechst-positive chromosome was observed and identified as a mouse chromosome with Giemsa banding. However, the mouse chromosome was different in the different cell lines. The Chinese hamster lines that expressed *TK* stably were then grown under conditions in which *TK* expression killed the cells. Again, chromosomes were isolated from each cell line and stained with both Hoechst and Giemsa. In some of the cell lines, the entire mouse chromosome had disappeared. In others, a mouse chromosome was present. However, some of the lines contained an intact chromosome while others contained only a portion of the mouse chromosome. When the different Chinese hamster cell lines were then switched again to conditions that required *TK* expression, only a few lines were capable of survival. No cell line that had lacked a mouse chromosome was capable of growing under conditions that required *TK* expression. No cell line that contained a portion of the mouse chromosome was capable of growing under conditions that required *TK* expression. Only some of the cell lines, but not all, that contained an intact mouse chromosome were capable of growing under conditions that required *TK* expression.

 a. What is implied if *TK*⁺ activity in the mouse cell line can be transferred along with a chromosome to the Chinese hamster cell line?

 b. Did the *Herpes simplex TK* gene insert in a specific site in the mouse genome?

 c. Is there any evidence that the *TK* gene was deleted in the Chinese hamster cell lines?

d. Is there any evidence of gene regulation in these experiments?

e. Do Chinese hamster cells contain any extended regions that are AT-rich?

15. Eukaryotic cells in culture have been demonstrated to phagocytose polystyrene balls, calcium phosphate crystals, and numerous other substances that cells are not normally exposed to. How can this ability to phagocytose large particles be exploited to study gene expression?

16. When mouse cells in culture are exposed to exogenous DNA in the presence of calcium phosphate precipitate, both the DNA and the crystals of calcium phosphate enter the cells by a process that looks like phagocytosis when viewed by electron microscopy. The DNA is intimately associated with the crystals. Within 24 hours, the phagocytic vesicles appear to be undergoing exocytosis, the secretion of the vesicle contents into the surrounding medium. If selection for an introduced marker is applied, a very small portion of the mouse cells will survive, and testing reveals that the selected marker is being expressed. Some cell lines will stably express the selected marker. The key event in this process seems to be the quality of the coprecipitate formed. Why is it that very few cells express the selected marker?

17. In the previous problem, it is stated that some cells stably express the selected marker. That means that expression continues in the absence of selection. When recipient chromosomes are examined in these stable expression lines, frequently one or more chromosomes have new bands that were not there before introduction of the exogenous DNA. The new band locations differ from line to line. What do these observations suggest?

Answers to Concept Map 15-1 Questions

1. A gene is cloned through transformation of bacterial cells.

2. *In vitro* literally means "in glass." In other words, something that occurs in vitro occurs in a test tube rather than in an organism.

3. *Mutagenesis* is a term that describes the deliberate causing of mutations, usually at a specific site and in a specific manner (base change, deletion, insertion, etc.).

4. Southern blotting is the name associated with probing for DNA, and Northern blotting is the name associated with probing for RNA.

5. If one probes chromosomes, the results can be seen on a microscope slide or, if pulsed field gel electrophoresis is used, on a gel.

6. If one probes cells, usually the results are seen in petri dishes where colonies of cells are grown.

7. Cloning allows the production of novel genotypes through the introduction of novel genes or mutated genes into an organism.

8. Reverse genetics is an analytic technique that begins with the isolation of a DNA sequence or a protein of unknown function in an organism, and ends with identifying the phenotype associated with the DNA sequence or protein. This is a process that is backwards from classical genetics, in which a phenotype leads to the identification of a gene.

9. The two types of viruses used for gene therapy are retroviruses and adenoviruses. Retroviruses can introduce genes only into dividing cells. The retroviral genome then integrates into host cell chromosomes, which could result in the disruption of an essential gene. Adenoviruses do not require that cells be dividing, which means that they can be used with almost any tissue. Furthermore, the adenovirus genome does not integrate into the host chromosome, which avoids the possibility of gene disruption.

10. Transformation is the introduction of naked DNA into a cell.

11. Therapeutic changes to somatic cells are not inherited because, except for some plants, somatic cells do not give rise to new individuals.

12. Techniques that are used for diagnosis include amniocentesis, chorionic villus sampling, restriction-site analysis, probing for specific sequences, and PCR.

13. a. The only chromosome that has the same pattern of "+" and "−" as the VSV infection is chromosome 19 in humans. Therefore, chromosome 19 in humans codes for a cell surface protein that can be used by the RD114 virus.

   b. The reverse transcriptase makes a DNA copy of the RNA viral genome so that the genome can insert into the human chromosome.

   c. Either the virus cannot infect humans or the cell receptor was not being expressed in the hybrid cells.

14. a. The *TK* gene was inserted into the structure of the mouse chromosome.

   b. Because the mouse genome has only one site for the thymidine kinase gene and the viral *TK* gene was inserted into several different mouse chromosomes, all but one integration must have occurred at a nonhomologous site.

   c. Those hamster cells that were first grown under conditions prohibiting *TK* expression and then were unable to grow under conditions requiring *TK* expression had lost either the entire mouse chromosome or a portion of it. The obvious interpretation was that at least the *TK* gene was deleted.

   d. The hamster cell lines that contained an intact mouse chromosome and were capable of growth, sequentially, under conditions that required *TK* expression, required *TK* nonexpression, and required *TK* expression appear to have been regulating the expression of the viral *TK* gene when it was inserted into the mouse chromosome and then transferred into the hamster cell line.

   e. No. The centromere of each mouse chromosome is AT-rich, but no Hoechst-positive centromeres exist among the Chinese hamster chromosomes.

15. The ability to take in large particles can be exploited to get cells to take in exogenous DNA. When isolated DNA is coprecipitated with calcium phosphate onto the surface of cells in culture, a small portion of those cells can later express a selectable gene that was contained in the DNA but not in the recipient cells. This is known as *eukaryotic transformation.*

16. Most of the DNA that enters the cells through phagocytosis leaves the cells through exocytosis. The DNA that passes into the nucleus must escape the phagocytic vesicle, perhaps by dissolving of the vesicle membranes by the calcium phosphate crystals.

17. The suggestion is that stable expression is correlated with the appearance of new chromosome bands. This, in turn, suggests that stable expression is brought about by integration, perhaps through a recombination-like process, of the introduced DNA in the chromosomes. The varying locations of the bands suggest that integration is random.

▶ SOLUTIONS TO PROBLEMS

Be sure that you have thoroughly read the entire chapter before you attempt any of the problems.

1. Plant 1 shows a typical heterozygous testcross result when crossed with the wild-type kanamycin-sensitive plant. This indicates that a single copy of the gene integrated. The cross is

    P $K^r K^s \times K^s K^s$

    F_1 1 $K^r K^s$ resistant

    1 $K^s K^s$ sensitive

    From this result, it can be concluded that resistance is dominant to sensitivity.

    Plant 2 results in a 3:1 ratio in a testcross. Assuming two separate integrations, the cross is

    P $K^{r1}/K^{s1}\ K^{r2}/K^{s2} \times K^{s1}/K^{s1}\ K^{s2}/K^{s2}$

    F_1 $K^{r1}/K^{s1}\ K^{r2}/K^{s2}$ 25% resistant

    $K^{s1}/K^{s1}\ K^{r2}/K^{s2}$ 25% resistant

    $K^{r1}/K^{s1}\ K^{r2}/K^{s2}$ 25% resistant

    $K^{s1}/K^{s1}\ K^{r2}/K^{s2}$ 25% sensitive

    The gene of interest should assort with kanamycin resistance.

2. Pulsed field gel electrophoresis involves the movement through a gel of chromosome-sized pieces of DNA. Unless the overall size of a chromosome is changed, it should always migrate to the same relative position.

    a.   7 bands identical with wild type.

    b.   7 bands identical with wild type.

    c.   The largest and smallest bands would be expected to disappear. Two new intermediate bands would appear, unless they happen to co-migrate with one of the other five wild-type bands.

    d.   One band would be larger than expected and one would be smaller than expected, when compared with the wild type.

    e.   7 bands identical with wild type.

    f.   7 bands.

    g.   The largest wild-type band would be missing and an even larger new band would be present.

3. The first task is to get the bacterial vector containing the gene of interest into the cell type you wish to test. This can be accomplished by one of several means such as literally shooting the vector into the cell or producing transformed plants.

   Assuming that you also have a resistance gene in the vector, you can select for that gene. Once the tissue is expressing the resistance gene, indicating possible integration of the vector into the plant chromosome, you can then assay for the protein of interest or for production of its mRNA.

   The quick way would be to do a Northern blot on the roots.

4. This is a cross between strains 1 and 2, with bands as the "genes." Because the sum of the bands from a single probe is much less heavy than the total molecular weight involved in the DNA spanning the two sites, you do not have to worry about the two probes detecting part of the same fragment. Before looking at the ascospores, consider the parents, using both probes separately and simultaneously:

Strain 1				Strain 2		
A	B	A+B		A	B	A+B
					5.5	—
	4.0	—				
					3.0	—
—	2.0	—				
	1.5	—				
—	1.0	—				

   If there is no recombination, a 4:4 ascus results (first-division segregation) with the A+B pattern of strain 1 (4) and the A+B pattern of strain 2 (4).

   If a crossover occurs between the marker and a centromere, a 2:2:2:2 pattern would be seen that is an alternation of the A+B patterns of strains 1 and 2.

   If a single crossover occurs between the two RFLP sites, there will also be a 2:2:2:2 ascus. Each strain pattern will be seen among the parentals. The recombinants will be reciprocal and have the following patterns, where the subscripts refer to the strain source of the pattern:

A_1+B_2		A_1+B_2
—	5.5	
	4.0	—
	3.0	—
—	2.0	
	1.5	—
—	1.0	

5. The missing EcoRI restriction site can be assayed for all members of the family, using a probe that spans the EcoRI site. DNA isolated from blood (white blood cells) of each individual would be digested with EcoRI, electrophoresed, transferred by Southern blot, and then probed.

   If the restriction site is missing within an individual, that person would then know that he carries the allele for cystic fibrosis. Those who have the restriction site could be relieved of worry about carrying the allele.

The two individuals carry within the same gene two different mutations that lead to cystic fibrosis. Because two copies of a cystic fibrosis gene, whether or not the defect is exactly the same, leads to expression of the disorder, 25% of the children from this mating would have cystic fibrosis.

6. One approach would be to "knock out" wild-type function and then observe the phenotype. To do this, follow the one-step protocol described for Figure 15-13. Insert a selectable gene into the cloned gene of interest. Linearize the plasmid and use it for transformation of yeast cells. Select for the selectable gene. Another possibility is site-directed mutagenesis (pp. 461–462).

7. After electrophoresis, Southern blot the gel and probe with radioactive copies of the cloned gene.

8. Insert cloned glucuronidase with a plant promoter in the correct orientation into either a seed or numerous different tissues. If the former approach is used, grow mature plants in the presence of X-gluc and assay the tissues for pigment. If the latter approach is used, expose the tissues to X-gluc and assay.

9. a. Let B = bent tail, b = normal tail, r_1 = 3.8 kb, and r_2 = 1.7 kb. The cross is $Bb\ r_1r_2 \times bb\ r_1r_1$. Among the bent progeny are

    40% $Bb\ r_1r_1$

    60% $Bb\ r_1r_2$

    Eliminating the contribution of the wild-type parent, the progeny are

    40% $B\ r_1$

    60% $B\ r_2$

    If independent assortment exists, a 1:1 ratio of r_1:r_2 will be seen; r_1 and r_2 are clearly not assorting independently. Therefore, the two genes are linked with 40 m.u. between them.

    b. The original cross is $B\ r_2/b\ r_1 \times b\ r_1/b\ r_1$. The wild-type progeny for tail conformation are

    60% $b\ r_1/b\ r_1$

    40% $b\ r_2/b\ r_1$

10. a. 1: *A1, B2,* recombinant

    2: *A2, B1,* recombinant

    3: *A1, B1,* parental

    4: *A2, B2,* parental

    b. <u>*A*   30   *B*</u>   or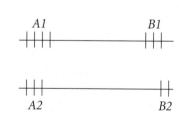

11. a. and b. Kanamycin resistance is dominant to sensitivity because it is a gain of function for this species. The kanamycin-resistance gene is integrated into the host chromosome in each case. If it were not integrated, then the specific ratios would not be observed. The ratios indicate that one gene is involved in plant A while two are involved in plant B. The two genes in plant B are independently assorting.

Below are diagrams of the relevant chromosomes for the two plants. The assumption used to make them is that the T1 plasmid linearized at some other region than the T-DNA region.

Plant A:

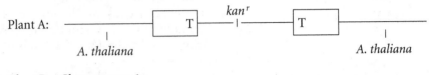

A. thaliana   *A. thaliana*

Plant B, Chromosome 1:

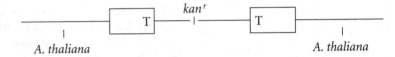

A. thaliana   *A. thaliana*

Chromosome 2 (not homologous with chromosome 1):

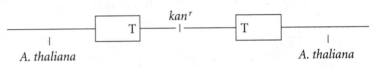

A. thaliana   *A. thaliana*

12. a. Recall that yeast plasmid vectors can exist free in the cytoplasm, producing multiple copies that are distributed randomly to daughter cells, or be integrated into yeast chromosomes, in which case they replicate with the host chromosome and are distributed as any other yeast gene during mitosis and meiosis.

YP1 produces 100% *leu*⁺ progeny, suggesting that the *leu*⁺ gene is not integrated into a chromosome and that enough copies exist of the plasmid so that all progeny obtained at least one copy.

YP2 produces 50% *leu*⁺ progeny, which is indicative of orderly distribution and suggests that the *leu*⁺ gene is integrated into a yeast chromosome.

b. The "insert," which is not cut with the enzyme, is the *leu*⁺ gene inserted into the plasmid.

Following digestion, electrophoresis, blotting, and probing, YP1 will produce one band that is somewhat heavier than the plasmid without an insert (that is, it migrates less distance than the plasmid during electrophoresis). YP2 will produce two bands that contain a portion of the integrated plasmid, plus genomic DNA. The location of these bands relative to the plasmid cannot be predicted.

13. a. If the plasmid never integrates, then *Xba*I digestion will simply cut once within it. Because the plasmid is linear, the *Bgl*II probe will pick up two bands, both of which are lighter than the intact linear plasmid (that is, they will migrate a greater distance than the intact linear plasmid).

b. If the plasmid occasionally integrates, then unintegrated copies will produce the two bands seen in part a and the integrated copies will also produce two bands, which consist of the two plasmid fragments plus genomic DNA. These two latter bands would most likely be heavier than the two former bands and thus migrate less distance through the gel.

14. The electrophoretic patterns are aligned with each individual who was tested, including the two fetuses. There are two alleles that need to be tracked: the 4.9-kb and the (3.7 + 1.2)-kb alleles. Neither of these alleles contains the Huntington disease allele. Instead, the alleles are located at a distance of 4 map units from the Huntington locus. Let the Huntington locus be designated *A* and the locus being tracked by Southern blotting *B*. Huntington disease is dominant and caused by *A*. Arbitrarily, let the (3.7 + 1.2)-kb fragments be *B* and the 4.9-kb fragment be *b*.

   Focus first on the III-1 fetus. Because its mother (II-2) is homozygous for the 4.9-kb fragment, the fetus had to have received the (3.7 + 1.2)-kb fragments from its father (II-1). The father, in turn, received those fragments from his father (I-1), who was homozygous for them and had Huntington disease.

   Individual I-1 is then *A B/a B*. His son, II-1, received −*B* from him and *a b* from his mother (I-2). The probability that II-1 is *A B/a b*, rather than *a B/a b*, is 1/2. If he is *A B/a b*, he produces the following gametes:

   0.48 *A B*

   0.48 *a b*

   0.02 *A b*

   0.02 *a B*

   The fetus received *B*. Therefore, the probability that the fetus has the Huntington disease allele is the probability that its father received it (1/2) times the probability that the father, having received it, passed it to his child, which is 100%(0.48)/(0.48 + 0.02). The total probability is 100%(1/2)(0.48)/(0.48 + 0.02) = 48%.

   The second fetus, III-2, received the (3.7 + 1.2)-kb fragments from its father, II-3. That means the fetus received the 4.9-kb fragment from its mother, II-4. That mother, in turn, must have received the 4.9-kb fragment from her mother and the (3.7 + 1.2)-kb fragments from her father. The mother of the fetus is −*B/a b*. The probability that she is *A B/a b*, rather than *a B/a b*, is 1/2. If she is *A B/a b*, then she makes the following gametes:

   0.48 *A B*

   0.48 *a b*

   0.02 *A b*

   0.02 *a B*

   The fetus received *b* from her. Therefore, the probability that the fetus has the Huntington disease allele is the probability that its mother received it (1/2) times the probability that the mother, if she received it, passed it to her child, which is 100%(0.02)/(0.48 + 0.02). The total probability is 100%(1/2)(0.02)/(0.48 + 0.02) = 2%.

15. *Unpacking the Problem*

a. Hyphae in *Neurospora* are the filaments that grow out from the original ascospore. Therefore, hyphal extension refers to the pattern of these filaments and the distance that they grow. Whatever exists in the world will be found to be interesting by at least one person; in this case, hyphal extension was interesting to a geneticist. On a more serious note, hyphal extension is a type of development that could be analogized to finger extension or leg extension. Therefore, it is likely that someone in the area of developmental genetics might find this gene to be of great interest.

b. Mutational dissection is the analysis of a biological trait through matings of individuals who are phenotypically different for the trait. The general approach of this experiment is the same.

c. *Neurospora* is haploid throughout the vast majority of its life cycle. This is relevant because the experimenter does not have to deal with dominance and recessiveness.

d. Yes, because the experiment will work.

e. Transformation, as originally discovered by Griffith, is the uptake and expression of naked DNA from one organism by another organism. In this situation, *Neurospora* has been pretreated in such a way as to cause the uptake of the bacterial plasmid. This is a well-used technique in molecular genetics as a way of introducing genes into organisms.

f. Plant cells are generally prepared for transformation by removal of their cell walls. The cell membranes are then exposed to a high salt concentration and the exogenous DNA. Studies indicate that the DNA enters the cell in two ways: (1) phagocytosis and (2) localized temporary dissolving of the membrane by the high salt concentration.

g. With successful transformation, the exogenous DNA passes through the cytoplasm and enters the nucleus, where it becomes integrated into a host chromosome.

h.

<u>Entry into Cell</u>

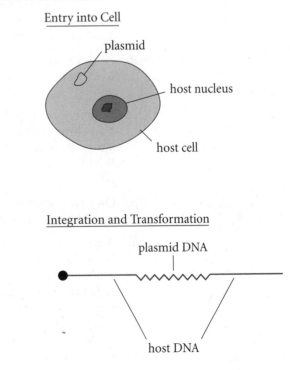

plasmid

host nucleus

host cell

<u>Integration and Transformation</u>

plasmid DNA

host DNA

i. It is completely unnecessary to know what benomyl is. Its use simply allows for the selection of cells that received and integrated some exogenous DNA. Virtually any resistance marker could have been used. The choice of a resistance marker usually depends on what is easily available to the researcher, although questions of toxicity to humans may play a role in the choice.

j. Because hyphal extension occurs in colonies, not at the one-cell stage, the researcher must look for mutants that are expressed by a clone or colony. Therefore, he is looking for mutants that are "colonial." Mutations that produce an aberrant colony in size or shape are, by definition, involved with the extension of hyphae.

k. The "previous mutational analysis" studies most likely involved crosses using a naturally produced (not engineered) mutant.

l.

nonresistant cells

noncolonial transformants

colonial transformants

m. Tagging is the process whereby the identification of a gene of interest is detected through the insertion of DNA within it so that a mutant phenotype is detected. In this case, the benomyl resistance is being inserted randomly within the genome, but insertion within a gene involved with hyphal extension should produce aberrant colonies because the hyphal extension gene has been disrupted by the insertion of the benomyl-resistance gene. The disruption causes a mutation within the gene of interest.

n. Refer to Chapters 3 and 20 for a detailed description of the life cycle of *Neurospora* and the ways in which experimenters have worked with it.

o. In order for the benomyl-resistance gene to be integrated within the host chromosome, it recombines into it. There is no point in calculating the recombination frequency, however, because it is meaningless in this context. For type 1, the RF = 0, but for type 2 the RF = 50%.

p. Only two types are possible: integration into the hyphal gene, which should be relatively rare, and ectopic integration.

q. A probe in experiments such as this one is usually a sequence of DNA that can be used to identify a specific DNA sequence within a genome or colony. The probe is labeled in some way to indicate its presence. The probe may be radioactive, it may have attached to it sequences that cause a color change in a dye, or it may cause a specific chemical reaction that is used to indicate its presence.

In this experiment, the probe was probably the bacterial plasmid (although it might have been benomyl-resistance gene), most likely radioactively labeled. After ascospores were dissected out and grown into colonies, dot blotting identified colonies.

r. A probe specific to the bacterial plasmid could be made by growing bacteria with the plasmid. The plasmids could be isolated through cesium chloride centrifugation and then labeled.

Solution to the Problem

a. Type 1 isolates are parental. In these, the two markers *col* and *ben–R* are from the mutant, and the *col*⁺ and *ben–S* are from the other. Recall, however, that *col* arises because *ben–R* is inserted within the *col*⁺ gene.

   Type 2 isolates are tetratypes that arise through "intragenic" crossing-over.

b. Only type 2 asci have an intact hyphal extension gene, and they should be used for cloning.

c. Tagging in this experiment consists of identifying colonies that grow on benomyl and have aberrant colonial morphology.

d. Recall that transformation was conducted using a bacterial plasmid into which was inserted the benomyl-resistance gene. Any cell that is benomyl-resistant will grow. It becomes benomyl-resistant through the integration of the bacterial plasmid into a host chromosome. The integration carries into the host chromosome the bacterial sequence also. Therefore, any colony that is benomyl-resistant should react with the probe that consists solely of bacterial plasmid sequences, even if intragenic crossing-over occurs. Refer to the diagram in part a above.

The Structure and Function of Eukaryotic Chromosomes

▶ IMPORTANT TERMS
AND CONCEPTS

A **eukaryotic chromosome** is composed of a single, continuous molecule of DNA complexed with proteins. One class of proteins associated with the DNA is the **histones**. There are four major histones: **H2A, H2B, H3,** and **H4.** Two molecules of each of the histones form a structure called a **nucleosome.** The DNA wraps twice around a nucleosome, and there is a nucleosome approximately every 200 base pairs of DNA. Nucleosomes associate to form the **solenoid** structure with the help of another histone, **H1.** The solenoid structure is arranged in loops from the **scaffold.** The scaffold contains **topoisomerase II,** which probably functions to prevent problems caused when DNA must unwind during replication and transcription.

DNA complexed with protein is called **chromatin. Euchromatin,** which is light staining, contains genetically active regions. **Heterochromatin,** which is darkly staining, is thought to be mostly genetically nonfunctional, although some genes may exist in these regions. Some portions of the genome are always heterochromatic (constitutive heterochromatin). Frequently, the DNA contained in the heterochromatic region is **satellite DNA.** Constitutive heterochromatin may play a structural or functional role, or both, in the chromosome. Some portions of the genome are heterochromatic in some cells and not in others (facultative heterochromatin). Faculative heterochromatin consists of genes that are nonfunctional in the specific cell in which it is seen. One example is the **inactivated X chromosome** constituting the **Barr body** (Chapter 3). Another example is an "eye gene" in a liver cell.

A **position effect** is seen when genes are translocated from euchromatin into or near heterochromatin.

Chromosome bands, whether naturally occurring as in *Drosophila* or induced by staining techniques such as the Giemsa banding procedures, are distinctive for each species. The Giemsa light bands are relatively GC-rich, tend to replicate early in S, and contain mainly the "housekeeping" genes (those genes that are active in all cells). The Giemsa dark bands are AT-rich, tend to replicate late in S, and contain mostly tissue-specific genes.

Eukaryotic DNA can be classified as **single-copy, multiple-copy,** or **repetitive,** and **spacer.** Single-copy DNA is composed of protein-encoding genes that are present in only one copy. The repetitive DNA contains both functional and nonfunctional sequences, and the multiple copies can be either in tandem or dispersed throughout the genome. Functional sequences include the rRNA genes, the tRNA genes, families of coding genes (the human globins, as an example) along with their associated pseudo-genes, and telomeric DNA. Such genes may or may not encode a protein. The repetitive sequences with no known function include the DNA in the centromere region, variable-number tandem repeats (VNTRs), and transposable elements.

The **nucleolus** forms in the **nucleolar organizing region** (NO or NOR), which is located at the **secondary constriction** of chromosomes. It is the site of rRNA synthesis and ribosome assembly.

▶ CONCEPT MAP 16-1: EUKARYOTIC CHROMOSOME STRUCTURE

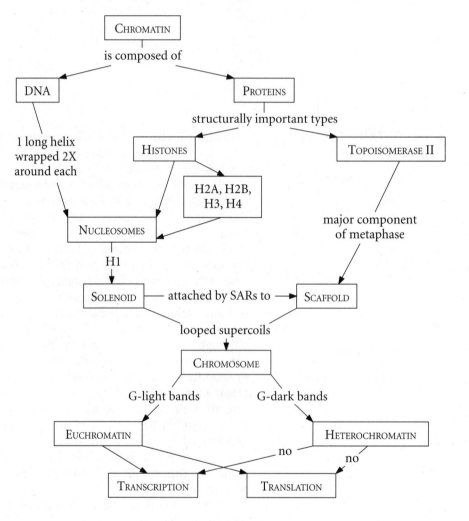

Questions for Concept Map 16-1

1. From where is chromatin isolated?

2. What is the diameter of the DNA helix?

3. What is the diameter of the DNA wrapped around a nucleosome?

4. Is the DNA wrapped around a single nucleosome a gene?

5. What is the relationship between a gene and nucleosomes?

6. When are nucleosomes replicated?

7. Does one DNA helix get all the old nucleosomes and the other get all new ones?

8. Which histones compose the octomer?

9. What is the role of the H1 histone?

10. What is the definition of a solenoid?

11. Why is the term *solenoid* used in DNA compaction?

12. Why is it necessary to compact DNA?

13. Is DNA always compacted?

14. What is a topoisomerase?

15. What does topoisomerase II do?

16. Do you think that the full scaffold structure is also present in interphase?

17. If so, do you think it is still composed of topoisomerase II?

18. If yes, do you think that the topoisomerase II has an enzymatic function during interphase?

19. What is the definition of a supercoil?

20. Give a common example of a supercoil found in virtually every home and office.

21. Distinguish between euchromatin and heterochromatin by staining properties, location in a chromosome, genetic activity (transcription), cesium chloride separation by centrifugation, and location at interphase.

22. Is all euchromatin constantly being transcribed? Explain.

23. Some heterochromatin is always heterochromatic in every cell type within a species. What is its most probable function?

24. Some DNA may be heterochromatic in some cells and not others. It may also be heterochromatic at some stages in the life cycle but not others. Why?

25. Is the variable heterochromatic DNA in question 24 capable of being transcribed in the heterochromatic state?

26. Why is heterochromatin darker-staining than euchromatin?

27. Transcription and replication require the separation of base-paired DNA strands. How might this be accomplished while the DNA is bound to nucleosomes?

28. What does *SAR* mean?

29. What is the function of the SAR?

30. During the transition from interphase to metaphase, what must occur with regard to supercoils?

31. Prokaryotes do not have histones. Does this mean that there are no nucleosomes? No compaction of the DNA?

32. Consider radiation passing through a nucleus. Would heterochromatin or euchromatin be more damaged by it?

Answers to Concept Map 16-1 Questions

1. Chromatin is isolated from the nuclei of cells.

2. The diameter of the DNA helix is 2 nanometers (nm).

3. The diameter of the DNA wrapped around a nucleosome is, in total, approximately 5.7 nanometers (nm) by 11 nm. It is often referred to as being 10 nm for simplicity's sake.

4. The DNA wrapped around a single nucleosome is not a gene. It takes approximately 140 base pairs (bp) of DNA to wrap twice (actually, one and three quarter times) around one nucleosome core particle. The amount of DNA from a specific point on one nucleosome core particle to the same point on the next nucleosome core particle is approximately 200 bp. In contrast, protein-encoding genes are an average of 125 amino acids, which is a minimum of 3×125, or 375, bp of DNA.

5. There is no relationship between a gene and nucleosomes, other than the fact that actively transcribed genes have very loosely bound nucleosomes while inactive genes have tightly bound nucleosomes. The nucleosome is simply a packing structure.

6. Nucleosomes replicate as the DNA is replicated.

7. The most current model is that each DNA helix has both old and new nucleosomes after DNA replication. For that reason, the model is called *dispersive*.

8. Two molecules each of histones H2A, H2B, H3, and H4 compose the octomer.

9. The H1 histone molecules bind to the stretch of DNA between nucleosome core particles, one molecule for each 60 bp of DNA, and to each other to form the solenoid structure.

10. The dictionary definition of a solenoid is a coil of wire commonly in the form of a long cylinder that when carrying a current resembles a bar magnet so that the movable core is drawn into the coil when the current flows. In genetics, a solenoid is a 30-nanometer chromatin fibril formed of nucleosomes that are packed approximately six per turn, all in a cylindrical shape. The H1 histone molecules and the DNA to which they are attached are within the center of the cylinder.

11. The term *solenoid* is used in DNA compaction because the structure formed resembles a solenoid.

12. It is necessary to compact DNA because the DNA is very long in comparison with the cell that contains it and because it could become impossibly entangled if not compacted to some degree. The compaction also allows for the partial sequestering of genes that may be nonfunctional in a particular cell type.

13. DNA is always compacted to some extent, even during transcription or replication.

14. Topoisomerase is a class of enzymes that allows one DNA strand to pass through another DNA strand.

15. Topoisomerase II is the major component of the scaffold structure in metaphase chromosomes. It apparently manipulates the chromatin to

allow for replication and transcription without entanglement of strands.

16. The full scaffold structure is also present in interphase.

17. The full scaffold structure is still composed of topoisomerase II at interphase.

18. Topoisomerase II is probably most active during interphase, when the chromatin is far less condensed and is actively engaged in replication and transcription.

19. A supercoil is the result of the double helix's being twisted in space around its own axis. For circular DNA, the structure becomes supercoiled in one region when it uncoils in another region for either replication or transcription. For linear DNA, no supercoiling will exist as the helix unwinds for replication or transcription unless there are two fixed points flanking the unwound region. The fixed points are thought to be provided by the scaffold attachment points.

20. A common example of a supercoil found in virtually every home and office is an entangled telephone cord. The supercoils are the tightly wound regions that stick out from the rest of the cord at almost a right angle.

21. Euchromatin stains lightly, is dispersed throughout the chromosome except at the centromeres and telomeres, is actively transcribed, appears in the main band of a cesium chloride gradient, and is dispersed throughout the interphase nucleus. Heterochromatin is darkly staining, is located primarily at the centromeres and telomeres of chromosomes, is usually not transcribed, in some species can be found in bands above or below the main band in a cesium chloride gradient, and tends to be located at the periphery of the interphase nuclear membrane.

22. Not all euchromatin is constantly being transcribed, although some forms are (genes coding for rRNA and tRNA, for example). Whether or not a euchromatic gene is being transcribed depends on the stage of the cell cycle and the current concentration of the gene product, along with the current needs of the cell.

23. The heterochromatin that is always heterochromatic in every cell type within a species is most likely structural. Possible examples include centromeric, telomeric, and SAR heterochromatin.

24. The DNA that is heterochromatic in some cells and not others, or heterochromatic at some stages in the life cycle but not others, is DNA encoding for tissue-specific genes or genes that are required in certain stages of development.

25. The variable heterochromatic DNA in question 24 is generally not capable of being transcribed in the heterochromatic state.

26. Heterochromatin is darker-staining than euchromatin simply because it is more condensed.

27. The separation of base-paired strands required for both transcription and replication might be accomplished while the DNA is bound to nucleosomes through a partial unfolding of the nucleosome structure.

28. *SAR* means "scaffold attachment region," which is a segment of DNA that appears to attach to the scaffold protein structure.

29. The SAR appears to attach to the scaffold, which serves to maintain DNA in a supercoiled state within the interphase nucleus and also allows for chromosome condensation with the onset of mitosis.

30. During the transition from interphase to metaphase, supercoils must become more compact.

31. Prokaryotes do not have histones or nucleosomes. While the DNA is not compacted in the same way that euchromatin is compacted, it is maintained in a supercoiled looped-domain state by attachment of the loops at their base. The mechanism of this attachment may be through proteins to the cell membrane, but this has not yet been demonstrated.

32. Radiation passing through a nucleus would damage any DNA in its path. Because heterochromatin is more tightly compacted than euchromatin, heterochromatin may be more damaged than euchromatin. Additionally, it would be more difficult for the enzymes involved in repair to reach the DNA damage when the DNA itself is more tightly compacted.

▶ CONCEPT MAP 16-2:
DNA TYPES

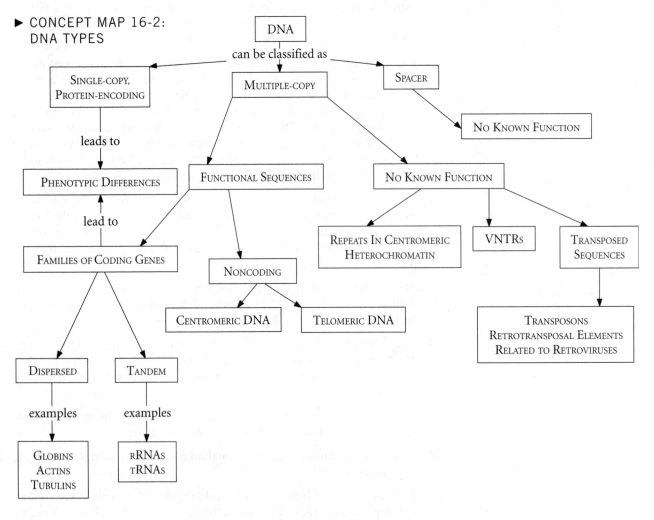

Questions for Concept Map 16-2

1. What are the processes that result in phenotypic differences from single-copy, protein-encoding genes?

2. Are they the same processes that result in phenotypic differences from dispersed families of coding genes??

3. Are they the same processes that result in phenotypic differences from tandem families of coding genes?

4. Discuss the structure and function of centromeric DNA.

5. Discuss the structure and function of telomeric DNA.

6. What is the role of the centromere?

7. What is the role of telomeres?

8. Do all chromosomes have centromeres and telomeres?

9. What happens to a chromosome that does not have a centromere?

10. What happens to a chromosome that lacks a telomere?

11. How many telomeres would be found in a chromosome that has a centrally located centromere?

12. Why are special sequences required within telomeres?

13. What do the special sequences within telomeres do during replication?

14. Discuss aging and telomeres.

15. What is the definition of a spacer?

16. What is spacer DNA?

17. Name at least two possible roles for spacer DNA.

18. What does *VNTR* mean?

19. How are VNTRs used in forensics?

20. What does *transposed* mean?

21. What are transposons?

22. What are retrotransposal elements?

23. What are retroviruses?

24. What does reverse transcriptase do?

25. What is cDNA?

26. What is the relationship between cDNA and retrotransposal elements?

Answers to Concept Map 16-2 Questions

1. Phenotypic differences from single-copy, protein-encoding genes arise through point mutation, complementation, epistatic suppression, chromosomal rearrangement, and environmental effects.

2. Phenotypic differences from coding genes that are in dispersed families have seldom been detected unless the genes begin to diverge and assume new functions, such as has happened with the globin genes. Obviously, genes in a low copy number are more likely to exhibit phenotypic differences than genes in a high copy number.

3. Basically, there are no phenotypic differences from tandem families of coding genes, because the genes apparently have a rectification process

that keeps them from accumulating mutations. The only possible difference that could arise in this case is through deletion or inversion, which is seen in both *Xenopus* in the case of deletion of the NO region and *Drosophila* in the case of the scute phenotype.

4. Centromeric DNA is highly repetitive, consisting of short tandem repeats, and the repeats have no known function.

5. Telomeric DNA is noncoding and consists of tandem arrays of short, simple sequences. They exist to prevent chromosome shortening at each round of replication. This is accomplished through the enzyme telomerase, which adds the repeats to the ends of the chromosome.

6. The centromere has no known function. However, the region is the site of the organization of the kinetochore, which is required for proper chromosome distribution during division. It may also block DNA synthesis of the underlying DNA during the first meiotic division.

7. The telomeres are required for preventing the shortening of chromosomes with each replication, which would occur without them.

8. All chromosomes have centromeres and telomeres.

9. A chromosome that does not have a centromere does not proceed to one of the two poles during cell division and usually is rapidly lost from a cell line.

10. A chromosome that lacks a telomere will shorten with each replication, leading to cell death eventually. It may also be that the lack of a telomere is what causes broken chromosome ends to behave as if they are "sticky" during replication. Such chromosomes then enter the breakage-fusion-bridge cycle as described initially by McClintock.

11. Two telomeres would be found in a chromosome that has a centrally located centromere.

12. Special sequences are required within telomeres because DNA polymerase requires an RNA primer and there is no room for one at the tip of the lagging strand.

13. The special sequences within telomeres appear to bend back upon themselves and form a hairpin structure which allows for a 3′ end which in turn allows replication at the tip of the lagging strand.

14. As cells replicate in culture, their telomeres shorten. Furthermore, there appears to be an age-dependent shortening of telomeres in at least some human tissues.

15. A spacer is something that simply occupies space between two objects that have a function.

16. A spacer is a segment of DNA that has no known function and cannot be classified as centromeric DNA, for instance, because of its location.

17. Spacer DNA could exist to provide bulk for the rest of the DNA in the cell, which might increase the rate of proper distribution at cell division (YAC chromosomes that are too short have a high rate of nondisjunction). Alternatively, the spacer may be required to allow for proper and efficient gene regulation.

18. *VNTR* means "variable-number tandem repeats."

19. VNTR polymorphisms are used to classify crime scene evidence.

20. *Transposed* means "moved to a different place."

21. Transposons are genetic elements that have the ability to move within the genome.

22. Retrotransposal elements are genetic elements that ultimately are derived from the reverse transcription (through a virally encoded gene product) of RNA.

23. Retroviruses are RNA viruses that carry into the cell a copy of reverse transcriptase. This enzyme makes a DNA copy of the RNA genome, which is then capable of inserting into a host chromosome.

24. Reverse transcriptase makes a DNA copy of an RNA segment.

25. cDNA is the DNA made by reverse transcriptase.

26. Retrotransposal elements appear to be derived from cDNA.

▶ SOLUTIONS TO PROBLEMS

Be sure that you have thoroughly read the entire chapter before you attempt any of the problems.

1. a. Half the bands in the child should be derived from the mother and half from the father. A child cannot contain a band that is not seen in either parent, unless a germ cell mutation has occurred. F2 is among the group of men who could be a father to the child in question.

   b. If the bands in the child are numbered from the top to the bottom, there are a total of 22 bands. The following bands are from the mother: 1, 4, 5, 10, 12, 13, 14, 15, 16, 17, 18, 19, 20, and 21. The remaining bands are from the presumed father.

   c. No.

   d. A mutation occurred that altered, eliminated, or made a restriction site.

   e. The cellular DNA is cut with a restriction enzyme and electrophoresed. This results in a continuous "smear" of DNA along the length of the gel. Discrete bands are seen because the probe is detecting regions that are complementary to it. The multiple bands on the gel indicate a repetitive sequence that is of variable distance from a restriction site from repeat to repeat. The probe most likely is detecting a subsection of the piece of DNA that is seen in the band.

2. In order to answer this problem, assume that the wild-type allele makes a product which is necessary for life and that the recessive lethal lacks that product.

   a. The *Su(var)* mutant would suppress the variable expression of the wild-type allele, allowing full wild-type phenotype. If the gene product is diffusible, the organism would be wild-type. If the gene product is not diffusible, then the organism would still have a wild-type phenotype because each cell makes the gene product.

b. The *E(var)* mutant would enhance the heterochromatin effect, leading to much less wild-type expression. This would, in turn, allow expression of the recessive lethal allele. If the gene product is diffusible, a few cells may be capable of producing it, which would probably not allow viability. If the product is not diffusible, then most cells would die, leading to organismal death.

c. Without either suppression or enhancement, variegation would result. If the product is diffusible, the organism would be partially wild-type, but would still probably die. If it is not diffusible, then most cells would die, resulting in organismal death.

3. a. One locus on each of the homologous chromosomes would be indicated.

b. Both ends of all chromatids would be indicated.

c. The chromosomal satellite region on all chromosomes containing NORs would be indicated.

d. Many small regions on many chromosomes would be indicated.

e. Centromeres, telomeres, and NORs would be indicated.

4. On average, each DNA fragment contains 2 SINE sequences. Assuming that all SINEs have the same orientation, then each fragment of DNA should be able to undergo complementary pairing with one or more fragments.

5. a. The pattern of suspect 1 is compatible with including him among the group of individuals who could have committed the rape.

b. Suspects 2 and 3 could not have committed the rape.

6. To work this problem, number bands in the two parents, then use those numbers to identify bands in their children. If the mother has bands 1–4, marked with an asterisk, and the father has bands 5–8, not marked, the children have the following set of bands:

CHILD 1	CHILD 2	CHILD 3	CHILD 4
1*	6	5	1*
6	2*	2*	5
7	3*	4*	7
3*	8	8	4*

Note that four bands are present in each person but that a total of eight different bands exist in this family. Each child receives two bands from each parent. There are, therefore, two loci that are being detected with the probe. The loci are not linked. Four alleles are being detected for each locus. The genotypes are (1*/2* 3*/4* × 5/6 7/8.)

7. Recall that exactly such an experiment was used to determine whether the formation and the extent of heterochromatin is under genetic control. For both white and rough, the wild-type phenotype should be present, which is red and smooth, unless the inversion causes some gene inactivation.

a. For the first inversion, red, which is closer to the heterochromatic region, is not expressed approximately 60% of the time, and sometimes (10%) smooth is also not expressed. As the breakpoint is moved farther away from the heterochromatic region, as in the second inversion, far less inactivation occurs. When just smooth is moved very close to the heterochromatic region, as in the third inversion, only that gene is inactivated. Inversion 4 shows lower inactivation of both genes because neither is close to the heterochromatic region. Comparing inversions 2 and 4, only distance seems to matter, not the source of the distance [whether from the DNA normally next to the heterochromatic region (inversion 2) or from the DNA brought into that region (inversion 4)].

b. Inversion 3 separates the two genes, yielding the expected result of 0% inactivation for the red allele. Inversion 4 brings both genes nearer the heterochromatic region, and the 0% observed for smooth alone is expected under the hypothesis of a spreading effect from the heterochromatic region: smooth cannot be inactivated unless the red gene on the same chromosome also is inactivated.

c. In a double inversion heterozygote of inversions 1 and 2, with all alleles wild-type on both chromosomes, only wild-type alleles are present to be expressed. Assuming that the absence of wild-type expression leads to whie eyes, the probability of both w^+ alleles being inactivated is 60% × 12% = 7.2%. The probability of the r^+ alleles and the w^+ alleles being inactivated in the same facets is 10% × 2% = 0.2%.

8. a. Isolate DNA from the cells, cut it with a restriction enzyme, and electrophorese the fragments. After Southern blotting, probe with radioactive transposon sequences.

b. You could disrupt the transposon and look for a loss of the abnormal phenotype. Also, you could do a full linkage analysis: those progeny with a mutant phenotype should react when probed with the transposon; normals should not react when probed with the transposon.

c. The transposon could have inserted in spacer DNA or in nonessential genes.

d. The 30:70 split may reflect the basic mutability of the genome, with 30% of the genome composed of spacer DNA and 70% composed of genes. If this is true, then more mutagenesis should lead to more detection of genes.

9. *Unpacking the Problem*

a. *Variegation* means that a trait is variable within an organism. It is expressed in some cells and not in others. The fly eye is a compound eye, and each segment is called a *facet*. In a *reciprocal translocation*, two nonhomologous chromosomes exchange genetic material. *Wild-type* is the most prevalent form observed in a specific population, whether in the laboratory or in nature. *Heterozygous* means that two different alleles are present in one organism for the gene in question. With regard to translocation, a heterozygous translocation indicates one normal chromosome and one translocated homologous chromosome.

b. *Mixed ancestry* refers to nonlaboratory-bred stock, perhaps, in which some genes are homozygous and some are heterozygous.

c. A freebie for the writer! Go to it.

d. White eyes is on the tip of the X chromosome. It is recessive and not wild-type.

e. The female is XX; the male is XY.

f. The male is hemizygous white, *w*/Y.

g. The female is homozygous wild-type red, w^+w^+.

h. The expectation is that all females and males would be red.

i. Some causes of variegation are X-inactivation, which doesn't occur in the fruit fly; chromosome loss, which is highly unlikely here; somatic crossing-over, which is highly unlikely to occur in 50% of the progeny; and a translocation in one of the parents that leads to gene inactivation because the gene is moved close to a heterochromatic region (position effect). The last is the most likely explanation in this situation.

j. Chromosome 3 is not specifically relevant to the problem.

k. Fifty percent of the progeny received their mother's normal X chromosome with a red allele on it.

l. The females are all w^+w, with inactivation occurring for the red allele, when it occurs. Inactivation occurs in all the eyes in which it is expected, 50% of all the female progeny. What needs to be explained is why there are equal numbers of 60% and 20% inactivation of facets.

   The most likely possibility is that there is another gene that directly affects the degree of inactivation, for which the male was heterozygous. Assume that the distance between the white gene and the heterochromatic region was such that 20% of the facets would be white owing to gene inactivation. Now assume that the male is heterozygous for an autosomal gene in which the wild-type allele $E(var)^+$ allows the normal 20% inactivation but the mutant allele $E(var)$ increases inactivation by an additional 40%. Half of all females would receive each allele. In the noninverted heterozygous female progeny, the mutant allele from their father for this gene would have no effect because the white allele is not near a heterochromatic region. In the inverted heterozygous female progeny, half would show 20% inactivation and half would show 60% inactivation of their facets.

m. The phenotype of the other facets is red.

n. Three distinct classes are seen in a heterozygous-by-heterozygous cross that has a distinct phenotype for each genotype.

o. It is significant that it was the proposed heterozygous female who gave rise to even lower percent facet inactivation in her progeny. All the homozygous dominant females (60% facet inactivation) would give rise to progeny with 20% of the facets inactivated because they would be *Aa*.

p. This has already been discussed in question l above.

Solution to the Problem

a. and b. The original cross and progeny were as below, with $E(var)$ dominant to $E(var)^+$.

P X^{red}/(tX-3) normal-3/(tX-3)^{red} $E(var)^+$/$E(var)^+$ ×
     X^{white}/Y normal-3/normal-3 $E(var)^+$/$E(var)$

F_1 1/2 X^{red}/X^{white}

      1/2 normal-3/normal-3 $E(var)^+$/$E(var)^+$ red

      1/2 normal-3/normal-3 $E(var)^+$/$E(var)$ red

   1/2 (tX-3)/X^{white}

      1/2 normal-3/(tX-3)^{red} $E(var)^+$/$E(var)^+$ 20% white facets

      1/2 normal-3/(tX-3)^{red} $E(var)^+$/$E(var)$ 60% white facets

c. At this point, the backcross needs to be considered. This is one of those classic situations in genetics that causes beginners, especially, to think they've made a mistake. The explanation provided for a and b above does not take the backcross into consideration. Now that it must be considered, the explanation needs to be revised. However, it was *not* incorrect; it simply failed to take into account all the data.

Assume that the original female was $E(var)^+$/$E(var)^+$ and the male was $E(var)^+$/$E(var)$ for a gene that, under normal circumstances, does not affect eye color, as was done above. In order to account for the backcross result of only 2% white facets in some of the female progeny, those progeny must be $E(var)$ $E(var)$ because that is the only genotype class that did not appear in the F_1. If $E(var)$ $E(var)$ leads to 2% inactivation (or the alternative, dosage effect), then the $E(var)$ allele must reduce the effects of the inversion-induced variegation. Alternatively, the $E(var)^+$ allele increases the inversion-induced variegation. This is exactly the reverse of the way the alleles were said to work, above. Given this change in thinking, the original cross then becomes

P X^{red}/(tX-3) normal-3/(tX-3)^{red} $E(var)^+$/$E(var)^+$ ×
     X^{white}/(Y normal-3/normal-3 $E(var)^+$/$E(var)$

F_1 1/2 X^{red} X^{white}

      1/2 normal-3/normal-3 $E(var)^+$/$E(var)^+$ red

      1/2 normal-3/normal-3 $E(var)^+$/$E(var)$ red

   1/2 (tX-3)/X^{white}

      1/2 normal-3/(tX-3)^{red} $E(var)^+$ $E(var)^+$ 60%* white facets

      1/2 normal-3/(tX-3)^{red} $E(var)^+$ $E(var)$ 20%* white facets

The asterisks indicate the change from the previous solution to this problem. Very simply, what we've done is provide an explanation that changes the 20% class to heterozygous for the $E(var)$ locus so that it can now provide an $E(var)$ allele, and thus yield an $E(var)$ $E(var)$ homozygote out of the backcross to the $E(var)$ $E(var)^+$ father.

d. There is a 1:1 proportion of 60% and 20% white-faceted females because there is equal probability they will receive *E(var)* or *E(var)*⁺ from the father.

10. **a. and b.** One explanation is that the gene is surrounded by some type or types of repeating sequences and that the probe contained at least a portion of that repeat. Alternatively, the *sp* locus is not a single, unique locus. Although the three rare cutters each produced different multiple cuts, the repeating sequence, or the *sp* gene family, was located in at least four different regions that were separated by the cuts in each case, producing four bands.

c. It is unlikely that each enzyme cut the gene in three places to produce four bands because the enzymes all recognized rare target sequences. Rather, each enzyme cut in the flanking regions.

Genomics

Genomics is the molecular characterization of entire genomes in different species. A large number of techniques have been developed to facilitate this characterization. The first task is to localize a specific gene to a specific chromosome. After this has been done, the gene is localized to smaller and smaller regions of the chromosome, until its specific location is determined. DNA sequence is also determined, along with the transcription product and its function within the organism. Ultimately, base changes leading to specific phenotypic variants are determined. The entire process does not necessarily proceed from the "large" to the "small" as described here, however. For instance, the DNA sequence may be known before any knowledge of the existence of the gene for which the DNA codes. Alternatively, the transcript may be characterized before the chromosomal location is known. The end point in genomic analysis is to know the specific location, product, and function of a gene; to know how that gene product interacts with the products of other genes to produce the mature organism; and to know how the functioning of the gene is regulated by or helps to regulate other genes. Ultimately, there may exist a chart or map, similar to a metabolic chart, that describes gene functioning of an organism in minute detail.

A number of different techniques are used to assign genes and markers to specific chromosomes. Using **meiotic recombination** studies, a gene may be found to be linked to a locus with a known location. *In situ hybridization* is used to visualize the location of a gene on chromosome metaphase spreads using either radioactivity or fluorescent dyes. Using **rearrangement breakpoints** that occur with translocations, inversions, and deletions, genes can be assigned to specific chromosomes if their function is inactivated by the chromosome breakage. **Pulsed field gel electrophoresis, PFGE,** allows for the separation of small chromosomes by size and shape; the specific chromosomes can then be probed for a gene or marker through **Southern blotting. Somatic cell hybrid** cell lines allow for gene and marker mapping to specific chromosomes.

Once a gene or marker is localized to a chromosome, it then needs to be localized within the chromosome. Again, a number of techniques exist that allow this level of mapping. **Meiotic mapping** is conducted using classical recombination studies. **Restriction fragment length polymorphisms, RFLPs,** are mapped using Southern blotting. Likewise, **simple sequence length polymorphisms, SSLPs,** and **minisatellite markers** are mapped through Southern blotting and probing. **Microsatellite markers** are mapped through the use of clone screening with probes made with **PCR** (polymerase chain reaction). Through **molecular genotyping of sperm,** individual sperm can be genotyped and recombination frequencies determined. **Homozygosity analysis** is a type of recombination study conducted in families with inbreeding. **Restriction landmark genomic scanning, RLGS,** combined with other molecular techniques, leads to the mapping of rare restriction sites. **RAPD, randomly amplified polymorphic DNAs,** can be mapped using most of the molecular techniques. **Chromosome-mediated gene transfer** is a method of introducing a small number of chromosomes from one species into cells of a second species. Cotransfer of two genes often leads to the detection of linkage relationships. Using the same basic idea as chromosome-mediated gene transfer, **irradiation and fusion gene transfer, IFGT,** introduces broken pieces of chromosomes from one species into the cells of a second species. This method allows for the detection of very small distances between genes.

Physical mapping of DNA fragments allows for very fine mapping of chromosomes. Most of the techniques described above can be utilized in the mapping of DNA fragments. An additional technique is **optical mapping,** which allows for the visualization of restriction sites. A variant on that technique is **chromosome microsurgery. Yeast artificial chromosomes, YACs,** can be constructed and used to study specific DNA fragments.

► CONCEPT MAP 17-1:
  GENOMICS

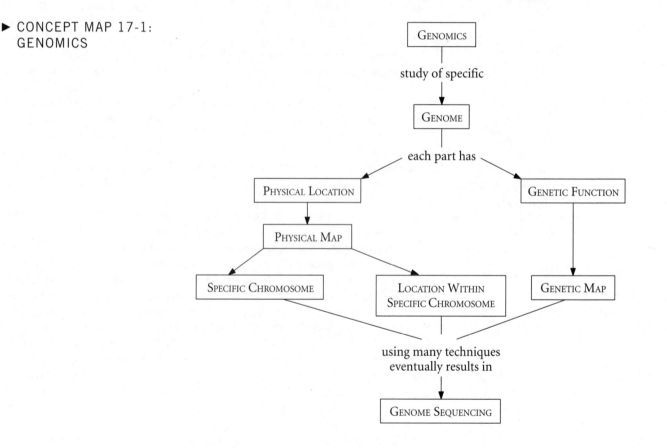

Questions for Concept Map 17-1

1. Make sure that you understand each technique that is utilized.

2. What is the definition of a *genome?*

3. How many genomes are present in humans?

4. Are the genomes within two humans the same?

5. If genomes differ between people, will differences be detected by genome sequencing?

6. How can you distinguish between real differences and errors made in sequencing?

7. What could cause two different YAC alignments for the same chromosome within a species?

8. Once a genome has been completely sequenced, what types of understanding can be gained from that information?

9. Will sequencing reveal epistasis?

10. Will sequencing explain why an allele is dominant or recessive?

11. Will sequencing explain why a genotype is codominant?

12. Will sequencing explain why dosage effects exist?

13. Will sequencing explain how transcription is initiated?

14. Will sequencing explain how replication is initiated?

15. Will sequencing explain the length of a cell cycle or how aging and development occur?

16. What is the difference between a physical map and a genetic map?

17. What is the relationship between a map unit and physical distance?

18. In a number of species, it has been shown that the genetic distance between two genes differs in males and females. How accurate is the relationship between a map unit and physical distance that you gave in question 17?

19. Could the relationship between a map unit and physical distance differ between species?

Answers to Concept Map 17-1 Questions

1. No question, no answer, just a big task.

2. A genome is the set of genetic information received from one parent.

3. Humans have two genomes.

4. The two sets of genomes in each person usually differ.

5. Genome sequencing in its present form utilizes one allele for each gene. If only one copy of each gene is sequenced, the differences cannot be recognized.

6. The way to distinguish between real differences and errors made in sequencing is through repeated sequencing.

7. A chromosome rearrangement could cause two different YAC alignments for the same chromosome within a species.

8. Once a genome has been completely sequenced, genes can be identified, especially those genes that cannot vary in an organism without being lethal. Classical genetic analysis cannot identify such genes, because of the lethality involved. Other types of information to be obtained are the so-called silent differences between people, variants that cause no phenotypic effects, and eventually information regarding both development and regulation.

9. Sequencing probably will not provide information on epistasis.

10. Sequencing may explain why an allele is dominant or recessive.

11. Sequencing possibly will explain why a genotype is codominant.

12. Sequencing possibly will explain why dosage effects exist.

13. Sequencing should help to explain how transcription is initiated.

14. Sequencing should help to explain how replication is initiated.

15. Sequencing may possibly help to explain the length of a cell cycle or how aging and development occur.

16. A physical map is a map of the physical location of a gene, while a genetic map is a map of the percentage of recombination between genes.

17. A map unit is approximately 1000 kb of DNA.

18. The estimate that a map unit is approximately 1000 kb of DNA is exactly that: an estimate. It is not very accurate.

19. It would be expected that the relationship between map units and physical distance would vary between species.

▶ SOLUTIONS TO PROBLEMS

Be sure that you have thoroughly read the entire chapter before you attempt any of the problems.

1. *Unpacking the problem*

    a. Two types of hybridizations that have already been studied are hybridizations between strains of a species and hybridizations between species. A third type of hybridization is referred to in this problem: molecular hybridization. Molecular hybridization can involve either DNA-DNA hybridization or DNA-RNA hybridization. In both instances, it relies on the specificity of complementary pairing and can take place in solution, on a gel, on a blot, or on a slide

$$
\begin{array}{ccccccccc}
5'\text{-U} & A & C & G & G & G & A & U\text{-}3' & \text{RNA} \\
\vdots & \vdots & \vdots & \vdots & \vdots & \vdots & \vdots & \vdots & \\
3'\text{-A} & T & G & C & C & C & T & A\text{-}5' & \text{DNA}
\end{array}
$$

    b. In situ hybridization usually is conducted on a slide so that the stained chromosomes can be observed and the specific portion of a chromosome to which the probe hybridizes can be identified.

    c. A YAC is a yeast artificial chromosome. It contains a yeast centromere (ORI), telomeres, and DNA that has been attached between them.

    d. Chromosome bands are dark regions along the length of a chromosome that occur in a characteristic pattern for each chromosome within an organism. They can occur naturally, as with *Drosophila* polytene chromosomes, or they can be induced by a number of chemical and physical agents, combined with staining to accentuate the bands and interbands.

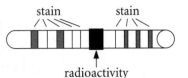

stain   stain

radioactivity

e. The five YACS could have been hybridized sequentially to the same chromosome preparation, which is, however, unlikely. Alternatively, the information could have been determined in five separate experiments. In either case, a YAC labeled with either radioactivity or fluorescence, and including the DNA of interest, was hybridized to a chromosome preparation, the chromosomes were properly treated to reveal the banding pattern, and the YACs were determined to hybridize to the same band.

f. A genomic fragment, by definition, contains a subportion of the genome being studied. In most instances, it actually contains a subportion of one chromosome. Five randomly chosen YACs would not be expected to contain the same genomic fragment or even fragments from the same chromosome. The fragments could have been produced by either physical (X-irradiation, shearing) or chemical (digestion, restriction) means, but it does not matter how they were produced.

g. A restriction enzyme is a naturally occurring bacterial enzyme that is capable of causing either single- or double-stranded breaks in DNA at specific DNA sequences.

h. A long cutter is a restriction enzyme that produces very long fragments of DNA because the sequence it recognizes occurs infrequently within the genome.

i. The YACS were radioactively labeled so that their location after hybridization could be detected through autoradiography. To *radioactively label* is to attach an isotope that emits energy through decay to a specific probe for a sequence. Commonly used radioactive labels are tritium (^{3}H) in place of normal hydrogen and ^{32}P in place of phosphorus.

j. An autoradiogram is a "self-picture" taken through radioactive decay from a labeled probe. When a gel or blot is used, the radioactive decay is captured by a piece of X-ray film. When in situ hybridization is used on slides, the emulsion coats the slide directly.

k. Free choice. Be sure you truly know the meaning of each term.

l. We are given a diagram of the composite autoradiographic results. The DNA from humans was isolated and subjected to digestion by a restriction enzyme that cuts very infrequently. Once the DNA was electrophoresed, it was Southern blotted and then probed sequentially with radioactively labeled YACs, followed by sequential exposure to X-ray film. Between probings, the previous YAC hybrid was removed through denaturing of the DNA-DNA hybrid. Alternatively, five separate Southern blottings were done.

m. The haploid human genome is thought to contain approximately 3.3×10^6 kilobases of DNA (Chapter 16, Table 16-1).

n. Restriction digestion of human genomic DNA would be expected to produce hundreds of thousands of fragments.

o. The fragments produced by restriction of human genomic DNA would be expected to be mostly different.

p. When subjected to electrophoresis and then stained with a DNA stain, the digested human genome would produce a continuous "smear" of DNA from very large fragments (in excess of tens of thousands of base pairs in length) to fragments that are very small (under a hundred bases in length).

q. In this question, only two distinct bands are produced, at most, in any one probing. The difference between what is seen with a DNA stain and what is seen with probing lies in the specificity of the agent being used. DNA stain will detect any DNA, while a DNA probe will detect only DNA that is complementary to the probe.

r. Number them from top to bottom, 1–3, across the gel. Thus, YACs A–C contain band 1, YACs C–D contain band 2, and YACs A and E contain band 3.

s. There are no restriction fragments on the autoradiogram. The fragments are on the substance (nitrocellulose, nylon) used to blot the gel. The radioactivity of the probes is captured by the X-ray film as it decays, producing energy that impinges on the film.

t. YACs B, D, and E hybridize to one fragment, and YACs A and C hybridize to two fragments.

u. A YAC can hybridize to two fragments if the YAC contains continuous DNA and there is a restriction site within that region. A YAC can also hybridize to two fragments if it contains discontinuous DNA from two locations in the genome that either are on different chromosomes (this is analogous to a translocation) or are separated by at least two restriction sites if they are on the same chromosome (this is analogous to a deletion). In this case, the former makes more sense. Because the YACS were selected for their binding to one specific chromosome band, it is unlikely that the YACS are composed of discontinuous DNA sequences. A YAC could hybridize to more than two fragments because the continuous DNA could contain many restriction sites or the discontinuous DNA could be composed of DNA from a number of regions in the genome.

v. Cytogeneticists use *band* to designate a region of a chromosome that is dark-staining. Molecular biologists use *band* to designate a region of dark appearing on an autoradiogram which is produced by radioactive decay from a specific probe which reacted with a population of molecules localized by gel electrophoresis. In both cases, *band* refers to a localization.

Solution to the Problem

a. Note that fragments 1 and 3 occur together and fragments 1 and 2 occur together, but that fragments 2 and 3 do not occur together. This suggests that the sequence is 2 1 3 (or 3 1 2).

b. If the sequence of the fragments is 2 1 3, then the YACS can be shown in relation to these fragments. YAC A spans at least a portion of both 1 and 3. YAC B is within region 1. YAC C spans at least a portion of regions 1 and 2. YAC D is contained within region 2. YAC E is contained within region 3. A diagram of these results is shown below. In the diagram, there is no way to know the exact location of the ends of each YAC.

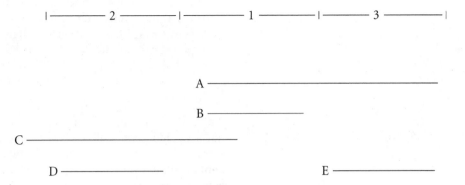

2. PFGE results in separation of the chromosomes, and the gel indicates the discreteness obtained by the separation. Because the *MATa* probe does not produce a band, there must not be a *MATa* gene in *Neurospora*. *LEU2* is contained within chromosome 4, and *ADE3* is contained within chromosome 7.

3. The cross is

$$cys\text{-}1\ RFLP\text{-}1^O\ RFLP\text{-}2^O \times cys\text{-}1^+\ RFLP\text{-}1^M\ RFLP\text{-}2^M$$

A parental type will have the genotype of either strain and be in a majority, while a recombinant type will have a mixed genotype and be in a minority of the ascospores. Clearly, the first two ascospore types are parental, with the remaining being recombinant.

a. The *cys-1* locus is in this region of chromosome 5. If it were not in this region, linkage to either of the RFLP loci would not be observed.

b. The entire region is approximately 17 map units in length (100% − % parentals = 100% − 100% × 83).

To calculate specific distances, you may need to review Chapters 5 and 6. Here, it is assumed that you recall that material.

cys-1 to *RFLP-1* = 100%(2 + 3)/100 = 5 map units

cys-1 to *RFLP-2* = 100%(7 + 5)/100 = 12 map units

RFLP-1 to *RFLP-2* = 100%(2 + 3 + 7 + 5)/100 = 17 map units

$$\vdash\!\!-5\ \text{m.u.}\!-\!\!\dashv\!\!-\!\!-\!\!-\!\!-12\ \text{m.u.}\!-\!\!-\!\!-\!\!-\!\dashv$$
RFLP-1   *cys-1*   *RFLP-2*

c. A suitable next step in cloning *cys-1* might be to do PFGE to obtain a pure sample of chromosome 5, followed by cloning of chromosome fragments. The clones could be screened by the *RFLP* probes.

4. a. and b. Compare each translocation gel to the wild-type gel and note the difference in bands that were obtained. Together, the translocations involve chromosomes 1, 2, and 4. The top band in all three gels is constant and must reflect chromosome 3, which is not involved in either translocation. Focusing on the three remaining bands in the wild-type gel, the last band is not altered in the 1/4 translocation but is altered in the 2/4 translocation. This means that it must reflect chromosome 2. The band second from the bottom in the wild-type

gel is altered in the 1/4 translocation but is not altered in the 2/4 translocation; it must reflect chromosome 1. The remaining band in the wild-type gel is altered in both the 1/4 and the 2/4 translocations; it must reflect chromosome 4.

In the 1/4 gel, the two new bands in comparison with the wild-type gel are due to the 1/4 translocation, while in the 2/4 gel, they are due to the 2/4 translocation. In both translocation gels, the *P* probe is associated with new-appearing bands, indicating that *P* is located on chromosome 4. This is confirmed in the wild-type gel.

5. Remember that a gene is one small region of a long strand of DNA and that a cloned gene will contain the entire sequence of the gene under normal circumstances. If there are two cuts within a gene, three fragments will be produced, all of which will interact with the probe, as was seen with enzyme 1. Cuts external to a gene will produce one fragment that will interact with the probe, which was seen with enzyme 2. One cut within a gene will produce two fragments that will interact with the probe, as was seen with enzyme 3.

6. This type of problem is analogous to the problems in Chapter 10.

   a. To determine the physical map showing the STS order, simply list the STSs that are positive, using parentheses if the order is unknown.

   YAC A: 1   4   3

   YAC B: 5   1

   YAC C: 4   3   7

   YAC D: (6 2)   5

   YAC E: 3   7

   b. Once the sequence of STSs is known, the YACs can be aligned as follows, although precise details of overlapping and the locations of ends are unknown:

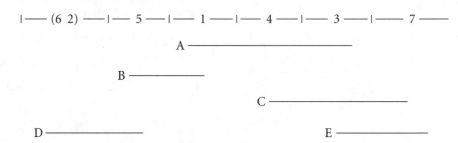

7. **a. and b.** There are four patterns that can be observed in the 15 2X2 comparisons that can be made between these six markers: ++, −−, +−, and −+. The first two indicate concordance and the second two indicate a lack of concordance. Ideally, data would show either 100% concordance for the seven hybrids, indicating linkage, or 100% discordance for the seven hybrids, indicating a lack of linkage.

   Because radiation hybrids involve chromosome breakage, two genes that are located very close together on the same chromosome may show some discordance despite the close linkage. Two genes that are located on different chromosomes may also show some concordance due to chance alone because two fragments may become established within a single hybrid line. Therefore, the problem is how to distinguish between reduced concordance due to chromosome fragmentation and chance concordance due to two fragments from different chromosomes. Obviously, a statistical solution is needed, but there are not enough data in this problem for a statistical analysis.

Sort the data into three groups: 100% concordance, 100% discordance, and mixed (concordance/discordance). This follows below:

100% CONCORDANCE	100% DISCORDANCE	MIXED
E-F	None	A-B 2/5
		A-C 2/5
		A-D 6/1
		A-E 2/5
		A-F 2/5
		B-C 5/2
		B-D 1/6
		B-E 3/4
		B-F 3/4
		C-D 3/4
		C-E 3/4
		C-F 3/4
		D-E 3/4
		D-F 3/4

Markers E and F are most likely located on the same chromosome. Markers B and D may be located on different chromosomes.

In the absence of statistical analysis, with so few total hybrids, it is important to pay more attention to the ++ patterns than the − − patterns simply because − − can arise either from linkage, with the specific chromosome missing in two hybrids, or from lack of linkage, with the two chromosomes lacking in the two hybrids. Therefore, going back to the mixed category and focusing on those marker pairs that had a high degree, but not 100%, of concordance, one sees that the 6/1 pattern of A-D and the 5/2 pattern of B-C stand out. For the A-D pair, 3 of the 6 concordances are ++, while only 2 of the 5 concordances for B-C are ++. It is unclear from the data whether this is a significant difference, and significance cannot be determined in any fashion. Therefore, it would be important to collect more data before drawing any conclusions at all, with the possible exception of linkage between markers E and F.

8. a. RAPDs are formed by being bracketed by two inverted copies of the primer sequence. Below, the primer is indicated by X's, and the amplified region appears in brackets. For convenience, the two amplified regions are shown on the same lengthy piece of DNA for strain 1.

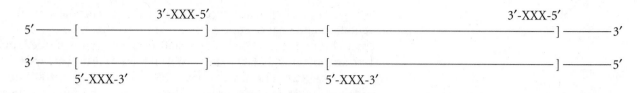

3'-XXX-5'   3'-XXX-5'

5' —— [————————————] ———————— [————————————————] —— 3'

3' —— [————————————] ———————— [————————————————] —— 5'

5'-XXX-3'   5'-XXX-3'

Strain 2 lacks one or two regions complementary to the primer.

b. Progeny 1 and 6 are identical with the strain 1 parent. Progeny 4 and 7 are identical with the strain 2 parent. Progeny 2 and 5 received the chromosome holding the upper band from the strain 1 parent and the chromosome holding the lower band from the strain 2 parent (resulting in no second band). Progeny 3 received the opposite: the chromosome holding the lower band from the strain 1 parent and the chromosome holding the upper band from the strain 2 parent (resulting in no second band). Therefore, bands 1 and 2 are unlinked.

c. Recall that a nonparental ditype has two types only, both of which are recombinant. Therefore, the tetrad would be composed of four progeny like progeny 2 and four progeny like progeny 3.

9. a. Look for the region of overlap among cosmids C, D, and E; that shows the location of gene *x*. Gene *x* is located in region 5.

b. The common region of cosmids E and F, or the location of gene *y*, is region 8.

c. Both probes are able to hybridize with cosmid E because the cosmid is long enough to contain part of genes *x* and *y*.

10. α: The only chromosome missing in A and B and present in C is 7.

β: The only chromosome present in all colonies is 1.

χ: The only chromosome missing in A and present in B and C is 5.

δ: The only chromosome present only in B is 6.

ε: Not on chromosomes 1 through 7.

11. Steroid sulfatase: Xp; phosphoglucomutase-3: 6q; esterase D: 13q; phosphofructokinase: 21; amylase: 1p; galactokinase: 17q.

12. a.

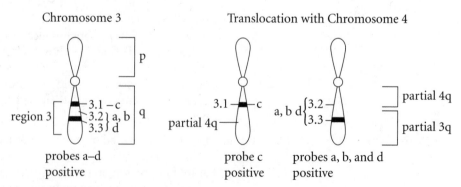

Chromosome 3   Translocation with Chromosome 4

region 3 [3.1 — c / 3.2 } a, b / 3.3 } d] q   p

probes a–d positive

partial 4q   3.1 — c   probe c positive

a, b d { 3.2 / 3.3 —   partial 4q / partial 3q   probes a, b, and d positive

b. The normal allele, *n*, has been localized to band 3q3.1. Furthermore, probes a–d are known to hybridize to this band. Probe c specifically appears to be closer to the centromere than the other probes because it is retained on the chromosome 3 centromere in reciprocal translocations.

In order to isolate and characterize the normal allele, a restriction digest of DNA from individuals who do not have the disease could be conducted. After blotting, the probes can be used to produce a restriction map of the region. A physical map of the region could be made using the probes to identify fragments that then would be cloned. In both cases, normal DNA and DNA from individuals with the disease should be run alongside each other to verify that the region in the normal individuals is homologous to the region from the individuals with the disease. Ultimately, the cloned region could be sequenced.

Alternatively, the clone giving rise to probe c could be used as the start of a chromosome walk. New clones would be probed with a, b, and d. A clone positive for a, b, and d, indicates gene N has been walked past. Intermediate clones for gene N would then be examined.

c. Once n is cloned, it can be used to identify the DNA region in individuals who have the disease but not a translocation. At this point, the mutant sequences can be compared with the normal sequence.

13. a. DNA from each individual was obtained. It was restricted, electrophoresed, blotted, and then probed with the five probes. After each probing, an autoradiograph was produced.

b. First compare the two parents and identify differences between them. Looking at the affected parent, she has the following differences from the unaffected parent: 1′, 2°, 3′, 4°, and 5′.

Next compare affected offspring with the unaffected parent to determine which chromosome came from that parent. The other chromosome in affected progeny had to have come from the affected parent and must contain the RFLP locus closest to the allele causing the disorder. Compare the two "affected" chromosomes with each other. The regions that the two chromosomes have in common are 2°, 3′ and 4°.

Unaffected progeny cannot have obtained the affected allele from the affected parent. Looking at them, identify which chromosome came from the affected parent, and then identify which, if any, of the three regions that must contain the affected allele were inherited from the affected parent. Both unaffected progeny contain regions 2° and 3′, and neither unaffected progeny has region 4°. Therefore, the RFLP locus closest to the disease allele is 4°.

c. Now that the RFLP locus closest to the gene has been identified, the disease gene can be cloned using probe 4° as an indicator of the region of interest. Shotgun cloning could be done first, and clones then screened with the probe. Transcripts from the clones positive for 4° could be compared and used to "fish out" the DNA region containing the gene of interest. Ultimately, DNA sequencing could be done, and the mutations causing the disease could be identified.

14. a. Breaks in different regions of 17R result in deletion of all genes from the breakpoint.

b. Because there is only 17R from humans, all the human genes expressed must be on 17R.

Notice that only gene *c* is expressed by itself. This means that gene *c* is closest to the mouse material. Next notice that if *c* and one other gene are expressed, that other gene is always *b*. This means *b* is closer to the mouse material than *a* is. The gene order is mouse –*c* –*b* –*a*.

The probability of a break between two genes is a function of the distance between them. Of the 200 lines tested, 48 expressed no human activity. Thus, the *c* gene is no more than (100%) (48)/200 = 24 relative m.u. from the mouse material. A break between *c* and *b* (cells express *c* only) occurred in 12 lines, placing these genes (100%) (12)/200 = 6 relative m.u. apart. A break between *b* and *a* (cells express *c* and *b*) occurred in 80 lines, placing these genes (100%) (80)/200 = 40 relative m.u. apart. Finally, 60 clones expressed all three genes, placing gene *a* (100%) (60)/200 = 30 relative m.u. from the end of the chromosome:

```
mouse          c            b            a           end
  |----------|-----------|-----------|-----------|
    24 m.u.      6 m.u.      40 m.u.     30 m.u.
```

    c. The dye could be used to correlate band presence with gene presence.

18

Control of
Gene Expression

Gene regulation is the regulation of transcription for specific genes. In prokaryotes, it is most often mediated by proteins that react to environmental signals by raising or lowering the rate of transcription.

An **operon** consists of two or more cistrons plus the regulatory signals that affect their transcription. Regulation occurs through the **promoter**, the *I* **locus**, and the **operator**. The **coordinately controlled genes** are transcribed in a **polycistronic** message. **Polar mutations** affect the gene within which they map and also reduce or eliminate the expression of all genes distal to the site of mutation in the polycistronic message.

Negative control is the blocking of transcription by a repressor protein that binds to an **operator.** The relief of **repression** is called **induction**.

The *lac* operon is under negative control. In the *lac* operon, **constitutive mutants** result in continuous expression in an unregulated fashion. Some are mutants of the *I* locus, *I⁻*. The *I* locus determines the synthesis of a **repressor molecule,** which blocks activation of a gene or genes. An **operator constitutive mutation**, *O*c, also results in unrepressed synthesis.

Positive control is the activation of transcription by a protein factor. **Catabolic repression** of the *lac* operon is an example of positive control. Here, the operon is activated by the presence of a large amount of CAP-cAMP.

Some operons are under both positive and negative control. The arabinose operon is an example.

Genes involved in the same metabolic pathway are frequently tightly clustered on prokaryotic chromosomes.

Feedback inhibition is the inhibition of the first enzyme in a metabolic pathway by the end product of that pathway, as exemplified in tryptophan biosynthesis. This is achieved by a process known as **attenuation**, an alteration of the secondary structure of the newly formed mRNA in the leader region.

Eukaryotic regulation is controlled by the **promoter, enhancers,** and **upstream activating sequences.** Several proteins have been identified that interact with these sites. Some steroid hormones also bind at these sites.

Genetic redundancy and **gene amplification** are cellular mechanisms that ensure an adequate supply of vital gene products.

► CONCEPT MAP 18-1:
GENERALIZED
PROKARYOTIC GENE
REGULATION

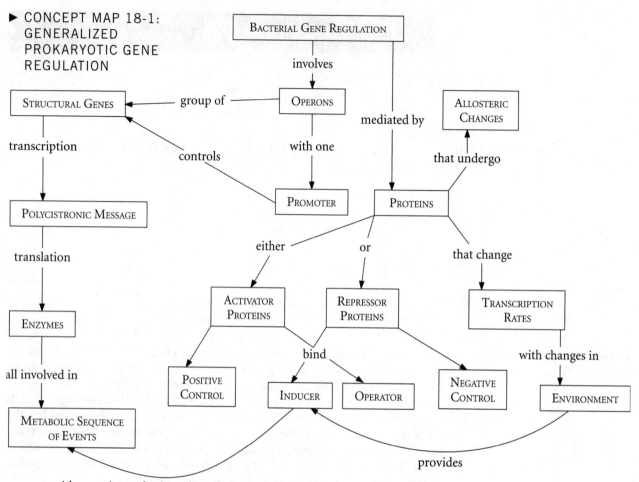

Questions for Concept Map 18-1

1. What does *regulation* mean?

2. What is the definition of an operon?

3. What is the definition of a promoter?

4. Do structural genes code for proteins? For rRNA? For tRNA?

5. Where does transcription occur in prokaryotes?

6. What is a cistronic message?

7. What is a polycistronic message?

8. Where does translation occur in prokaryotes?

9. What protein-rRNA structure is involved in translation?

10. If a gene codes for an enzyme, why is it called a *structural gene*?

11. How does a promoter control a structural gene?

12. What can bind to promoters?

13. Of what chemical is a promoter composed? An operon?

14. From where do the proteins that regulate bacterial genes come?

15. What is an allosteric change?

16. What is an activator protein?

17. What is a repressor protein?

18. What does *positive control* mean?

19. What does *negative control* mean?

20. What is the source of an inducer?

21. Where is the operator located?

22. Of what chemical is an operator composed?

23. Which system would be inducible: one that is regulated by an activator protein or one that is regulated by a repressor protein?

24. Which system would be repressible: one that is regulated by an activator protein or one that is regulated by a repressor protein?

25. If lactose is an inducer, do the enzymes whose synthesis it induces participate in anabolism or catabolism?

26. If histidine is an inducer, do the enzymes whose synthesis it induces participate in anabolism or catabolism? Can histidine actually be an inducer of the *his* operon?

27. During repression, do transcription rates increase?

28. During induction, do transcription rates increase?

29. Can any cellular process be governed by positive feedback?

Answers to Concept Map 18-1 Questions

1. *Regulation* means "controlling."

2. An operon is a cluster of cistrons (genes) that is controlled by a single promoter and is transcribed as a unit.

3. A promoter is a portion of an operon (or gene, in eukaryotes) that controls transcription from downstream cistrons.

4. Structural genes code for proteins but not for rRNA or tRNA.

5. In prokaryotes, transcription occurs in the nucleoid, which is not segregated by a nuclear membrane.

6. A cistronic message is composed of mRNA and contains the information for one polypeptide.

7. A polycistronic message is composed of mRNA and contains the information for more than one polypeptide.

8. Translation occurs in the nucleoid in prokaryotes.

9. The protein-rRNA structure that is involved in translation is the ribosome.

10. A structural gene is defined by mutation. If, for example, no cell wall protein is made because of a mutation, then the gene is called *cell wall*⁻, and it is a structural gene even though the actual code that has mutated is for one or more enzymes that are involved in the synthesis of a cell wall protein.

11. A promoter controls a structural gene by regulating RNA polymerase binding that in turn regulates transcription.

12. Proteins bind to promoters.

13. A promoter is composed of DNA, as is an operon.

14. The proteins that regulate bacterial genes come from other operons.

15. An allosteric change is a reversible change in structure of a protein that alters its metabolic activities.

16. An activator protein is a protein that binds to a promoter and thereby allows transcription to occur. An activator protein "activates" an operon.

17. A repressor protein is a protein that binds to a promoter and thereby blocks transcription from occurring. A repressor protein "represses" an operon.

18. *Positive control* means that the binding of a gene product promotes transcription.

19. *Negative control* means that the binding of a gene product blocks transcription.

20. The environment is the source of an inducer.

21. The operator is a region of a cistron.

22. An operator is composed of DNA.

23. An inducible system is one in which the synthesis of an enzyme occurs because of the presence of its substrate. Systems that are regulated by activator proteins or by repressor proteins can be inducible.

24. A repressible system is one in which the excess of its end product halts the synthesis of the enzyme which makes that end product. Systems that are regulated by either activator proteins or by repressor proteins can be repressible.

25. If lactose is an inducer, the enzymes whose synthesis it induces participate in catabolism. Lactose is used as an energy source.

26. If histidine is an inducer, the enzymes whose synthesis it induces participate in anabolism. However, histidine is not actually an inducer of the *his* operon.

27. During repression, transcription rates decrease.

28. During induction, transcription rates increase.

29. No cellular process can be governed by positive feedback, because positive feedback is inherently unstable.

Questions for Concept Map 18-2

1. Does the gene that makes the repressor protein have to be trans?

2. Do the promoter and operator have to be cis?

3. In the *lac* operon, what type of mutants does not make repressor protein? How does this affect the operon?

► CONCEPT MAP 18-2:
NEGATIVE CONTROL
IN PROKARYOTES

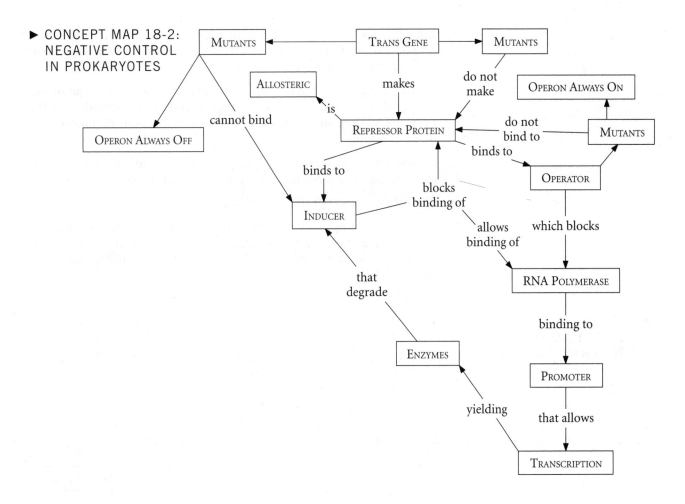

4. In the *lac* operon, what type of mutants makes a repressor protein that cannot bind the inducer? How does this affect the operon?

5. In the *lac* operon, what type of mutant has an operator to which the repressor protein cannot bind? How does this affect the operon?

6. What happens to the repressor protein when it binds an inducer?

7. How does a repressor protein block transcription?

8. Is the system diagrammed inducible or repressible?

9. Why is this called *negative control?*

Answers to Concept Map 18-2 Questions

1. The gene that makes the repressor protein does not have to be trans.

2. The promoter and operator have to be cis in order to regulate access of RNA polymerase to the cistrons associated with them.

3. In the *lac* operon, mutants that do not make repressor protein are I^- mutants. The operon is always turned on because repression does not occur.

4. In the lac operon, mutants that make a repressor protein that cannot bind the inducer are I^s mutants. The operon is always turned off because the repressor cannot be pulled off the operator by the inducer.

5. In the *lac* operon, mutants that have an operator to which the repressor protein cannot bind are O^c mutants. The operon is always turned on because the repressor protein cannot repress it.

6. When the repressor protein binds an inducer, it undergoes a conformational (allosteric) change. This change results in an alteration of the operator-binding site so that the repressor protein can no longer bind to the operator.

7. A repressor protein blocks transcription by binding the operator, thus blocking access of RNA polymerase to the structural genes.

8. The system diagrammed is inducible.

9. This is called *negative control* because the operon is turned off in the absence of a controlling substance (inducer). A positively controlled operon would be turned on in the absence of a controlling substance (repressor).

► CONCEPT MAP 18-3: POSITIVE CONTROL IN PROKARYOTES

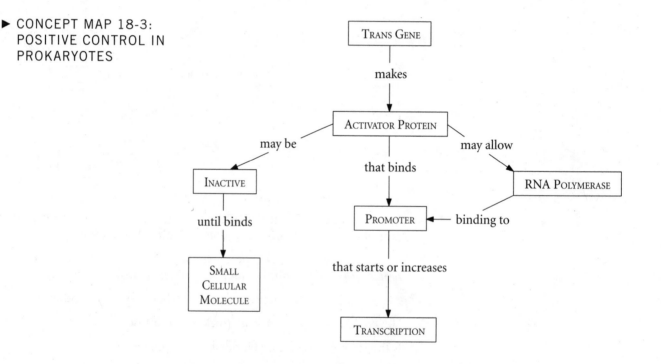

Questions for Concept Map 18-3

1. Must the activator protein always be trans?

2. If an activator protein is the only regulator of a cistron, and it is always present, what is the effect on the cistron? What is another name for this process?

3. If an activator protein is the only regulator of a cistron and it is inactive in the absence of some other regulatory molecule, is the operon inducible?

4. If an activator protein is the only regulator of a cistron and the operon is not always turned on, is something regulating either the production or the activity of the activator protein? If yes, what could regulate the production of the activator protein?

5. What is possibly happening to the DNA that is bound by an activator protein?

Answers to Concept Map 18-3 Questions

1. The activator protein does not have to be trans.

2. If an activator protein is the only regulator of a cistron, and it is always present, the cistron is permanently turned on. Another name for this is *positive feedback.*

3. If an activator protein is the only regulator of a cistron and it is inactive in the absence of some other regulatory molecule, the operon is inducible.

4. If an activator protein is the only regulator of a cistron and the operon is not always turned on, something is regulating either the production or the activity of the activator protein. A repressor protein could regulate the production of the activator protein.

5. The DNA that is bound by an activator protein may be unpairing in the region of the promoter so that RNA polymerase can bind to it.

▶ CONCEPT MAP 18-4:
CATABOLITE
REPRESSION
(POSITIVE CONTROL)

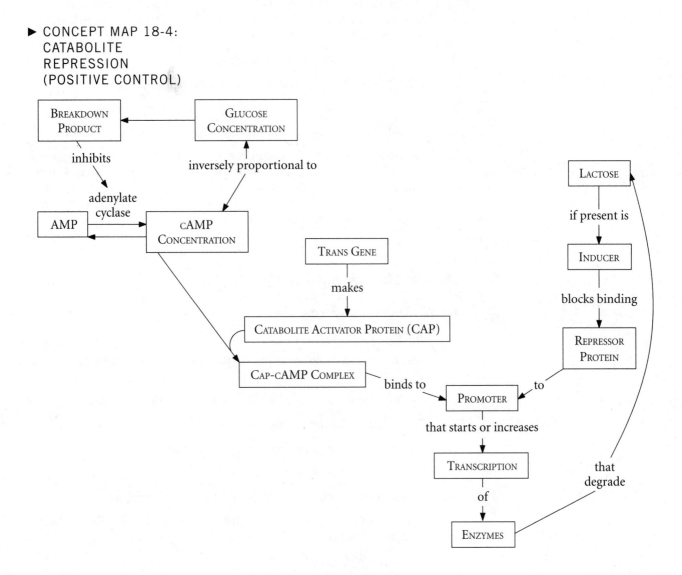

Questions for Concept Map 18-4

1. What are the two subunits of lactose?

2. Why does it make sense for cells to use glucose first if both glucose and lactose are present?

3. What does *AMP* stand for?

4. What does *cAMP* stand for?

5. If no cAMP is present, does any transcription at all occur?

6. Where does the CAP-cAMP complex bind?

7. How many control systems exist in this diagram? What type are they?

Answers to Concept Map 18-4 Questions

1. Lactose is composed of glucose and galactose.

2. The catabolism of lactose and glucose yields energy for the cell. In order to break down lactose to glucose and gaplactose, the cell must invest energy to break bonds. It is far better for the cell to invest as little energy as possible in order to obtain energy.

3. *AMP* stands for "adenosine monophosphate."

4. *cAMP* stands for "cyclic adenosine monophosphate."

5. If no cAMP is present, transcription occurs at a very low rate.

6. The CAP-cAMP complex binds to the CAP region of the promoter.

7. In a sense, the answer depends on how one counts. Two major control systems exist, however. The catabolite repression system constitutes positive control (binding results in transcription). The degradation of lactose is under the control of a repressor protein, which is negative control (binding halts transcription).

Questions for Concept Map 18-5

1. What does *upstream* mean?

2. What does *cis* mean?

3. Does the upstream cis gene that makes the control protein have to be cis?

4. How can a single molecule be both an activator and a repressor?

5. What does *positive control* mean?

6. What does *negative control* mean?

7. What is an initiator region?

8. How many different active sites are on the control protein?

9. What happens to the operon if the control protein is nonfunctional owing to a mutation?

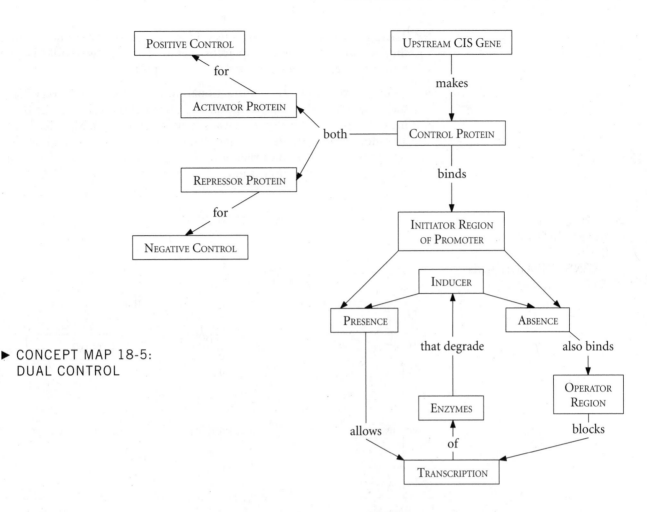

► CONCEPT MAP 18-5:
  DUAL CONTROL

Answers to Concept Map 18-5 Questions

1. The template strand runs from 3′ to 5′. Sequences that precede (are 3′ to) the gene on the template strand (or are 5′ to the gene on the complementary nontemplate strand) are referred to as being *upstream.* At the other end of the gene, the sequences are downstream. A flowing river analogy is used because RNA polymerase is attracted to upstream sequences and then seems to "float" downstream to the promoter.

2. *Cis* means "on the same chromosome." In bacterial genetics, it means that a partial diploid has been constructed so that two genes can be either cis or trans (on two physically separate pieces of DNA).

3. Because the gene product is a protein that can diffuse, it need not be cis (nor upstream).

4. A single molecule can be both an activator and a repressor if it has at least two binding sites.

5. *Positive control* means that binding by the control protein initiates transcription.

6. *Negative control* means that binding by the control protein stops transcription.

7. An initiator region is a portion of the promoter where the control protein has a binding site.

8. There is an initiator-binding site, a binding site for the inducer, and a third binding site for the operator region. The last two could be the same or different, depending on the control protein.

9. If the control protein is nonfunctional owing to a mutation, very likely there would be some transcription at a low level because it is the binding of the operator region that leads to no transcription. Most likely, the activator protein dramatically increases transcription over the low level by opening the promoter region to RNA polymerase, rather than truly being required for transcription.

► CONCEPT MAP 18-6: ATTENUATION IN COUPLED TRANSCRIPTION AND TRANSLATION

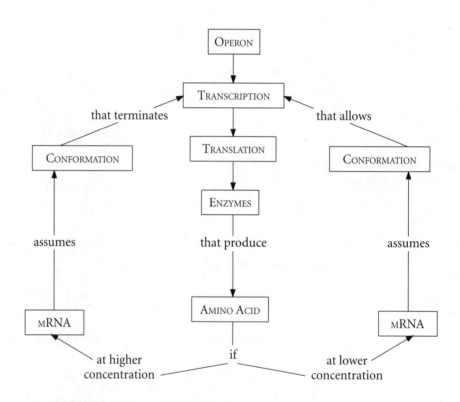

Questions for Concept Map 18-6

1. What does *attenuate* mean?

2. What does *coupled transcription and translation* mean?

3. What does *conformation* mean?

4. In attenuation, which molecule is undergoing changes in its secondary structure?

5. What leads to the change in conformation in the *trp* operon?

6. Could eukaryotes use this method of control?

Answers to Concept Map 18-6 Questions

1. *Attenuate* means "to weaken, reduce, make thin, to draw out and make thin."

2. *Coupled transcription and translation* means that a single mRNA molecule is being transcribed at one end while translation has already begun at the other end.

3. *Conformation* means "shape."

4. The mRNA molecule is undergoing changes in its secondary structure. It is forming alternative stem-loop structures by complementary pairing.

5. The rate of passage of the ribosome along the length of the mRNA molecule during translation leads to changes in secondary structure of the mRNA.

6. Eukaryotes could not use this method of control because transcription occurs in the nucleus and translation occurs in the cytoplasm. They are never coupled.

► CONCEPT MAP 18-7: GENERALIZED REGULATION IN EUKARYOTES

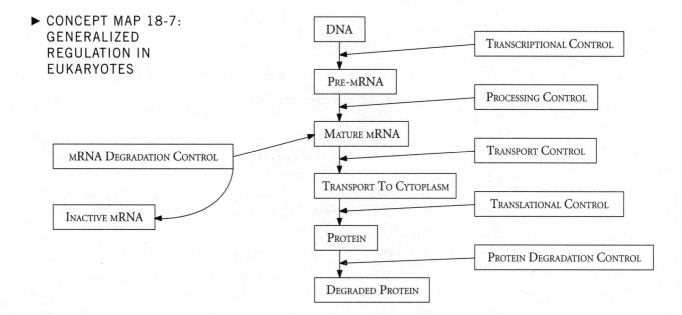

Questions for Concept Map 18-7

1. Can regulation occur before the DNA → pre-mRNA step?

2. What types of pre-transcriptional control could exist?

3. What types of transcriptional control could occur?

4. What types of processing control could occur?

5. Does processing control exist in prokaryotes?

6. If transport control exists, why would a cell make an mRNA molecule and then not transport it for translation?

7. What types of transport control could exist?

8. Do prokaryotes have any translational control?

9. What types of translational control could exist?

10. What types of protein degradation control could exist?

11. What types of mRNA degradation control could exist?

12. Is very much known about eukaryotic regulation?

13. Is Barr body formation a type of control?

14. Is methylation a type of control?

15. Is the formation of heterochromatin a type of control?

16. Do eukaryotes have promoters?

17. Do you think eukaryotes have regulatory proteins?

18. Do you think positive and negative controls exist in eukaryotes?

19. Do inducers exist in eukaryotes?

20. Are there inducible and repressible genes in eukaryotes?

21. Are nucleosomes a type of control?

Answers to Concept Map 18-7 Questions

1. Regulation occurs at the level of the structure of DNA before translation.

2. Repressor and activator proteins could exist as forms of pre-transcriptional control.

3. Most of the mechanisms that exist in prokaryotes could exist for transcriptional control in eukaryotes, including promoters, inducers, and regulatory proteins.

4. Splicing, the addition of poly(A) tails, the length of the poly(A) tails, and the addition of a 5′ CAP occur as forms of processing control in eukaryotes.

5. Processing control does not appear to exist in prokaryotes.

6. A cell might make an mRNA molecule and then not transport it for translation because of finely tuned adjustments to substrate concentrations.

7. Transport control could be at the level of rate of passage into the cytoplasm or could involve binding to "carrier" molecules.

8. Prokaryotes have translational control in attenuation.

9. The control during translation could involve availability of charged tRNA molecules, the binding of ribosomes to mRNA, the rate of polypeptide synthesis, and the types and rate of protein processing.

10. Protein degradation control might involve rate, types of proteins, and location of degradation enzymes.

11. The control of degradation of mRNA could involve selection of mRNAs, poly(A) tails, the presence of introns, the presence of a 5′ CAP, and alternative splicing.

12. In comparison with prokaryotic regulation, not very much is known about eukaryotic regulation. However, the greater complexity of eukaryotic cells suggests that there may be many more types and levels of control in eukaryotes than in prokaryotes.

13. Barr body formation is a type of control.

14. Methylation is a type of control.

15. The formation of heterochromatin is a type of control.

16. Eukaryotes have promoters.

17. Eukaryotes are known to have regulatory proteins, including steroid hormones, transcription factors, and translation factors.

18. Positive and negative controls exist in eukaryotes.

19. Inducers exist in eukaryotes. Drugs and alcohol, for instance, are inducers of genes in the liver cells.

20. There are inducible and repressible genes in eukaryotes.

21. Nucleosomes are a type of control.

▶ CONCEPT MAP 18-8: EUKARYOTIC GENE REGULATION

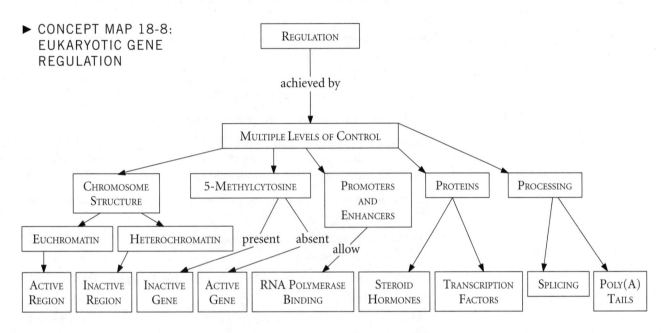

Questions for Concept Map 18-8

1. Name some of the multiple levels of control that either do or could exist.

2. Is the formation of constitutive heterochromatin a type of control?

3. Are all active regions or active genes always transcribed at maximal rates?

4. How do nucleosomes participate in control of gene expression?

5. Can methylation be reversed experimentally?

6. Can methylation be reversed in organisms?

7. What is an enhancer?

8. How do steroid hormones, which are transported from one cell type to another, control gene expression?

9. How do steroid hormones enter cells?

10. How do other regulatory proteins that are made in one cell type and function in another cell type enter the cells where they function?

11. Do all regulatory proteins need to be transported to a different cell in order to regulate gene expression?

12. Where does splicing occur?

13. Where does the addition of poly(A) tails occur?

14. Are all mRNAs spliced?

Answers to Concept Map 18-8 Questions

1. The multiple levels of control that either do or could exist are indicated on Concept Map 18-7.

2. The formation of constitutive heterochromatin is a type of control.

3. All active regions and active genes are not always transcribed at maximal rates. Many genes are in active regions but not transcribed continuously or at maximal rates.

4. Nucleosomes participate in control of gene expression by controlling access of the RNA polymerases to promoters.

5. Methylation can and has been reversed experimentally.

6. Under certain physiological conditions, the globin genes that have been inactivated during development can be expressed in the adult. This suggests that methylation can be reversed in organisms.

7. An enhancer is an upstream sequence that dramatically increases the rate of transcription of a gene.

8. Steroid hormones frequently bind directly to DNA as part of the control of gene expression.

9. Steroid hormones enter cells by being soluble in lipids. They literally dissolve through the cell membrane.

10. Other regulatory proteins enter the cells, where they function by binding to cell receptors.

11. Not all regulatory proteins need to be transported to a different cell in order to regulate gene expression; many are made within the cells where they function.

12. Splicing occurs on the spliceosome in the nucleus.

13. The addition of poly(A) tails occurs in the nucleus.

14. Not all mRNAs are spliced, because not all have introns.

▶ SOLUTIONS TO PROBLEMS

Be sure that you have thoroughly read the entire chapter before you attempt any of the problems.

1. The I gene determines the synthesis of a repressor molecule, which blocks expression of the *lac* operon and which is inactivated by the inducer. The presence of the repressor, I^+, will be dominant to the absence of a repressor, I^-. I^s mutants are unresponsive to an inducer. For this reason, the gene product cannot be stopped from interacting with the operator and blocking the *lac* operon. Therefore, I^s is dominant to I^+.

2. O^c mutants have impaired binding of the repressor product of the I gene to the operator, and therefore, the *lac* operon associated with the O^c operator cannot be turned off. Because an operator controls only the genes on the same DNA strand, it is *cis* (on the same strand) and dominant (cannot be turned off).

3. **a.** Comparing lines 1 and 2, *a* and *c* in a + or − state do not affect the expression of the *Z* gene. Therefore, *b* is the *Z* gene.

   In line 6, the *I* gene is functioning in a *trans* configuration, indicating that *c* is *I*. This leaves *a* as the *O* region, which is confirmed by line 7.

   b. *Line 1:* $a^- = O^c$

   Line 2: $c^- = I^-$

   Line 3: $c^- = I^-$ or I^s

   Line 4: $a^- = O^c$, and $c^- = I^-$ or I^s

   Line 5: $a^- = O^c$, and $c^- = I^-$

   Line 6: $a^- = O^c$, and $c^- = I^-$

   Line 7: $a^- = O^c$, and $c^- = I^-$ or I^s

4.

PART	β-GALACTOSIDASE		PERMEASE	
	NO LACTOSE	LACTOSE	NO LACTOSE	LACTOSE
a	−	+	−	+
b	+	+	−	−
c	−	−	−	−
d	−	−	−	−
e	+	+	+	+
f	+	+	−	−
g	−	+	−	+

5. **a.** A lack of only E_1 or only E_2 function indicates that both genes have enzyme products that are responsible for a conversion reaction. Because the two genes are in different linkage groups, they cannot be regulated by a single operator and promoter like the *Z* and *Y* genes of the *lac* operon. Type 3 mutants must be mutants of a site that produces a diffusible regulator of the E_1 and E_2 genes. The type 3 mutants identify a site that produces either a repressor (like *I* in the *lac* operon) or an activator (analogous to CAP) of the other two genes.

   b. Separate operator and promoter mutants might be found for each gene.

6. If there is an operon governing both genes, then a frameshift mutation could cause the stop codon separating the two genes to be read as a sense codon. Therefore, the second gene product will be incorrect for almost all amino acids. However, there are no known polycistronic messages in eukaryotes. The alternative, and better, explanation is that

both enzymatic functions are performed by the same gene product. Here, a frameshift mutation beyond the first function, carbamyl phosphate synthetase, will result in the second half of the protein molecule being nonfunctional.

7. Nonpolar Z^- mutants cannot convert lactose to allolactose, and thus, the operon is never induced.

8. Because very small amounts of the repressor are made, the system as a whole is quite responsive to changes in lactose concentration. In the heterodiploids, repressor tetramers may form by association of polypeptides encoded by I^- and I^+. The operator binding site binds two subunits at a time. Therefore, the repressors produced may reduce operator binding, which in turn would result in some expression of the *lac* genes in the absence of lactose.

9. An operon is turned off or inactivated by the mediator in negative control, and the mediator must be removed for transcription to occur. An operon is turned on by the mediator in positive control, and the mediator must be added or activated for transcription to occur.

10. The *lacY* gene produces a permease that transports lactose into the cell. A *lacY*$^-$ gene could not transport lactose into the cell, so β-galactosidase will not be induced.

11. Activation of gene expression by trans-acting factors occurs in both prokaryotes and eukaryotes. In both cases, the trans-acting factors interact with specific sequences that control expression of cis genes.

    In prokaryotes, proteins bind to a specific DNA sequence, which is regulated by the binding protein and which, in turn, regulates one or more downstream cistrons.

    In eukaryotes, highly conserved sequences such as CCAAT and enhancers increase transcription controlled by the downstream TATA box promoter. Several proteins have been found to bind to the CCAAT sequence, upstream GC boxes, and the TATA sequence in *Drosophila*, yeast, and other organisms. Specifically, the Sp1 protein recognizes the upstream GC boxes of the SV40 promoter and many other genes, GCN4 and GAL4 proteins recognize upstream sequences in yeast, and many hormones bind to specific sites on the DNA (e.g., estrogen binding to a sequence upstream of the ovalbumin gene in chicken oviduct cells). Additionally, the structure of some of these trans-acting DNA-binding proteins is quite similar to the structure of binding proteins seen in prokaryotes. Further, protein-protein interactions are important in both prokaryotes and eukaryotes. For the above reasons, eukaryotic regulation is now thought to be very close to the model for regulation of the bacterial *ara* operon.

12. The bacterial operon contains a promoter region that extends approximately 35 bases upstream of the site where transcription is initiated. Within this region is the promoter. Inducers and repressors, both of which are trans-acting proteins that bind to the promoter region, regulate transcription of associated cistrons in cis only.

    The eukaryotic cistron has the same basic organization. However, the promoter region is somewhat larger. Also, enhancers up to several thousand nucleotides upstream or downstream can influence the rate of transcription. A major difference is that eukaryotes have not been demonstrated to have polycistronic messages.

13. The *araC* product has two conformations, which are determined by the presence and absence of arabinose. When it has bound arabinose, the *araC* product can then bind to the initiator site (*araI*) and activate transcription. When it is not bound to arabinose, the *araC* product binds to both the initiator (*araI*) and the operator (*araO*) sites, forming a loop of the intermediary DNA. When both sites are bound to the *araC* product, transcription is inhibited. The *araC* product is trans-acting.

Many eukaryotic trans-acting protein factors also bind to promoters, enhancers, or both that are upstream from the protein-encoding gene. These factors are required for the initiation of transcription. Additionally, some bind to other proteins, such as RNA polymerase II, in order to initiate transcription. Like their counterparts in the *ara* operon, the eukaryotic trans-acting protein factors can bind DNA at two sites, with the intermediary DNA forming a loop between the binding sites.

14. A reasonable model is that one dimer portion of the tetramer binds O_1 and one dimer portion of the tetramer binds O_2. The two dimers then bend the DNA when forming the tetramer complex, which results in a blocking of transcription.

15. Normally, the repressor searches for the operator by rapidly binding and dissociating from nonoperator sequences. Even for sequences that mimic the true operator, the dissociation time is only a few seconds or less. Therefore, it is easy for the repressor to find new operators as new strands of DNA are synthesized. However, when the affinity of the repressor for DNA and operator is increased, it takes too long for the repressor to dissociate from sequences on the chromosome that mimic the true operator, and, as the cell divides and new operators are synthesized, the repressor never quite finds all of them in time, leading to a partial synthesis of β-galactosidase. This explains why in the absence of IPTG there is some elevated β-galactosidase synthesis. When IPTG binds to the repressors with increased affinity, it lowers the affinity back to that of the normal repressor (without IPTG bound). Then, the repressor can rapidly dissociate from sequences in the chromosome that mimic the operator and find the true operator. Thus, β-galactosidase is repressed in the presence of IPTG in strains with repressors that have greatly increased affinity for operator. In summary, because of a kinetic phenomenon, we see some type of reverse induction curve.

Mechanisms of Genetic Change I: Gene Mutation

A mutation consists of a change in DNA sequence. **Spontaneous** mutations occur at a low rate in all cells; the mechanisms for spontaneous mutations include errors in DNA replication and the action of transposable genetic elements. **Induced** mutations are caused by one or more environmental agents.

A **tautomeric shift** results in a **transition** mutation: a purine is substituted for a purine or a pyrimidine is substituted for a pyrimidine. **Transversion** mutations substitute a purine for a pyrimidine or a pyrimidine for a purine.

Frameshift mutations result in a change in the reading frame; they are caused by slipped mispairing during the replication of repeated sequences. **Deletion** and **duplication** mutations can be caused by replication errors or recombinational errors.

Depurination, the loss of the nitrogenous purine base from the nucleotide, can result in a mutation. **Deamination** of cytosine yields uracil, which pairs with adenine (a transition).

Base analogs induce mutation at a high rate; they are incorporated into the DNA and alter base pairing. **Alkylating agents** chemically modify the nitrogenous bases to cause mutations. **Intercalating agents** insert themselves between the nitrogenous bases, causing additions and deletions.

DNA is **repaired** by several different systems: the **SOS system**, **detoxification** by **superoxide dismutase**, the **photoreactivating enzyme**, **alkyltransferase**, **excision repair**, the **AP endonucleases**, the **DNA glycosylase repair pathway**, the **mismatch repair system**, and **recombinational repair**. A defect in one of these repair systems can produce a **mutator** phenotype; that is, a strain can have a very high rate of spontaneous mutation.

The **Ames test** is a screening test for possible mutagens.

Questions for Concept Map 19-1

1. What are some sources of genetic change?

2. Which has a higher rate of mutation: spontaneous or induced mutation?

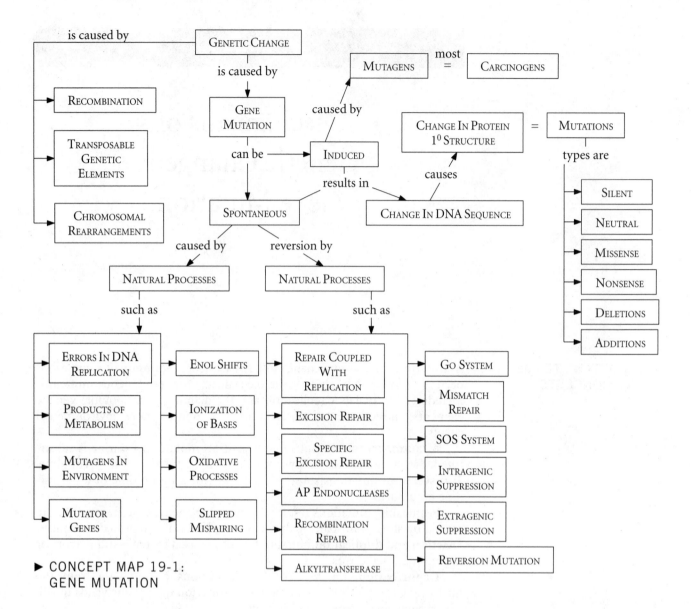

► CONCEPT MAP 19-1:
GENE MUTATION

3. Do the changes that occur in induced mutation differ from those that occur spontaneously?

4. What happens to the DNA sequence with all mutations?

5. How does the change of a DNA sequence lead to the expression of a mutant phenotype?

6. Are all genes that are mutated protein-encoding genes?

7. What types of changes occur in the DNA sequence?

8. If the DNA sequence of a protein-encoding gene changes, does this necessarily result in a change in the protein primary structure? Explain.

9. What is a silent mutation?

10. What is a neutral mutation?

11. What is a missense mutation?

12. What is a nonsense mutation?

13. What do additions and deletions of one base cause?

14. What kinds of errors occur during DNA replication?

15. Do you expect errors to occur at a low rate during DNA replication?

16. If the protein-encoding gene for DNA polymerase III were mutated, what would this do to the mutation rate in a cell?

17. Would you expect errors in DNA replication to occur at a higher frequency with aging?

18. If a person lives long enough, will that person ultimately develop a malignancy? Why?

19. What would happen in a person who inherited a defect in one or more repair systems?

20. What would happen in a person who experienced a mutation in one or more repair systems?

21. How do the products of metabolism cause mutation?

22. Are there natural mutagens in the environment or have all of them been placed there by humans?

23. Would you expect the mutation rate to be increasing?

24. Would you expect the malignancy rate to be increasing?

25. What is an enol shift?

26. What is ionization of bases?

27. What natural oxidative processes result in mutation?

28. What is slipped mispairing?

29. In what way is repair coupled with replication?

30. What is excision repair?

31. What is recombination repair?

32. What does alkyltransferase do?

33. What is the GO system?

34. What is mismatch repair?

35. What is the SOS system and how is it inducible?

36. What are intragenic and extragenic suppression?

37. What is reversion mutation?

Answers to Concept Map 19-1 Questions

1. Sources of genetic change include recombination, transposable genetic elements, chromosomal rearrangements, and gene mutation.

2. Induced mutations have a higher rate of mutation.

3. The changes that occur in induced mutation do not differ from those that occur spontaneously.

4. The DNA sequence changes with all mutations.

5. The change of a DNA sequence results in the expression of a mutant phenotype because the DNA sequence change leads to a change in protein function.

6. Not all genes that are mutated are protein-encoding genes. Any region of the DNA can mutate.

7. The types of changes that occur in the DNA sequence are transitions, transversions, additions, and deletions.

8. If the DNA sequence of a protein-encoding gene changes, that does not necessarily result in a change in the protein primary structure. The reason is that there is degeneracy in the code so that more than one codon can call for the same amino acid in a protein.

9. A silent mutation is a mutation that results in the same amino acid because of code degeneracy.

10. A neutral mutation is a mutation that results in a functionally equivalent amino acid replacement, so that the protein continues to work at the same level as before the mutation occurred.

11. A missense mutation is a mutation that results in an amino acid change that disrupts functioning of the protein.

12. A nonsense mutation is a mutation that leads to premature termination of translation, resulting in a shorter and nonfunctional protein.

13. Additions and deletions of one base cause frameshifts. This in turn results in virtually all the amino acids beyond the point of change being incorrect.

14. During DNA replication, an enol shift can occur, a nitrogenous base can ionize, slipped mispairing can occur, and simple mismatching can occur. In addition, depurination, deamination, oxidative damage, and dimer formation can occur. All these events lead to a change in DNA sequence.

15. Errors are expected to occur at a low rate during DNA replication.

16. If the protein-encoding gene for DNA polymerase III were mutated, it would drastically increase the mutation rate in a cell.

17. Errors in DNA replication would be expected to occur at a higher frequency with aging because the errors accumulate and, in effect, become compounded over time.

18. If a person lives long enough, that person will ultimately develop a malignancy because eventually an error will occur in a gene that regulates the cell cycle or is a tumor suppressor.

19. A person who inherited a defect in one or more repair systems will, at the minimum, suffer a much higher than normal rate of somatic mutation. Most likely, the person would develop a malignancy very early.

20. A person who experienced a mutation in one or more repair systems would, from that point on, suffer a much higher than normal rate of somatic mutation. Most likely, the person would develop a malignancy very soon after the mutation.

21. The products of metabolism cause mutation because detoxification can lead to mutagenic end products and because hydrogen peroxide and radicals are normally produced during metabolism.

22. There are many natural mutagens in the environment.

23. One would expect the mutation rate to be increasing because of the tremendous increase in pollution and because of exposure to so many nonnatural chemicals.

24. One would expect the malignancy rate to be increasing because of the expected increase in mutation rate

25. An enol, or tautomeric, shift is a change to an isomer of a base. The more common keto form has different pairing properties than the rare enol form.

26. The ionization of bases occurs when a nitrogenous base looses a hydrogen, changing the charge and the pairing ability of the base.

27. The formation of superoxide radicals, hydrogen radicals, and hydrogen peroxide during metabolism results in mutation.

28. Slipped mispairing occurs within repeats. It involves the DNA polymerase's getting out of register within the repeat segment, leading to either duplications or deletions of a repeat.

29. Repair is coupled with replication through the existence of excision repair, specific excision repair, mismatch repair, recombination repair, and the SOS system in bacteria.

30. Excision repair is the repair of a DNA lesion by removal of the faulty DNA segment and its replacement with a wild-type segment.

31. Recombination repair is the repair of DNA lesions through a process, similar to recombination, that uses recombination enzymes.

32. Alkyltransferase removes the alkyl group that was erroneously added to a base.

33. The GO system repairs the oxidative damage caused by 8-oxodG by removal of the lesion.

34. Mismatch repair corrects the damage caused by faulty pairing during replication by removal and replacement of the faulty base or bases.

35. The SOS system repairs lesions that completely stop DNA replication. The system does, however, lead to mutations. It is inducible by DNA damage, although the details of triggering are not understood.

36. Intragenic suppression is the blocking of the manifestations of a first mutation by a second mutation within the same gene. Extragenic suppression is the blocking of the manifestations of a first mutation by a second mutation within a different gene.

37. Reversion mutation is the restoration of the wild type by a second mutation at the same location.

▶ SOLUTIONS TO PROBLEMS

Be sure that you have thoroughly read the entire chapter before you attempt any of the problems.

1. **a.** A transition mutation is the substitution of a purine for a purine or the substitution of a pyrimidine for a pyrimidine. A transversion mutation is the substitution of a purine for a pyrimidine, or vice versa.

b. Both are base-pair substitutions. A silent mutation is one that does not alter the function of the protein product from the gene, because the new codon codes for the same amino acid as did the nonmutant codon. A neutral mutation results in a different amino acid that is functionally equivalent, and the mutation therefore has no adaptive significance.

c. A missense mutation results in a different amino acid in the protein product of the gene. A nonsense mutation causes premature termination of translation, resulting in a shortened protein.

d. Frameshift mutations arise from the addition or deletion of one or more bases which are not a multiple of three, thus altering the reading frame for translation and, therefore, the amino acid sequence from the site of the mutation to the end of the protein product of the gene. Frameshift mutations can result in nonsense (stop) mutations.

2. Frameshift mutations arise from the addition or deletion of one or more bases which are not a multiple of three, thus altering the reading frame for transition and, therefore, the amino acid sequence from the site of the mutation to the end of the protein product of the gene. Frameshift mutations can result in nonsense (stop) mutations.

3. The Streisinger model proposed that frameshifts arise when loops in single-stranded regions are stabilized by slipped mispairing of repeated sequences. In the *lac* gene of *E. coli*, a four-base-pair sequence is repeated three times in tandem, and this is the site of a hot spot.

   The sequence is 5'-CTGG CTGG CTGG-3'. During replication the DNA must become single-stranded in short stretches for replication to occur. As the new strand is synthesized and becomes hydrogen-bonded to the template strand, it can pair out of register with that strand by a total span of four bases. Depending on which strand, new or template, loops out with respect to the other, there will be an addition or deletion of four bases, as diagrammed below:

$$
\begin{array}{cc}
\text{T} & \text{G} \\
\text{C} & \text{G} \\
\end{array}
$$

5'-C T G G   C T G G-3' ⟶ DNA synthesis

3'-G A C C ———— G A C C G A C C-5'

   In this diagram, the upper strand looped out as replication was occurring. The loop is stabilized by base pairing on either strand. As replication continues at the 3' end, an additional copy of CTGG will be synthesized, leading to an addition of four bases. This will result in a frameshift mutation.

4. (1) In Problem 3, had the lower strand looped, the result would have been a deletion in the newly synthesized upper strand.

   (2) As with the bar-eye allele in *Drosophila*, misalignment of homologous chromosomes during recombination results in one chromosome with a duplication and the other with a deletion:

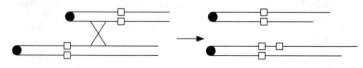

At the DNA level, recombination between two homologous repeats in a looped DNA molecule can lead to deletion. The models below are supported by DNA sequencing results.

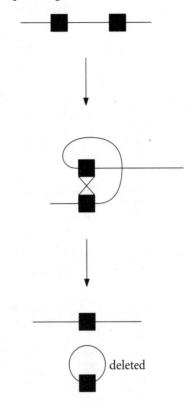

5. Depurination results in the loss of adenine or guanine from the DNA. Since the resulting apurinic site cannot specify a complementary base, replication is blocked. Under certain conditions, replication proceeds with a near random insertion of a base opposite the apurinic site. In three-fourths of these insertions, a mutation will result.

    Deamination of cytosine yields uracil. If left unrepaired, the uracil will be paired with adenine during replication, ultimately resulting in a transition mutation.

    8-OxodG (8-oxo-7-hydrodeoxyguanosine) can pair with adenine, resulting in a transversion.

6. 5-Bromouracil is an analog of thymine. It undergoes tautomeric shifts at a higher frequency than does thymine and, therefore, is more likely to pair with G than is thymine during replication. At the next replication this will lead to a GC pair rather than the original AT pair. On the other hand, 5-bromouracil can also be incorporated into DNA by mispairing with guanine. In this case it will convert a GC pair to an AT pair.

    Ethyl methanesulfonate is an alkylating agent that produces O-6-ethylguanine. This alkylated guanine will mispair with thymine, which leads from a GC pair to an AT pair at the next replication.

7. An AP site is an apurinic or apyrimidinic site. AP endonucleases introduce chain breaks by cleaving the phosphodiester bonds at the AP sites. Some exonuclease activity follows, so that a number of bases are removed. The resulting gap is filled by DNA pol I and then sealed by DNA ligase.

When aflatoxin B_1 damage occurs in *E. coli*, several compounds may bind to the damaged site, resulting in the so-called bulky adducts. The *uvrA*, *uvrB*, and *uvrC* gene products recognize a distortion in the DNA helix. Again, excision is followed by gap filling and ligation.

8. Mismatch repair occurs if a mismatched nucleotide is inserted during replication. The new, incorrect base is removed and the proper base is inserted. The enzymes involved can distinguish between new and old strands because, in *E. coli*, the old strand is methylated.

Recombination repair occurs if lesions such as AP sites and UV photodimers block replication (there is a gap in the new complementary strand). Recombination fills this gap with the corresponding segment from the sister DNA molecule, which is normal in both strands. This produces one DNA molecule with a gap across from a correct strand, which can then be filled by complementation, and one with a photodimer across from a correct strand (refer to Figure 19-40 a).

9. Leaky mutants are mutants with an altered protein product that retains a low level of function. Enzyme activity may, for instance, be reduced rather than abolished by a mutation.

10. The wild type contained a gene that increased the spontaneous mutation rate. This new gene seems to be unlinked to *ad-3*. Call the new gene *B*. Cross *A* (*ad-3 B+*) × wild type (*ad-3+ B*). The progeny should reflect independent assortment.

Progeny: 1/4 *ad-3*   *B*

1/4 *ad-3*   *B+*

1/4 *ad-3+*   *B*

1/4 *ad-3+*   *B+*

Further crosses should verify the above.

11. a. Because 5′-UAA-3′ does not contain G or C, a transition to a GC pair in the DNA cannot result in 5′-UAA-3′. 5′-UGA-3′ and 5′-UAG-3′ have the DNA antisense-strand sequence of 3′-ACT-5′ and 3′-ATC-5′, respectively. A transition to either of these stop codons occurs from the nonmutant 3′-ATT-5′, respectively. However, a DNA sequence of 3′-ATT-5′ results in an RNA sequence of UAA, itself a stop codon.

b. Yes. An example is 5′-UGG-3′, which codes for Trp, to 5′-UAG-3′.

c. No. In the three stop codons the only base that can be acted upon is G (in UAG, for instance). Replacing the G with an A would result in 5′-UAA-3′, a stop codon.

12. a. and b. *Mutant 1*: Most likely a deletion. It could be caused by radiation.

Mutant 2: Because proflavin causes either additions or deletions of bases and because spontaneous mutation can result in additions or deletions, the most probable cause was a frameshift mutation by an intercalating agent.

Mutant 3: 5-BU causes transitions, which means that the original mutation was most likely a transition. Because HA causes GC-to-AT transitions and HA cannot revert it, the original must have been a GC-to-AT transition. It could have been caused by base analogs.

Mutant 4: The chemical agents cause transitions or frameshift mutations. Because there is spontaneous reversion only, the original mutation must have been a transversion. X-irradiation or oxidizing agents could have caused the original mutation.

Mutant 5: HA causes transitions from GC to AT, as does 5-BU. The original mutation was most likely an AT-to-GC transition, which could be caused by base analogs.

c. The suggestion is a second-site reversion linked to the original mutant by 20 map units (m.u.) and therefore most likely in a second gene. Note that auxotrophs equal half the recombinants.

13. To understand these data, recall that half of the progeny should come from the wild-type parent.

a. A lack of revertants suggests either a deletion or an inversion within the gene.

b. *Prototroph A*: Because 100 percent of the progeny are prototrophic, a reversion at the original mutant site may have occurred.

Prototroph B: Half the progeny are parental prototrophs, and the remaining prototrophs, 28 percent, are the result of the new mutation. Notice that 28 percent is approximately equal to the 22 percent auxotrophs. The suggestion is that an unlinked suppressor mutation occurred, yielding independent assortment with the *nic* mutant.

Prototroph C: There are 496 "revertant" prototrophs (the other 500 are parental prototrophs) and 4 auxotrophs. This suggests that a suppressor mutation occurred in a site very close [$100\% (4 \times 2)/1000 = 0.8$ m.u.] to the original mutation.

14. a. To select for a nerve mutation that blocks flying, place *Drosophila* at the bottom of a cage and place a poisoned food source at the top of the cage.

b. Make antibodies against flagellar protein and expose mutagenized cultures to the antibodies.

c. Do filtration through membranes with various-sized pores.

d. Screen visually.

e. Go to a large shopping mall and set up a rotating polarized disk. Ask the passersby to look through the disk for a free evaluation of their vision and their need for sunglasses. People with normal vision will see light with a constant intensity through the disk. Those with polarized vision will see alternating dark and light.

f. Set up a Y tube (a tube with a fork giving the choice of two pathways) and observe whether the flies or unicellular algae crawl to the light or the dark pathway.

g. Set up replica cultures and expose one of the two plates to low doses of UV.

Mechanisms of Genetic Change II: Recombination

Crossing-over occurs during prophase I of meiosis. It involves the **breakage and reunion** of DNA molecules. **Chiasmata** are the sites of crossing-over.

The **Holliday model** involves the creation of **heteroduplex** DNA, which can undergo **branch migration**. A number of enzymes are postulated to function in this process. The model accounts for **gene conversion**, **polarity**, and **co-conversion**.

Site-specific recombination occurs between two specific sequences that need not be homologous.

Questions for Concept Map 20-1

1. What processes can give rise to genetic change?

2. What chromosomal rearrangements can result from recombination?

3. What is the cytological structure that is evidence of the molecular event of recombination?

4. When does crossing-over occur?

5. Why does DNA synthesis occur as part of crossing-over?

6. What is the chi structure?

7. What is the chi site?

8. How does λ phage integration occur?

9. What is site-specific recombination?

10. How does site-specific recombination differ from generalized recombination?

11. What is gene conversion?

12. What is mismatch repair?

13. What polarity is observed during recombination?

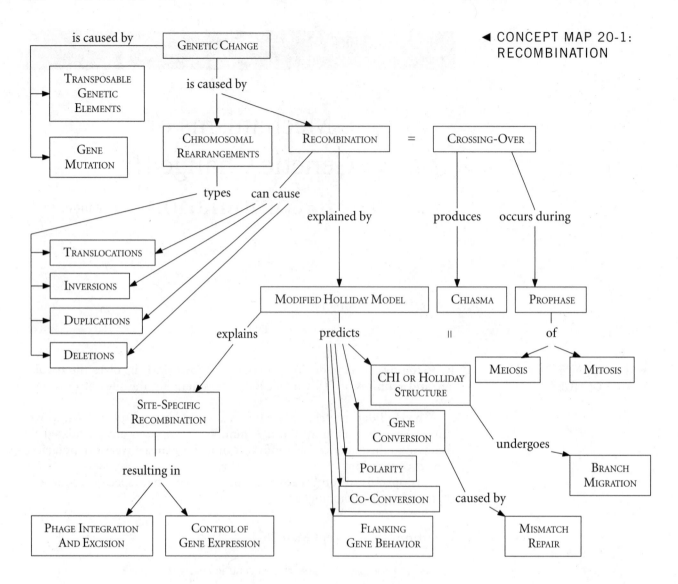

◀ CONCEPT MAP 20-1:
RECOMBINATION

14. What does *co-conversion* mean?

15. What is branch migration?

16. Can the Holliday model be applied to sister-chromatid exchange? If yes, what initial assumption must you make?

17. Some mutagens, in addition to causing sister-chromatid exchange, cause mitotic recombination between homologs. What does this suggest about the interphase location of homologous chromosomes?

18. Which of the following linear asci show gene conversion?

1	2	3	4	5	6
+	+	+	+	+	+
+	+	+	+	+	+
met	met	+	+	+	met
met	met	+	+	+	+
met	+	+	+	met	met

1	2	3	4	5	6
met	*met*	*met*	+	*met*	*met*
met	*met*	*met*	*met*	*met*	+
met	*met*	*met*	*met*	*met*	+

Answers to Concept Map 20-1 Questions

1. Recombination, chromosomal rearrangement, gene mutation, and transposition can give rise to genetic change.

2. Translocations, inversions, duplications, and deletions can result from recombination.

3. The chiasma is the cytological structure that is evidence of the molecular event of recombination.

4. Crossing-over occurs during prophase.

5. DNA synthesis occurs as part of crossing-over because of mismatch repair.

6. The chi structure is the heteroduplex structure formed during recombination that is capable of branch migration.

7. The chi site is the eight-base sequence that exists in *E. coli* where recombination takes places with the aid of the *rec* genes.

8. Integration of λ phage occurs at the attachment site, which contains a 15-base homology between λ and *E. coli*.

9. Site-specific recombination is recombination that occurs only at specific sites, such as the λ integration site.

10. Site-specific recombination is limited to special sites and involves regions of homology (14–15 bases) within regions that generally are not homologous. Generalized recombination occurs through the eukaryotic genome and involves homology of an unknown length.

11. Gene conversion is a meiotic process of directed change in which one allele directs the conversion of a partner allele to its own form. It is the result of mismatch repair in a heteroduplex region.

12. Mismatch repair occurs in heteroduplex regions in which complete complementarity does not exist.

13. There is a gradient of gene conversion that occurs, which is polarity.

14. *Co-conversion* means that a gene extremely close to another gene is also converted, but at a lower rate, during gene conversion.

15. Branch migration is the process whereby an invading strand extends its partial pairing with its complementary strand as it displaces the resident strand.

16. The Holliday model can account for sister-chromatid exchange. It must be assumed that a break caused by a mutagen signals an enzyme with endonuclease function to cause a break in the homologous site on the sister chromatid.

17. It suggests that homologous chromosomes are located near each other in mitotic interphase. Although it is generally assumed that the distribution of chromosomes during interphase is random, there are several studies that indirectly indicate that chromosome distribution is non-random during interphase.

18. All asci except ascus 5 show gene conversion.

▶ SOLUTIONS TO PROBLEMS

Be sure that you have thoroughly read the entire chapter before you attempt any of the problems.

1. Gene conversion may result in a deviation from a 4:4 ratio, with the order unimportant. The following asci show gene conversion: 3, 6.

   Ascus 4 is also produced by gene conversion. To recognize it as such, recall the sequence that gives rise to the eight meiotic products in *Neurospora*. The pattern generated could be produced only if the two DNA strands have a region of mismatch.

2. Note that the first printing of the text contains an error in the rare ascospore pattern: there should be a fourth "1'+" ascospore at the bottom of the column.

   In the first case 1' is being converted to 1'+. In the second case, 1" is being converted to 1''+. The difference in frequency is due to polarity.

3. A fixed break point is the point at which a DNA strand breaks and begins unwinding as the first step in recombination. The highest level of gene conversion is seen at this point.

   Gene conversion has occurred in cistrons 1, 2, and 3, and the conversions all are from mutant to wild type. Therefore, most likely one piece of heteroduplex DNA extended across the three cistrons:

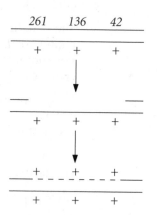

   These data are compatible with the idea, but do not prove, that the heteroduplex DNA initiates from a site to the left of cistron 1, corresponding to a promoter of a polycistronic message.

4. The actual mechanism of sister-chromatid exchange induction by mutagens is unknown but would be expected to vary with mutagen effects. The end point must be a break in a DNA strand, very likely as part of postreplication repair.

5. First notice that gene conversion is occurring. In the first cross, a_1 converts (1:3). In the second cross, a_3 converts. In the third cross, a_3 con-

verts. Polarity is obviously involved. The results can be explained by the following map, where hybrid DNA enters only from the left.

$$\underline{\qquad \overset{a_3}{\qquad} \qquad\qquad \overset{a_1}{\qquad} \qquad\qquad \overset{a_2}{\qquad} \qquad}$$

6. Rewrite the cross and results so that it is clear what they are.

P A $(m_1\ m_2\ m^+)$ $B \times a\ (m^+\ m^+\ m_3)\ b$

F_1 A $(m_1\ m_2\ m^+)$ B parental

   A $(m_1\ m^+\ m^+)$ b recombinant

   a $(m_1\ m_2\ m^+)$ B recombinant

   a $(m^+\ m^+\ m_3)$ b parental

Next, note the frequency of each allele:

$m_1{:}m_1^+ = 3{:}1$

$m_2{:}m_2^+ = 1{:}1$

$m_3{:}m_3^+ = 1{:}3$

Two gene conversion events have occurred, involving m_1 and m_3.

   To understand this at the molecular level, consider the following diagram:

A single excision-repair event changed $m^+\ m_3$ to $m_1\ m^+$, and the other mismatches remained unrepaired.

7. The ratios for a_1 and a_2 are both 3:1. There is no evidence of polarity, which indicates that gene conversion as part of recombination is occurring. The best explanation is that two separate excision-repair events occurred and, in both cases, the repair retained the mutant rather than the wild type.

8. **a. and b.** A heteroduplex that contains an unequal number of bases in the two strands has a larger distortion than does a simple mismatch. Therefore, the former would be more likely to be repaired. For such a case, both heteroduplex molecules are repaired (leading to 6:2 and 2:6) more often than one (leading to 5:3 or 3:5) or none (leading to 3:1:1:3). The preference in direction (i.e., adding a base rather than subtracting) is analogous to thymine dimer repair. In thymine dimer repair, the unpaired, bulged nucleotides are treated as correct and the strand with the thymine dimer is excised.

   A mismatch more often than not escapes repair, leading to a 3:1:1:3 ascus.

Transition mutations would not cause as large a distortion of the helix, and each strand of the heteroduplex should have an equal chance of repair. This would lead to 4:4 (two repairs each in the opposite direction), 5:3 (1 repair), 3:1:1:3 (no repairs or two repairs in opposite directions), and, less frequently, 6:2 (two repairs in the same direction).

c. Because excision repair excises the strand opposite the larger buckle (i.e., opposite the frameshift mutation), the *cis* transition mutation will also be retained. The nearby genes are converted because of the length of the excision repair.

9. The easiest way to handle these data is to construct the following table, which shows the repair rates for 0, 1, and 2 hybrid DNA molecules:

	0.5 +/+	0.3 +/g_1	0.2 g_1/g_1
0.5 +/+	0.25	0.15	0.1
	(6:2)	(5:3)	(4:4)
0.3 +/g_1	0.15	0.09	0.06
	(5:3)	(3:1:1:3)	(3:5)
0.2 g_1/g_1	0.1	0.06	0.04
	(4:4)	(3:5)	(2:6)

The aberrant asci are all those that are not 4:4. The 4:4 asci occur 20 percent of the time. Correcting for them,

a. 6:2 = 25%/0.8 = 31.25%

b. 2:6 = 4%/0.8 = 5%

c. 3:1:1:3 = 9%/0.8 = 11.25%

d. 5:3 = 30%/0.8 = 37.5%

e. 3:5 = 12%/0.8 = 15%

10. The map is

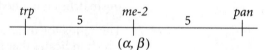

Rewrite the crosses and results to be sure that you understand them.

Cross 1: trp ($\alpha \beta^+$) pan$^+$ × trp+ ($\alpha^+\beta$) pan

Cross 2: trp ($\alpha^+\beta$) pan$^+$ × trp$^+$ ($\alpha \beta^+$) pan

Note that all progeny must be $\alpha^+ \beta^+$, which requires a crossover between them, and that the order of α and β is unknown.

Consider the first cross. If the sequence is *trp* α β *pan*, then one crossover should lead to a high frequency of ++++. The conventional double crossovers +++− and −+++ should be equally frequent and of lower frequency than ++++. The pattern −++− would result from a triple crossover and would be least frequent. This is summarized in the tabulation below, along with the results if the opposite gene order is true.

		NUMBER OF CROSSOVERS REQUIRED	
PATTERN	FREQUENCY	*TRP α β PAN*	*TRP β α PAN*
++++	56	1 CO	3 CO
−+++	26	DCO	DCO
+++−	59	DCO	DCO
−++−	16	3 CO	1 CO

For the second cross, the patterns and their interpretation are

		NUMBER OF CROSSOVERS REQUIRED	
PATTERN	FREQUENCY	*TRP α β PAN*	*TRP β α PAN*
++++	15	3 CO	1 CO
−+++	84	DCO	DCO
+++−	23	DCO	DCO
−++−	87	1 CO	3 CO

Both crosses indicate that the sequence is *trp α β pan*, but these results are, on the surface, confusing. We see that a double-crossover event does not lead to reciprocal results and, in fact, one double-crossover product occurs as frequently as the single-crossover product. The difficulty is not in the cross but in thinking of the results in terms of a conventional Mendelian cross rather than in terms of gene conversion. Double crossovers are not occurring in the Mendelian sense. In both crosses, "crossing-over" between *β* and the *pan* allele is occurring at a much higher frequency than expected, which means that *β* is being converted at a higher level than is *α*. By convention, that means the polarity runs from *pan* toward *trp*. The asymmetry due to polarity is also seen in the *trp* +:+ *pan* ratios in each cross.

11. Rewrite the original cross:

P $A \, x \, y^+ \times a \, x^+ \, y$

The progeny of parental genotypes will be like either of the two parents. The backcrosses are as follows, with the prime indicating progeny generation.

Cross 1: $a' \, x^+ y \times A \, x \, y^+ \longrightarrow 10^{-5}$ prototrophs

Cross 2a: $A' \, x \, y^+ \times a \, x^+ y \longrightarrow 10^{-5}$ prototrophs

Cross 2b: $A' \, x \, y^+ \times a \, x^+ y \longrightarrow 10^{-2}$ prototrophs

Recombination is allowing for the higher rate of appearance of prototrophs. Cross 2 is obviously a backcross for some gene affecting the rate of recombination. Whatever that gene is, the allele in the *A* parent blocks recombination (cross 1 and cross 2a), and the allele in the *a* parent allows recombination (cross 2b). It is unlinked to the *his* gene, since cross 2 yields results in a 1:1 ratio. The allele that blocks recombination (in A) is dominant, while the allele that allows recombination is recessive. This is demonstrated by the original cross, in which prototrophs occurred at the lower rate, and by cross 1.

To test this interpretation, one-fourth of the crosses between the A $x\ y^+$ and $a\ x^+\ y$ progeny should yield a high rate of recombination and therefore have a high frequency of prototrophs.

Mechanisms of Genetic Change III: Transposable Genetic Elements

► IMPORTANT TERMS AND CONCEPTS

Transposable genetic elements exist in both prokaryotes and eukaryotes. They move from location to location within the genome. Both the DNA sequence from which they are removed and the DNA sequence into which they are inserted are altered, potentially causing phenotypic changes that are scored as mutations.

Prokaryotic **insertion sequences**, termed **IS elements**, usually block the expression of all genes downstream in the operon from the site of insertion. These are **polar mutations.**

Prokaryotic **transposons** possess **inverted repeat (IR) sequences**, which are IS elements that flank one or more genes. If they are located on a plasmid, they can be passed during conjugation from organism to organism within a species or between closely related species. The transposon can move from plasmid to plasmid or between a plasmid and a bacterial chromosome. Transposition can be **replicative** or **conservative.**

There are several types of eukaryotic transposable genetic elements: the yeast **Ty elements**; the *Drosophila copia*-like elements, **fold-back elements** (**FB**), and **P elements**; and the maize **controlling elements.**

Retroviruses are RNA viruses that integrate into host chromosomes as a DNA copy of the viral genome made using **reverse transcriptase.** When integrated, they are termed **proviruses.** They possess long terminal repeats like those of prokaryotic transposons.

In eukaryotes the transposable elements use an RNA intermediate, while in prokaryotes transposition occurs at the DNA level.

Questions for Concept Map 21-1

1. How do transposable genetic elements use recombination?

2. How does a transposable genetic element cause gene mutation?

3. Are true revertants formed from a mutation caused by transposable genetic elements?

4. How do transposable genetic elements cause chromosomal rearrangements?

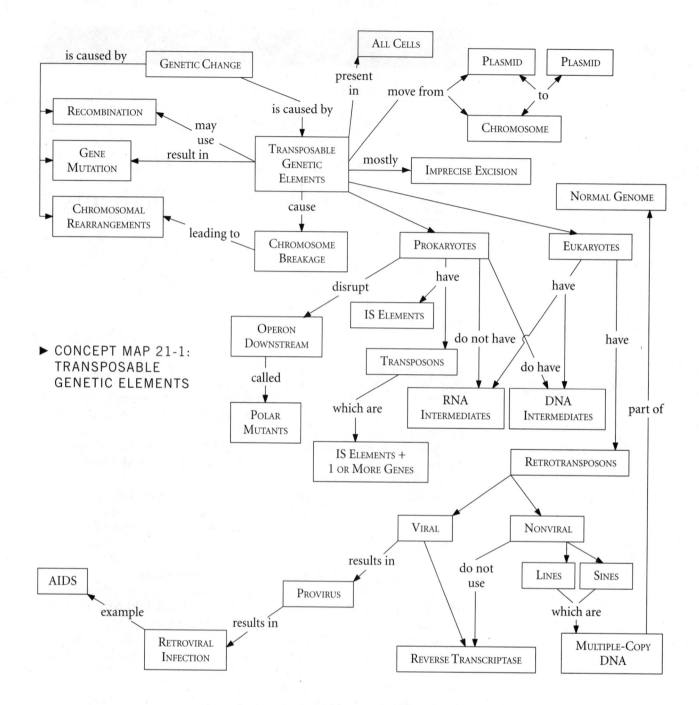

► CONCEPT MAP 21-1:
TRANSPOSABLE
GENETIC ELEMENTS

5. Are transposable genetic elements present in all cells?

6. To what site can transposable genetic elements move?

7. How do transposable genetic elements compare with episomes with regard to excision?

8. Can an episome have a transposable genetic element inserted?

9. What is a polar mutant?

10. How can transposable genetic elements pass from species to species?

11. What is the implication of the finding that nonviral retrotransposons are a part of the normal genome of eukaryotes?

12. What enzyme is used by viral retrotransposons for integration?

13. What evidence indicates that transposable elements in prokaryotes do not literally jump from one location to another?

14. A retrovirus has 27 percent G. When it is in the provirus state, what will be its percentage of G?

15. Consider the following genetic units: episome, lysogenic virus, plasmid, retrovirus, and lytic virus. Arrange them on a continuum and justify the arrangement. The answer to this question requires a creative approach.

16. The virus causing AIDS is a retrovirus. Suggest a way that it could be detected in humans.

17. Some retroviral infections are associated with a high rate of malignancy. List several mechanisms that could account for this.

Answers to Concept Map 21-1 Questions

1. Transposable genetic elements use recombination for integration.

2. A transposable genetic element causes gene mutation by integrating within a gene and disrupting its function.

3. True revertants are formed from a mutation caused by transposable genetic elements because, upon excision, the gene sequence is restored if excision is precise.

4. Transposable genetic elements cause chromosomal rearrangements by causing chromosome breakage.

5. Transposable genetic elements are present in all cells.

6. Transposable genetic elements move from plasmid to plasmid, or from plasmid to chromosome, or from chromosome to plasmid.

7. Transposable genetic elements have imprecise excision at a higher rate than episomes.

8. An episome like the F factor can have a transposable genetic element inserted.

9. A polar mutant occurs in prokaryotes. These mutants have a transposable genetic element inserted within an operon, and all genes downstream from the insertion are disrupted.

10. Transposable genetic elements pass from species to species through transformation.

11. The fact that nonviral retrotransposons are a part of the normal genome of eukaryotes implies, at a minimum, that mutations will occur and that malignancy is "built into" eukaryotes because of the potential disruption of genes involved in regulating the cell cycle.

12. Viral retrotransposons use reverse transcriptase for integration.

13. If transposable elements did jump, there would be a correlation between loss at one location and gain at another. This is not observed. Also, intermediate cointegrate structures have been observed during transposition of some elements.

14. There is no way to predict the percentage of G in the DNA copy of the retrovirus, because a retrovirus is single-stranded and the percentage of C is not known.

15. Several different schemes could be proposed. One follows.

plasmid	an episome that has lost the ability to integrate
episome	a virus that has lost the ability to exist extracellularly
lytic virus	a virus that has lost the ability to undergo lysogeny
retrovirus	a virus that has lost the ability to cause lysis
lysogenic virus	a complete virus

16. Restriction analysis should be able to detect the AIDS virus. Another possibility would be the detection of reverse transcriptase. A third would be the detection of viral antigens.

17. If the virus specifically integrates into cells of the immune system, the functioning of the immune system is impaired. Thus, the immune system is unable to destroy the malignant cells, which are always being formed in multicellular organisms.

    By integrating into specific genes that repair DNA lesions, the virus can eliminate repair functions. Thus, the genome can become more and more altered over time, eventually leading to malignancy.

    The virus may itself carry genes that cause malignancy.

    By causing chromosome breaks, a vital regulatory gene for such processes as control of cell division may be disrupted, leading to malignancy.

▶ SOLUTIONS TO PROBLEMS

Be sure that you have thoroughly read the entire chapter before you attempt any of the problems.

1. Isolate a λ*dgal* phage from the wild type and cross the mutant allele into it. Then determine the density of each phage in a CsCl gradient. If there is an insertion in the mutant phage, the phage will band at a higher density than for the wild type. Also, the phage can be used as a source of DNA for heteroduplex mapping, hybridizing separated strands of the wild type and mutant phage and then reannealing and observing the molecules under the electron microscope. As Figure 21-3 shows, a looped out segment will result at the position of the insertion.

    Another method will be to use restriction enzymes to cut the DNA from both the wild type and the mutant. One can do a Southern blot from each and probe with a sequence from the *gal* region. With some enzymes, particularly those that give large fragments, one should be able to detect a larger fragment in the mutant than in the wild type.

2. Polar mutations affect the transcription or translation of the gene or operon on only one side of the mutant site, usually described as *downstream*. Examples are nonsense mutations, frameshift mutations, and IS-induced mutations.

3. In replicative transposition a new copy of the transposable element is generated during transposition. The experiment by Ljungquist and Bukhari first demonstrated this by using restriction enzyme digests to

show that new mu-bacterial chromosome junctions appeared during mu transposition, while the presence of the original mu could still be detected .

In conservative transposition no replication occurs. The element is excised from its location and integrated into the new site. Kleckner and co-workers demonstrated this by constructing heteroduplexes of a transposon carrying different alleles of *lacZ* and using them to transpose. The resulting cells formed sectored colonies, indicating that they received heteroduplex DNA.

4. R plasmids are the main carriers of drug resistance. They acquire these genes by transposition of drug-resistance genes located between IR (inverted repeat) sequences. Once in a plasmid, the transposon carrying drug resistance can be transferred upon conjugation if it stays in the R plasmid, or it can insert into the host chromosome.

5. Boeke, Garfinkel, Fink, and their co-workers demonstrated that transposition of the Ty element in yeast involved an RNA intermediate. They constructed a plasmid using the Ty element. It had a promoter near the end of the element that could be activated by galactose, and it had an intron inserted into the Ty transposon–coding region. After transposition, they found that the new transposon lacked the intron sequence. Because intron splicing occurs only in RNA, there must have been an RNA intermediate.

6. P elements are transposons (genes flanked by inverted repeats, allowing for great mobility). Because they are transposons, they can insert into chromosomes. By inserting specific DNA between the inverted repeats of the P elements and injecting the altered transposons into cells, a high frequency of gene transfer will occur.

7. The a_1a_1 *DtDt* plant has the *A* (target gene) inactivated by the insertion of a receptor element into it. The regulator is at the *Dt* locus. When crossed with an a_1a_1 *dtdt* tester, the progeny will be a_1a_1 *Dtdt*, dotted.

Excision of the receptor element restores the *A* gene function.

Black kernels would arise if the receptor element excises from the chromosome prior to fertilization. Dotted kernels arise from sporadic excision at a later stage in development of the kernel.

8. The best explanation is that the mutation is due to an insertion of a transposable element.

9. The sn^+ patches in an *sn* background and the occurrence of sn^+ progeny from an *sn* × *sn* mating mean that the sn^+ allele is appearing at a fairly high frequency. The *sn* allele is unstable, suggesting that an insertion element in the sn^+ gene results in *sn*.

10. a. The expression of the tumor is blocked in plant B. This suggests either that plant B can suppress the functioning of the plasmid that causes the tumor, or that plant A provides something to the tissue with the tumor-causing plasmid that plant B does not provide.

    b. Tissue carrying the plasmid, when grafted to plant B, appears normal, but the graft produces tumor cells in synthetic medium. This indicates that the plasmid sequences are present and capable of functioning in the right environment. However, the production of normal type A plants from seeds from the graft suggests a permanent loss of the plasmid during meiosis.

11. *Cross 1:*

P $C/c^{Ds}\ Ac/Ac^{+} \times c/c\ Ac^{+}/Ac^{+}$

F_1 1 $C/c\ Ac/Ac^{+}$ (solid pigment)

   1 $C/c\ Ac^{+}/Ac^{+}$ (solid pigment)

   1 $c^{Ds}/c\ Ac/Ac^{+}$ (unstable colorless or spotted)

   1 $c^{Ds}/c\ Ac^{+}/Ac^{+}$ (colorless)

Cross 2:

P $C/c^{Ac} \times c/c$

F_1 1 C/c (solid pigment)

   1 c/c^{Ac} (spotted)

Cross 3:

P $C/c^{Ds}\ Ac/Ac^{+} \times C/c^{Ac}\ Ac^{+}/Ac^{+}$

F_1 1 $C/C\ Ac/Ac^{+}$ (solid pigment)

   1 $C/c^{Ac}\ Ac/Ac^{+}$ (solid pigment)

   1 $C/C\ Ac^{+}/Ac^{+}$ (solid pigment)

   1 $C/c^{Ac}\ Ac^{+}/Ac^{+}$ (solid pigment)

   1 $C/c^{Ds}\ Ac^{+}/Ac^{+}$ (solid pigment)

   1 $C/c^{Ds}\ Ac^{+}/Ac$ (solid pigment)

   1 $c^{Ds}/c^{Ac}\ Ac^{+}/Ac^{+}$ (spotted)

   1 $c^{Ds}/c^{Ac}\ Ac^{+}/Ac$ (spotted)

The Extranuclear Genome

Extranuclear genes exist in eukaryotes. These genes are located in organelles. They do not show Mendelian patterns of inheritance.

Organelles such as **chloroplasts** and **mitochondria** contain circular DNA. The phenotypes coded for by organelle DNA generally show a **maternal inheritance** pattern. **Segregation** is commonly seen among the progeny when two or more genetically different chloroplast or mitochondrial DNAs exist in the mother. **Recombination** and **extranuclear mutation** exist in organelle DNA. Organelle DNA can be mapped.

Most organelle-encoded polypeptides unite with nucleus-encoded polypeptides to form active proteins. These proteins function in the organelle.

Maternal inheritance does not always indicate extranuclear inheritance. For some characteristics, the maternal nuclear genome determines a progeny phenotype. This is called **maternal effect.**

Questions for Concept Map 22-1

1. What are the locations of the complete genome?

2. Do extranuclear genes follow the rules of Mendel?

3. If not, what types of inheritance patterns do extranuclear genes follow?

4. Can these genes be mapped using standard methods of mapping?

5. What are the locations of the extranuclear genes?

6. What is the function of the mitochondria?

7. What is the function of chloroplasts?

8. What types of cells have these two organelles?

9. What is uniparental inheritance?

10. What is maternal inheritance?

11. In humans, what besides the male pronucleus enters the egg from the sperm?

12. What is cytoplasmic segregation?

13. In protein synthesis, do mitochondria and chloroplasts use cytoplasmic ribosomes?

▶ CONCEPT MAP 22-1:
 EXTRANUCLEAR
 INHERITANCE

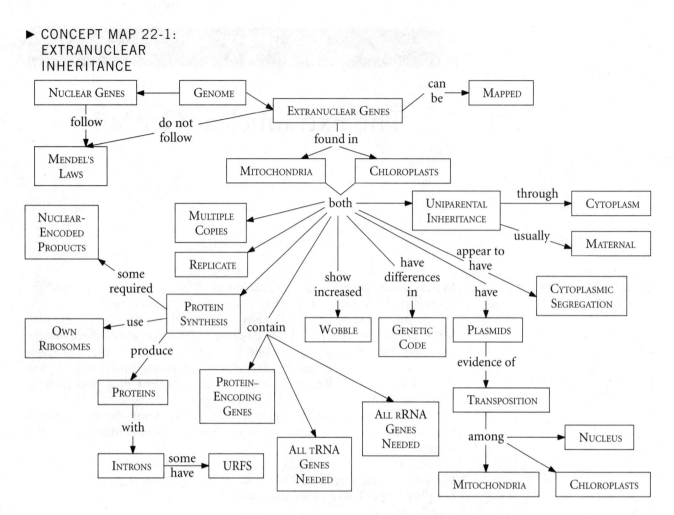

14. What are the sources of the proteins in the ribosomes used for protein synthesis within mitochondria and chloroplasts?

15. What is the source of the rRNA in the ribosomes used for protein synthesis within mitochondria and chloroplasts?

16. What is the source of the tRNAs used for protein synthesis within mitochondria and chloroplasts?

17. Are exons found in mitochondrial genes?

18. Are plasmids found in mitochondria?

19. Is the genetic code truly universal?

20. How can mitochondria have fewer tRNAs than the nucleus?

21. Is there evidence of transposition between mitochondria and the nucleus?

22. What are Mendel's laws?

23. How do mitochondria and chloroplasts reproduce?

Answers to Concept Map 22-1 Questions

1. The complete genome is found in the nucleus, mitochondria, and chloroplasts.

2. The extranuclear genes do not follow the rules of Mendel.

3. The extranuclear genes follow uniparental inheritance and cytoplasmic segregation.

4. These genes can be mapped using standard methods of mapping, including intragenic crossing-over.

5. The extranuclear genes are found in mitochondria and chloroplasts.

6. The mitochondria are involved with electron transport as part of ATP synthesis.

7. Chloroplasts use photosynthesis to capture energy through electron transport.

8. Cells that carry out the reactions involved in ATP synthesis and photosynthesis include prokaryotes and eukaryotes. However, only eukaryotes have both organelles.

9. Uniparental inheritance is the phenomenon whereby only one parent contributes genetic information regarding a trait.

10. Maternal inheritance is uniparental inheritance in which the mother is the source of genetic material for a trait.

11. In humans, only the centrioles and the male pronucleus enter the egg from the sperm.

12. Cytoplasmic segregation is the phenomenon whereby a heterogeneous mix of cytoplasmic organelles from two or more sources ultimately becomes homogeneous through segregation of like organelles during cell division.

13. In protein synthesis, mitochondria and chloroplasts use ribosomes that are found in each of these structures.

14. The proteins in the ribosomes used for protein synthesis within mitochondria and chloroplasts come from both the organelles and the nucleus.

15. The source of the rRNA in the ribosomes used for protein synthesis within mitochondria and chloroplasts is the organelle itself.

16. The source of the tRNAs used for protein synthesis within mitochondria and chloroplasts is the organelle itself.

17. Exons and introns are found in mitochondrial protein-encoding genes.

18. Plasmids are found in mitochondria.

19. The genetic code is not truly universal, because there are some codon differences between the nucleus and mitochondria.

20. Mitochondria can have fewer tRNAs than the nucleus because wobble is increased within them.

21. There is evidence of transposition between mitochondria and the nucleus.

22. Mendel's laws are the law of segregation of alleles and the law of independent assortment of genes.

23. Mitochondria and chloroplasts reproduce by fission.

Be sure that you have thoroughly read the entire chapter before you attempt any of the problems.

1. Most organelle-encoded polypeptides unite with nucleus-encoded polypeptides to produce active proteins, and these active proteins function in the organelle.

2. Reciprocal crosses reveal cytoplasmic inheritance. Cytoplasmic inheritance also can be demonstrated by doing a series of backcrosses, using hybrid females in each case, so that the nuclear genes of one strain are functioning in cytoplasm from the second strain. A heterokaryon test will also demonstrate cytoplasmic inheritance.

3. Maternal inheritance of chloroplasts results in the green-white color variegation observed in *Mirabilis*.

   Cross 1:   variegated female × green male $\longrightarrow$ variegated progeny

   Cross 2:   green female × variegated male $\longrightarrow$ green progeny

4. Stop-start growth for the first cross; normal growth for the reciprocal cross.

5. Both yeast parents contribute mitochondria to the cytoplasm of the resulting diploid cell. Subsequent meiosis shows uniparental inheritance for mitochondria. Therefore, 4:0 and 0:4 asci will be seen.

6. a. The ant^R gene may be mitochondrial.

   b. Some of the petites have a deleted ant^R gene. Other petites, in which the ant^R gene was retained, had a deletion in a different region.

7. The genetic determinants of R and S are cytoplasmic and are showing maternal inheritance.

8. The cpDNA from mt^- *Chlamydomonas* is lost. The results of the crosses are

   Cross 1:   all morph 1; 2-kb and 3-kb bands

   Cross 2:   all morph 2; 3-kb and 5-kb bands

9. Both yeast parents contribute mitochondria to the cytoplasm of the resulting diploid cell. Because of cytoplasmic segregation, the asci will be of four types:

   $oli^R\ cap^R$

   $oli^S\ cap^S$

   $oli^R\ cap^S$

   $oli^S\ cap^R$

10. Both crosses show maternal inheritance of a chloroplast gene. The rare variegated phenotype is probably due to a minor male contribution to the zygote. Variegation must result from a mixture of normal and prazinizan chloroplasts.

11. If the mutation is in the chloroplast, reciprocal crosses will give different results, while if it is in the nucleus and dominant, reciprocal crosses will give the same results.

12. This pattern is observed when a maternal recessive nuclear gene determines phenotype (called *maternal effect*). The crosses are

P   *dd* dwarf female × *DD* normal male

F$_1$ *Dd* dwarf (all dwarf because mother is *dd*)

F$_2$ 3/4 *D*–:1/4 *dd* normal (all normal because mother is *Dd*)

F$_3$ 3/4 normal (mother is *D*–):1/4 dwarf (mother is *dd*)

13. After the initial hybridization, a series of backcrosses using pollen from B will result in the desired combination of cytoplasm A and nucleus B. With each cross the female contributes all the cytoplasm and half the nuclear contents, while the male contributes half the nuclear contents.

14. **a.** Maternal inheritance is suggested.

    b. The net result was that a line carrying the nuclear genes of *E. hirsutum* contained the cytoplasm of *E. luteum*. This demonstrated that the very tall progeny were not the result of a hybrid nucleus.

15. Let male sterility be symbolized by MS.

    a. A line that was homozygous for *Rf* and contained the male-sterility factor would result in fertile males. When this line was crossed for two generations with females from a line not carrying the restorer gene, the male-sterility trait would reappear.

    b. The F$_1$ would carry the male-sterility factor and would be heterozygous for the *Rf* gene. Therefore, it would be fertile.

    c. The cross is *Rf rf* MS × *rf rf* (no cytoplasmic transmission). The progeny would be 1/2 *Rf rf* MS (fertile) and 1/2 *rf rf* MS (sterile).

    d. (i) P   *Rf-1 rf-1 Rf-2 rf-2* × *rf-1 rf-1 rf-2 rf-2* MS

           F$_1$ 1/4 *Rf-1 rf-1 Rf-2 rf-2* MS   fertile

                 1/4 *Rf-1 rf-1 rf-2 rf-2* MS   fertile

                 1/4 *rf-1 rf-1 Rf-2 rf-2* MS   fertile

                 1/4 *rf-1 rf-1 rf-2 rf-2* MS   male sterile

        (ii) P   *Rf-1 Rf-1 rf-2 rf-2* × *rf-1 rf-1 rf-2 rf-2* MS

           F$_1$ 100% *Rf-1 rf-1 rf-2 rf-2* MS   fertile

        (iii) P   *Rf-1 rf-1 rf-2 rf-2* × *rf-1 rf-1 rf-2 rf-2* MS

           F$_1$ 1/2 *Rf-1 rf-1 rf-2 rf-2* MS   fertile

                 1/2 *rf-1 rf-1 rf-2 rf-2* MS   male sterile

        (iv) P   *Rf-1 rf-1 Rf-2 Rf-2* × *rf-1 rf-1 rf-2 rf-2* MS

           F$_1$ 1/2 *Rf-1 rf-1 Rf-2 rf-2* MS   fertile

                 1/2 *rf-1 rf-1 Rf-2 rf-2* MS   fertile

16. The suggestion is that the mutants were cytoheterozygotes for streptomycin resistance.

17. Realize that the closer two genes are, the higher the rate of cosegregation. A rough map of the results is as follows:

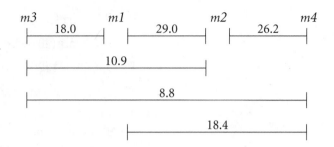

18. The red phenotype in the heterokaryon indicates that the red phenotype is caused by a cytoplasmic organelle allele.

19. (1) Injection of normal mitochondria into wild-type *Neurospora*. This should result in the wild-type phenotype in several generations unless there is an injection effect.

    (2) Injection of buffer solution into wild-type *Neurospora*. This also should result in the wild-type phenotype in several generations unless there is an injection effect.

    (3) In order to be sure that a cytoplasmic factor rather than a nuclear gene is being transferred, all donor material should be derived from organisms with many alleles variant from the wild type.

20. **a., b., and c.** Notice that the results are not reciprocal, which indicates an extranuclear gene. Notice also that in the first cross, there is a 1:1 ratio of phenotypes, indicating a nuclear gene. This may be a case where the same function has a different location in different species. The two genes are incompatible in the hybrid. Let cy = cytoplasmic factor in *N. sitophila*, A^c = the nuclear allele in *N. crassa*, and A = the nuclear allele in *N. sitophila*. Aconidial is then A^c cy. The crosses are

    A $cy \times A^c \longrightarrow$ 1/2 A cy (normal) : 1/2 A^c cy (aconidial)

    $A^c \times A$ (no cy contribution from male) $\longrightarrow$ all normal (A and A^c)

    Neither parent was aconidial, because the sporeless phenotype requires the interaction of a nuclear allele A^c from one species with a cytoplasmic factor (cy) from the other species.

21. Let poky be symbolized by (c). Let the nuclear suppressor of poky be symbolized by n. To do these problems, you cannot simply do the crosses in sequence. For instance, the parental genotypes in cross a must be written taking cross c into consideration.

CROSS	PROGENY
a. (+) + × (c) +	all (+) +
b. (+) n × (c) +	1/2 (+) n : 1/2 (+) +
c. (c) + × (+) +	all (c) +
d. (c) + × (+) n	1/2 (c) + (= D) : 1/2 (c) n (= E)
e. (c) n × (+) n	all (c) n
f. (c) n × (+) +	1/2 (c) n : 1/2 (c) +

22. *Unpacking the Problem*

   a.   mtDNA is mitochondrial DNA.

   b.

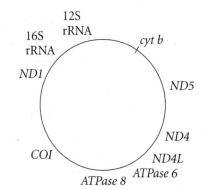

   c.   A cytoplasmic petite is a small, mutant yeast colony that uses fermentation to grow rather than oxidative phosphorylation because of defective mitochondrial DNA. It is "cytoplasmic" because the defect is in the cytoplasmically located mitochondria rather than in a nuclear gene, and it is "petite" because it grows slowly and, in comparison with wild type, produces smaller colonies per unit time.

   d.   At the mitochondrial level there is no protein synthesis because the mtDNA is either completely missing or contains a large deletion.

   e.   The wild-type DNA of mitochondria is circular. Petite DNA is represented as an arc to emphasize that, although circular DNA is found within the mitochondria of petites, most of the information has been lost owing to deletion and that the DNA present consists of tandem repeats of a small segment of the wild-type sequence.

   f.   Compare restriction fragments of a petite with wild-type DNA.

   g.   An *mit⁻* mutant is equivalent to a point mutation. Like petites, the *mit⁻* mutant grows by fermentation, producing small colonies. Unlike petites, they can form true revertants.

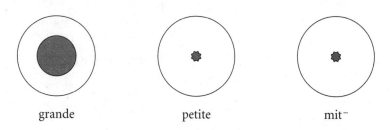

           grande   petite   mit⁻

   h.   Drug susceptibility is frequently due to interference with ribosomal function, specifically the ribosomes within mitochondria in this question. Because the yeast cells grow by fermentation, cytoplasmic ribosomes are needed, not mitochondrial ones. Therefore, drug sensitivity or resistance is a moot point until the ribosomes of the mitochondria have the ability to function, which they do not in the *mit⁻* strains. However, *mit⁻* strains are not drug-resistant.

   i.   *Cytohet* stands for "cytoplasmic heterozygote," or "heteroplasmon," which means that the cytoplasm contains mitochondria from a grande and a petite.

j. In this problem, cytohets were made by fusion of the *mit–* and petite strains. Auxotrophic markers would be very useful to distinguish revertants from cytohets. They actually force the formation of prototrophic cytohets.

k. This question cannot actually be answered from what is presented in the text. Membranes of two separate cells fuse spontaneously to form one cell.

l. The yeast cells bud, and colonies eventually form on growth medium.

m.

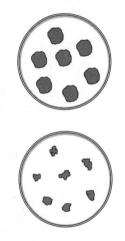

n.

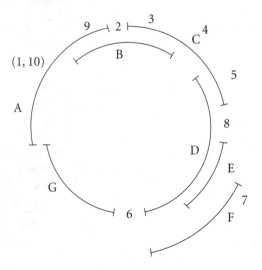

o. The "+" indicates grande growth, and the "–" indicates petite growth. Recombination produces wild-type mtDNA, and this allows grande growth.

p. No; 1 and 10 are the same.

q. If two *mit⁻* mutants show the same behavior in these crosses, they have mutations in the same region of the mitochondrial chromosome.

r. Petites E and F show the same behavior with the *mit⁻* mutants. If E and F can produce the same results with the same *mit⁻* mutant, then the two petites have retained overlapping regions.

Solution to the Problem

23. a. and b. Each meiosis shows uniparental inheritance, suggesting cyto-plasmic inheritance.

    c. Because ant^R is probably mitochondrial and because petites have been shown to result from deletions in the mitochondrial genome, ant^R may be lost in some petites.

24. a. The first tetrad shows a 2:2 pattern, indicating a nuclear gene, while the second tetrad shows a 0:4 pattern, indicating maternal inheritance and a cytoplasmic factor (mitochondrial).

    b. The nuclear gene should always show a 1:1 segregation pattern. The mitochondrial gene could produce an ascus that is all $cyt2^+$.

    c. Both produce proteins involved with the cytochromes. Either the products of the two genes affect different steps in mitochondrial function or they affect the same step if the two proteins interact to form one enzyme.

25. Consider the following hypothetical situation of three genes:

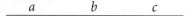

$$a \qquad\qquad b \qquad\qquad c$$

If a deletion of one of them occurred, b by itself would be least likely to be the only gene deleted if the genes were closely linked. The closer a is to b, the less likely it is that a will be deleted if b is retained.

    Applying the above logic to the data, the *apt-cob* pair had the lowest rate of loss (45 total), and the *apt-bar* pair had the highest rate of loss (207 total). This puts *cob* between *apt* and *bar*. The relative rate of loss is

apt-cob:cob-bar, or $45:117 = 1:2.6$.

$$\underline{\qquad apt \qquad\qquad\qquad cob \qquad\qquad\qquad bar \qquad}$$
$$\qquad\quad 1 \qquad\qquad\qquad\qquad\quad 2.6$$

26. Remember that petites arise by deletion. Hybridization with any fragment means that the gene being tested is on that fragment.

    Culture 1:   *cap* and rRNA$_{large}$ are on the same fragment

    Culture 2:   tRNA$_4$ is not next to *cap*, *ery*, *oli*, or *par*

    Culture 3:   tRNA$_2$ and tRNA$_5$ are on the same fragment

    Culture 4:   *oli* and rRNA$_1$ are on the same fragment

    Culture 5:   *ery*, rRNA$_{large}$, and rRNA$_{small}$ are on the same fragment

    Culture 6:   *cap* and tRNA$_3$ are on the same fragment

    Culture 7:   *oli*, *par*, and tRNA$_2$ are on the same fragment

    Culture 8:   rRNA$_{small}$, tRNA$_4$, and tRNA$_5$ are not next to *cap*, *ery*, *oli*, or *par*

    Culture 9:   *ery* and rRNA$_{large}$ are on the same fragment

    Culture 10:   tRNA$_1$ and tRNA$_3$ are on the same fragment

    Culture 11:   *ery* and rRNA$_{large}$ are not on the same fragment that includes all other genes

    Culture 12:   *ery*, rRNA$_{large}$, and rRNA$_{small}$ are on the same fragment

    Once you have identified each fragment, begin arranging them in overlapping order. For example,

Culture 1:   *cap*   rRNA$_{large}$

Culture 9:   rRNA$_{large}$ *ery*

Culture 5:   rRNA$_{large}$ *ery*   rRNA$_{small}$

gives you *cap*–rRNA$_{large}$–*ery*–rRNA$_{small}$. The entire sequence is a circle:

cap–rRNA$_{large}$–*ery*–rRNA$_{small}$–tRNA$_4$–tRNA$_5$–tRNA$_2$–*par*–*oli*–tRNA$_1$–tRNA$_3$–*cap*

27. Some tetrads will show strain-1 type, some will show strain-2 type, and some will be recombinant.

28. **a.** No; during diploid budding all the progeny receive one type of mtDNA. Most likely it is a plasmid or episome.

b. One *Eco*RI site in a circular molecule will produce one linear piece of DNA that migrates to the 2-μm position. With two *Eco*RI sites, two pieces will be produced, and their sum will be 2μm. Therefore, ascospores from the diploid buds will show three bands on a Southern blot.

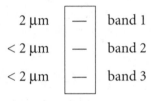

2 μm — band 1

< 2 μm — band 2

< 2 μm — band 3

Bands 2 and 3 total the size of band 1.

29.

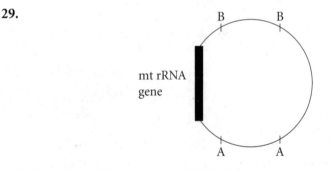

mt rRNA gene

30. First, prepare a restriction map of the mtDNA using various restriction enzymes. Using the assumption of evolutionary conservation, on a Southern blot hybridize equivalent fragments from yeast or another organism in which the genes have already been identified.

31. **a.** The cytoplasm from senescent cultures is a mixture of normal and abnormal mitochondria. The mitochondrial types are distributed in different ratios to different spores. The abnormal mitochondria appear to have a replicative advantage over the normal, since senescence seems, ultimately, to "win out" over normal nonsenescence. The rapidity of the onset of senescence seems to be related to the ratio of normal-to-abnormal mitochondria.

b. The mutation is a 10-kb insertion into fragment 20. It carries bands E and G and splits the original fragment into two fragments, B and C. It has three *Eco*RI sites, and also adds a new *Hin*dIII site. In the accompanying figure, the divided parts of *Hin*dIII fragment 20 are labeled in parentheses.

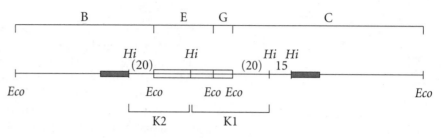

32. a. The plant is a mosaic: one cell line is normal and the other cell line has both cpDNAs. Assume that the plant is self-fertilizing.

Recall endosperm formation from Chapter 3. The endosperm is derived from two identical haploid female nuclei and one haploid male nucleus. In order for four bands to appear on the gel, lane 3, the cytoplasm of the meiocyte giving rise to the seed must have contained a mixture of both cpDNAs. This could occur only if a progenitor cell contained the original mutant and normal cpDNA, and segregation of the two types of chloroplasts did not occur.

In order to get homozygous cpDNA, seen in lanes 1 and 2, segregation of chloroplasts had to occur. Lane 1 is derived from normal cpDNA, which could have come from the normal cell line or the mutant line, through segregation. Lane 2 is derived from mutant cpDNA, through segregation.

b. Both *Gryllus* and *Drosophila* would be expected to have segregation of mitochondria, but it is not being seen in these rare females. Therefore, it must be hypothesized that their mitochondria all contain two genomes, one of which carries the mutation and one of which is normal. Segregation of mitochondria may not occur in these species.

33. The two repeats, center to center, are separated by 83 kb. If the two direct repeats lined up as below and then experienced a recombination event, the two smaller circles would be 83 kb and 135 kb, each containing a single copy of the direct repeat:

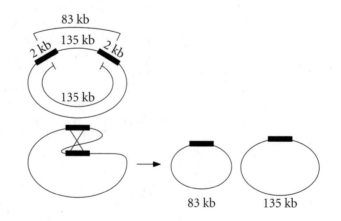

34. a. Both the gametophyte and the sporophyte are closer in shape to the mother than the father. Note that a size increase occurs in each type of cross.

b. Gametophyte and sporophyte morphology are affected by extranuclear factors. Leaf size may be a function of the interplay between nuclear genome contributions.

c. If extranuclear factors are affecting morphology while nuclear factors are affecting leaf size, then repeated backcrosses could be conducted, using the hybrid as the female. This would result in the cytoplasmic information remaining constant while the nuclear information becomes increasingly like that of the backcross parent. Leaf morphology should therefore remain constant while leaf size would decrease toward the size of the backcross parent.

35. a. The complete absence of males is the unusual aspect of this pedigree. In addition, all progeny that mate carry the trait for lack of male offspring. If the male lethality factor were nuclear, the male parent would be expected to alter this pattern. Therefore, cytoplasmic inheritance is suggested.

b. If all females resulted from chance alone, then the probability of this result is $(1/2)^n$, where n = the number of female births. In this case n is 72. Chance is an unlikely explanation for the observations.

The observations can be explained by cytoplasmic factors by assuming that a new mutation in mitochondria is lethal only in males.

Mendelian inheritance cannot explain the observations, because all the fathers would have had to carry the male-lethal mutation in order to observe such a pattern. This would be highly unlikely.

36. a. The pedigree clearly shows maternal inheritance.

b. Most likely, the mutant DNA is mitochondrial.

37. Recall that each cell has many mitochondria, each with numerous genomes. Also recall that cytoplasmic segregation is routinely found in mitochondrial mixtures within the same cell.

The best explanation for this pedigree is that the mother in generation I experienced a mutation in a single cell that was a progenitor of her egg cells (primordial germ cell). By chance alone, the two males with the disorder in the second generation were from egg cells that had experienced a great deal of cytoplasmic segregation prior to fertilization, while the two females in that generation received a mixture.

The abortions that occurred for the first woman in generation II were the result of extensive cytoplasmic segregation in her primordial germ cells: aberrant mitochondria were retained. The abortions of the second woman in generation II also came from such cells. The normal children of this woman were the result of extensive segregation in the opposite direction: normal mitochondria were retained. The affected children of this woman were from egg cells that had undergone less cytoplasmic segregation by the time of fertilization so that they were viable.

23

Gene Regulation during Development

The types and relative amounts of **proteins** in a cell determine the characteristics of that cell. Because both transcription and translation are involved, the **regulation of gene expression** through transcription and translation is ultimately responsible for the differentiated state of each cell. In turn, the coordinated development of each differentiated cell gives rise to the final organism.

The regulation of protein synthesis occurs by a number of processes: **modification of gene structure, modulation of transcription rate, modulation of processing of the primary transcript into mature RNA, modulation of translation rate,** and **modulation of protein modifications.**

Modification of gene structure occurs by a number of processes, including **gene amplification** (histones), **DNA rearrangement** (the immune system), and **methylation** of genes.

Modulation of transcription rate occurs through **tissue-specific enhancers** controlled by **transcription factors** and coordinated by **hormones.**

The modulation of processing of the primary transcript into mature RNA can occur through regulation of the splicing of introns.

Sex determination can occur through environmental, chromosomal, or gene differences in those species that have two sexes. A combination of genetic and environmental control determines sex in some species. In *Drosophila*, the state of sexual differentiation is established and maintained through regulation of transcription and splicing. At the cellular level, each cell is autonomous for sex. In mammals, each cell is not autonomous for sex, and sexual differentiation is established and maintained by the transcription of androgens.

Questions for Concept Map 23-1

1. How does the environment impose limits on a developing organism?

2. How do epigenetic phenomena impose limits on a developing organism?

3. What is the meaning of *epigenetic*?

4. What is parental imprinting?

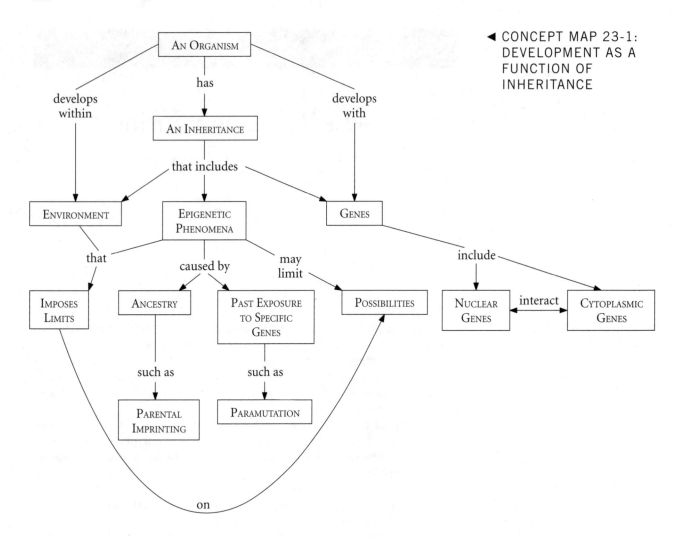

5. What is paramutation?

6. Do epigenetic phenomena alter the DNA sequence?

7. Do epigenetic phenomena modify the genetic sequence?

8. How do genes define the possibilities of a developing organism?

9. How do nuclear and cytoplasmic genes interact?

10. How is X-inactivation involved in parental imprinting?

11. How does the physical development of an organism parallel the social development of an individual human?

Answers to Concept Map 23-1 Questions

1. The environment imposes limits on a developing organism through the nutrition that is available, the possibilities for safe development that exist, the mutagens present that can alter the genetic information and in a host of other ways.

2. Epigenetic phenomena impose limits on a developing organism by restricting the expression of the genetic information that is present.

3. *Epigenetic* is a combination of *epi*, which means "over" in this situation,

and *genetic*. Thus, *epigenetic* means "over genetics," suggesting that something beyond just the genetics of the situation guides development. No mysticism is implied as part of the definition.

4. *Parental imprinting* refers to the changes that exist that mark a chromosome as from a particular source. These changes are not changes in DNA sequence but in expression, and they appear to involve methylation. The phenomenon suggests that the right information must be present in the right amount and from the right source to obtain normal development. Also recall that the information must be in the right location (position effect).

5. Paramutation is the permanent change of one gene by another gene without any change occurring in the DNA sequence.

6. Epigenetic phenomena do not alter the DNA sequence.

7. Epigentic phenomena appear to modify the genetic sequence through methylation.

8. Genes define the possibilities of a developing organism through the information contained within them.

9. Nuclear and cytoplasmic genes interact in that some nuclear proteins are required for the expression of cytoplasmic genes. On another level, cytoplasmic genes ultimately provide the energy needed by nuclear genes for eventual expression.

10. The paternal X is inactivated in the extraembryonic membranes (amnion, chorion) of all mammals. Also, in some mammals, the paternal X is always inactivated in females. These are clear examples of parental imprinting.

11. Individual humans develop within a specific familial, social, and physical environment that affects the expression of the possibilities within each human. The past history of each component in the environment also places limits on the individual. The physical development of an organism occurs within an environment that affects the expression of the genetic possibilities of the organism. The past history of the elements of the environment (including epigenetic history) also places limits on the development of an individual.

Questions for Concept Map 23-2

1. What is a cascade?

2. At what level does regulation occur?

3. What is an autoregulatory loop?

4. What is differentiation?

5. The presence or absence of what gene makes a major difference in mammals?

6. What is site-specific recombination?

7. What is a major difference between *Drosophila* and human sex determination?

8. With on/off switches in the *Drosophila* system of sex determination, what do the proteins that are produced do?

▶ CONCEPT MAP 23-2:
DEVELOPMENT AS A
CASCADE

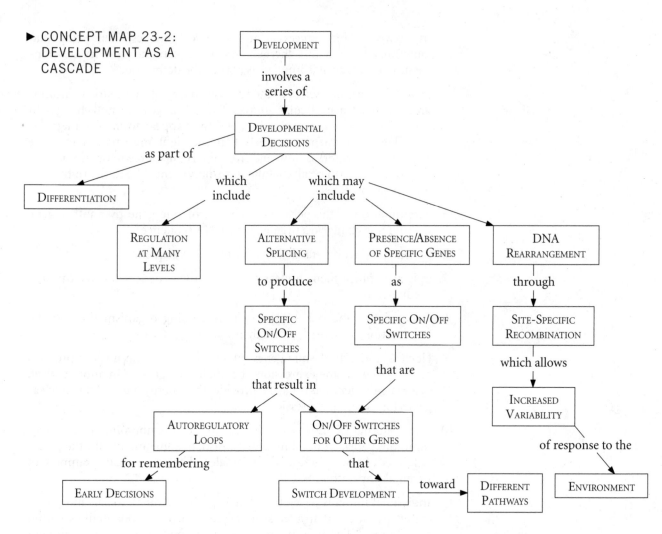

9. How does competition for polypeptide subunits work to determine sex?

10. In *Drosophila*, what happens if the master switch is nonfunctional because of mutation?

11. In *Drosophila*, what happens if the *tra* gene is not functioning?

12. In humans, what happens if the *TDF* gene is not present?

13. In humans, what happens if an androgen receptor is nonfunctional?

14. What is sex reversal in humans?

Answers to Concept Map 23-2 Questions

1. A cascade is a series of events such that each event derives from and leads to another event.

2. Regulation occurs at virtually every level. See Chapters 18 and 22 for a partial review.

3. An autoregulatory loop is a system in which a substance controls its own production.

4. Differentiation is the set of cumulative changes that occur in cells during development.

5. The *TDF*, testes-determining factor, gene makes a major difference in mammals because its presence determines maleness and its absence determines femaleness.

6. Site-specific recombination is recombination that occurs only in regions of homology.

7. In *Drosophila* sex determination is a cell-by-cell process, and in humans sex determination occurs on a global scale because it is caused by circulating hormones.

8. The proteins that are produced alter splicing and are helix-loop-helix binding proteins in the *Drosophila* system of sex determination.

9. Competition for polypeptide subunits functions to determine whether the dimer protein complex is capable of repressing female or male genes.

10. In *Drosophila*, if the master switch is nonfunctional because of mutation, the organism develops along the male line.

11. In *Drosophila*, if the *tra* gene is not functioning, the organism develops along the male line.

12. In humans, if the *TDF* gene is not present, the organism develops along the female line.

13. In humans, if an androgen receptor is nonfunctional, the organism develops along the female line.

14. Sex reversal in humans is the development of a phenotypic male with two X chromosomes. This occurs because improper recombination has occurred between an X and Y chromosome, transferring the *TDF* gene to the X. The Y chromosome, which now lacks that gene, can give rise to a phenotypic female with XY chromosomes.

Questions for Concept Map 23-3

1. Why must germ cell DNA remain unchanged during development?

2. Are liver genes expressed in eye cells? If not, what happens to the liver genes?

3. What is being amplified in "tissue-specific amplification"?

4. Don't the rearrangements of DNA and the amplification of DNA contradict the common understanding that all cells in an organism have the same DNA?

5. What is the role of promoters?

6. What is the role of enhancers?

7. What is the role of silencers?

8. What is meant by "splicing variations?"

9. What is competitive binding of proteins?

10. What is enzymatic modification of proteins?

11. What is the 3' end of mRNA?

12. Can double-stranded RNA exist?

► CONCEPT MAP 23-3: MECHANISMS OF DEVELOPMENTAL CASCADES

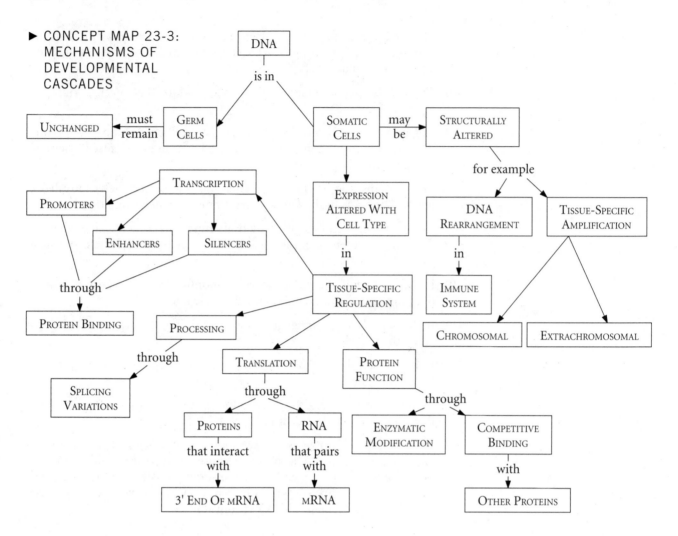

13. Can double-stranded RNA be translated?

14. When genes are introduced into recipient eukaryotic cells through one of several techniques available, how likely is it that these genes will be regulated properly?

15. Muscular dystrophy in mice has been observed. When normal and mutant embryos are fused, wild-type muscle cells enervated by nerve cells from the dystrophic strain have muscular dystrophy. If adult mice from the two strains are surgically linked (parabiosis), wild-type muscle cells enervated by nerve cells from the dystrophic strain are normal. What do these results indicate?

16. How would you select for the following behavioral mutants in *Drosophila*? *Note:* flies taste with their feet.

   a. blindness

   b. inability to fly

   c. abnormal feeding responses

Answers to Concept Map 23-3 Questions

1. Germ cell DNA must remain unchanged during development in order

to ensure that each generation receives the same DNA with regard to amount, organization, and location. Only minor variations, mutations, are tolerated.

2. Liver genes are not expressed in eye cells. They are turned off through many levels of gene regulation, beginning perhaps with chromosome structure (heterochromatin).

3. DNA is being amplified in "tissue-specific amplification."

4. The rearrangements of DNA and the amplification of DNA contradict the common understanding that all cells in an organism have the same DNA, which just goes to show that one should closely evaluate common understandings of anything.

5. Promoters bind transcription factors that facilitate transcription of the downstream gene.

6. Enhancers bind transcription factors that facilitate the binding of RNA polymerase to the downstream gene.

7. Silencers bind transcription factors that block the binding of transcription factors to enhancers.

8. Splicing variations occur as part of regulation, and they shift gene expression into alternative pathways.

9. The competitive binding of proteins can occur as part of regulation to turn on or off a pathway during development.

10. Proteins can be enzymatically modified as part of developmental regulation.

11. The 3′ end of mRNA is the untranslated region downstream from a gene.

12. Double-stranded RNA can and does exist.

13. Double-stranded RNA cannot be translated and is one mechanism used for gene regulation.

14. The answer to this question could go one of several ways, and it might vary with the gene being introduced. It is known that position effect does occur in translocated material. A similar effect would most likely occur for introduced genes. Thus, a gene normally functioning within the context of euchromatin would be expected to function at a lower rate when located within heterochromatin, and vice versa. Genes that normally are regulated in cis are unlikely to be regulated in a normal fashion simply because the chance that they will be located close to their cis regulator in the new cell is low. Genes that normally are regulated by trans-acting factors are more likely to be properly regulated in their new locations.

15. The studies suggest that muscular dystrophy is caused by defects in the nerves that subsequently cause muscular defects. The parabiosis studies show that the muscular defects cannot be induced in adult muscle. Therefore, the effects of the nerves on the muscles must occur during development.

16. a. The wild-type flies are attracted to light. Select flies that do not move toward a light at the end of a tube.

    b. Select for flies that do not fly by placing a poisoned food source high enough in a fly cage so that the flies can only reach it by flying.

Alternatively, place mutagenized flies on a window ledge and scare them. The ones that do not fly away are very brave, are very stupid, or are the mutants that you want.

c. Add a tasteless toxic compound to a sugar solution on filter paper in the bottom of the cage. Those flies that cannot taste sugar will not eat it, and those that can taste sugar will eat it and die.

▶ SOLUTIONS TO
PROBLEMS

Be sure that you have thoroughly read the entire chapter before you attempt any of the problems.

1. a. Recall that in early *Drosophila* development the *Sxl* gene is expressed in females but not in males. That means that female embryos will react to both antibodies but males embryos will not react to either.

   b. In late development, the female produces the full protein, whereas the male produces a vastly truncated version of it that is not functional. As in early development, the female will react to both probes. The male will react to the probe for the amino end only.

2. a. If only one gene coded for antibodies, then the number of different alleles would limit antibody production to two per organism. The multimeric nature of antibodies, combined with DNA rearrangements, allows for a vast number of different antibodies to be produced. Without either aspect, production of different antibodies would be greatly reduced.

   b. The enhancer elements that contribute to antibody production are located near the C, or constant, region because that is one region that will always exist in the antibody gene after rearrangement. With the enhancer near the constant region, transcription will occur only after a successful rearrangment brings a promoter to the vicinity of the enhancer. If the enhancer were located near the V segment farthest from the C segment, then an additional deletion would have to occur in order to bring the enhancer close to the rearranged gene.

3. Sex determination in *Drosophila* is autonomous at the cellular level. The *Sxl* gene is permanently turned on or remains off early in development in response to the concentration of X:A transcription factor. Because the X:A ratio is established by the interaction of gene products made in the ovary and in the early zygote, a chemical gradient would be expected to exist that would be sufficiently high in some cells to result in femaleness but low enough in other cells to result in maleness, making the individual an intersex.

4. The first task is to identify wing-development genes. This can be accomplished by first using a reporter gene to find the enhancer for the wing development gene and then using enhancer trapping. Once this has been done, the transposase gene can be linked to a copy of the enhancer for the gene for wing development and inserted into an egg.

5. a. Allelic exclusion must precede clonal expansion, which ensures a rapid response to the specific antigen. If allelic exclusion did not occur, then a cell that has already rearranged so that it can respond

to one antigen might subsequently rearrange to respond to a second antigen. This would leave the organism with no cell programmed to respond to the first antigen.

  b. Because all combinations can occur, the answer is approximately $300 \times 4 \times 1000 \times 4 \times 12 = 57,600,000$ different antibody molecules.

6. Gain-of-function mutations can result from an inversion that causes a new fused gene, such as the *Tab* mutant. Reversion would have to consist of inactivating the fused gene. The inactivation could occur by a mutation that affects the enhancer, the actual protein-encoding portion, or any transcription factor that interacts with either component. The high rate of reversion is due to the fact that any inactivation of the gene product will revert the phenotype.

7. **a.** The most likely explanation is that the intron has a stop codon for translation.

  b. Assuming that the cDNA-encoded protein is capable of function in somatic cells, it can be fused with an enhancer and then introduced into *Drosophila* eggs.

  c. Because the probe detects all eight fragments, each fragment contains some exon material, and no intron can contain more than one enzyme recognition site.

8. Construct a set of reporter genes with the promoter region, the introns, and the region 3′ to the transcription unit of the gene in question containing different alterations that do not disrupt transcription or processing. Use these reporter genes to make transgenic animals by germ-line transformation. Assay for expression of the reporter gene in various tissues and the kidney of both sexes.

9. Normally, the *tra* gene in the female is active, while in the male it is not active. The active *tra* form of the gene product results in a change in the *dsx* product, shifting development toward the female. If the *dsx* product is not altered, development proceeds along the male line. A mutation in the gene that results in chromosomal females developing as phenotypic, but sterile, males must involve an inactive *tra* product. Homozygotes for the *tra* mutation could be transplanted with male germ cells very early in development, which should result in normal gonad development.

10. In humans, a single copy of the Y chromosome is sufficient to shift development toward normal male phenotype. The extra copy of the X chromosome is simply inactivated. Both mechanisms seem to be all-or-none rather than to be based on concentration levels.

11. Because maleness is based on the presence of androgens produced by the developing testes and femaleness is based on the absence of those androgens, what seems to be crucial here is whether the migrating germ cells organize a testes. Although the determinator is unknown, it may be that a minimal number of XY cells are required to organize a testes. If, in the mosaic, not enough exist, then development will be female. If a sufficient number exist, development will be male.

12. The concentration of *Sxl* is crucial for female development and dispensable for male development. The dominant Sxl^M male-lethal mutations may not actually kill all males but simply produce an excessive amount of gene product so that only females (fertile XX and XY) result. The reversions may eliminate all gene product, resulting in XX (sterile)

and XY males. The reversions would be recessive because, presumably, a single normal copy of the gene may produce enough gene product to "toggle the switch" in development to female.

24

Genetics and Cellular Differentiation

▶ IMPORTANT TERMS
AND CONCEPTS

The **cytoskeleton** is responsible for cellular shape; the mechanical properties of a cell; the movement of chromosomes, organelles, and molecules through the cell; the location of the cleavage plate; and morphogenesis. Three proteins form the cytoskeleton. The **microtubules** are composed of **tubulin,** the **microfilaments** are composed of **actin,** and the **intermediate filaments** are composed of **vimentin.**

Cell division, and more generally the cell cycle, is very closely regulated by a series of positive and negative regulators. Each step in the cascade leading to cell division involves the functioning of a heterodimer composed of **cyclin** and a **cyclin-dependent kinase (CDK).** The complex phosphorylates specific amino acids, thereby activating transcription factors for the next step. There are also negative regulators, most of which detect DNA damage. One, **p53,** detects DNA mismatches and activates **p21,** which then inhibits the cyclin-CDK complex. **Cell division cycle (cdc) mutants** have been linked to the development of cancer.

Intercellular communication is necessary for coordinate development and function in multicellular organisms. This communication is carried out primarily by chemicals. A **ligand** is released from a signaling cell. It binds to a **receptor** on the target cell and either activates or inhibits transcription. Ligands include **proteins, steroid hormones,** and some **vitamins.** They may bind to transmembrane proteins, receptors within the cytoplasm, or receptors within the nucleus. Some bind directly to DNA. Most induce a **signal transduction cascade** that consists of phosphorylation of transcription factors, thereby activating them.

Cancer, defined as unregulated growth by cells that have uncoupled from the normal regulatory mechanisms of cell proliferation, results from multiple mutations within a single cell. The mutations occur in two classes of genes: dominant mutations in **proto-oncogenes** and recessive mutations in **tumor suppressor genes.** Proto-oncogenes are positive regulators of cell proliferation. Tumor suppressor genes are negative regulators of cell proliferation.

Oncogenes are genes that are also found in tumor viruses that cause cancer in multicellular organisms. The oncogenes are mutated cellular **proto-oncogenes** that entered the viruses through improper excision. Many tumors of the same type possess the same oncogene. Different types of tumors contain different oncogenes.

The normal cellular functions of the proto-oncogenes involve transcription factors, parts of the system that convert signals received by a cell into a change in gene activity, and parts of the system that involve communication between cells. Dominant oncogenic mutations activate the proto-oncogene. They result in a protein's being expressed on a continuous basis, at too high a level, or in the wrong tissue. Recessive tumor suppressor mutations inactivate the inhibiting function of the gene.

▶ CONCEPT MAP 24-1:
   THE CYTOSKELETON

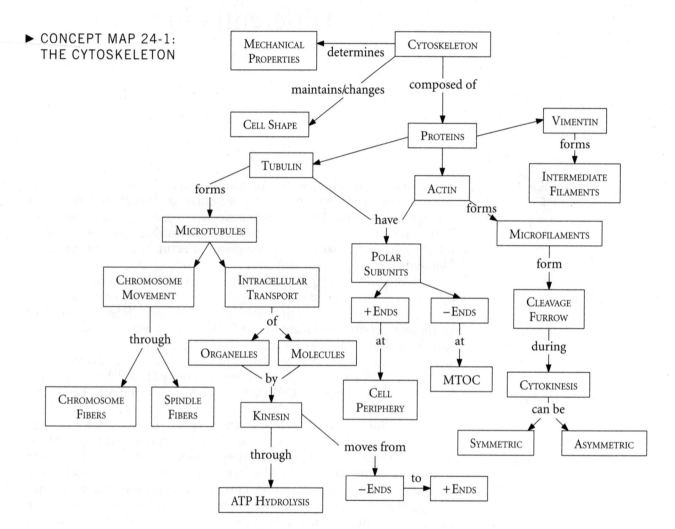

Questions for Concept Map 24-1

1. What is the cytoskeleton?

2. Which proteins are part of the cytoskeleton?

3. Which filament does each type of protein form?

4. Which filaments in the cytoskeleton are known to have polar subunits?

5. Which filament type moves chromosomes?

6. Which filament type helps form the cleavage furrow?

7. What is the cleavage furrow?

8. What is the MTOC?

9. Which protein actually moves organelles and molecules along micro-tubules as part of intracellular transport?

10. How does the protein in Question 9 obtain energy for the movement?

11. What is the role of actin in the cell?

12. What is vimentin and what does it do?

13. What chemical disrupts microtubules?

14. What chemical disrupts microfilaments?

Answers to Concept Map 24-1 Questions

1. The cytoskeleton is the protein matrix of a cell that determines the cell's mechanical properties and shape and moves organelles and molecules through the cytoplasm. It is also responsible for chromosome movement and location of cytokinesis within the cell.

2. Tubulin, actin, and vimentin are the proteins of the cytoskeleton.

3. Tubulin forms microtubules, actin forms microfilaments, and vimentin forms the intermediate filaments.

4. The microtubules and the microfilaments have polar subunits.

5. Microtubules move chromosomes.

6. Microfilaments form the cleavage furrow.

7. The cleavage furrow is the midpoint of the metaphase plate where cytokinesis occurs following telophase.

8. The MTOC is the microtubule organizing center.

9. Kinesin actually moves organelles and molecules along microtubules as part of intracellular transport.

10. Kinesin obtains energy for the movement through the hydrolysis of ATP.

11. Actin is involved in forming the cleavage furrow and in maintaining cell shape and internal organization.

12. Vimentin is a protein of the cytoskeleton that forms the intermediate filaments.

13. Colchicine, or its less toxic derivative colcemid, disrupts microtubules.

14. Cytochalasin disrupts microfilaments.

Questions for Concept Map 24-2

1. What are the four major divisions of the cell cycle?

2. Does each division have a series of stages?

3. How does positive regulation of the cell cycle occur?

4. How does negative regulation of the cell cycle occur?

5. What is the role of p53?

6. What is the role of p21?

► CONCEPT MAP 24-2:
CONTROL OF THE
CELL CYCLE

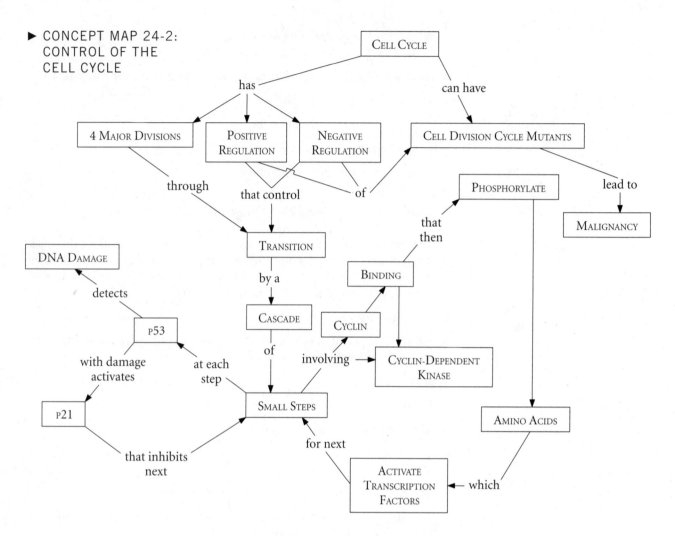

7. What does CDK protein do to cyclin?

8. What does cyclin do for CDK protein?

9. How are transcription factors activated?

10. How are transcription factors inhibited?

11. What would happen to a cell that lost its receptor for a necessary signal to inhibit transcription?

12. What would happen to a cell that lost its receptor for a necessary signal to activate transcription?

Answers to Concept Map 24-2 Questions

1. The four major divisions of the cell cycle are mitosis, G_1, S, and G_2.

2. Each major division has a series of stages.

3. Recall that positive regulation is the activation of transcription through protein binding. In cell cycle regulation, transcription factors are activated when phosphorylated by a cyclin–cyclin-dependent kinase complex.

4. Recall that negative regulation is the inhibition of transcription by bind-

ing of a protein. In cell cycle regulation, DNA damage triggers p53 to activate p21, which inhibits the cyclin–cyclin-dependent kinase complex.

5. The p53 detects mismatches caused by DNA damage and activates p21.

6. The p21 blocks the cyclin-CDK complex from phosphorylating a transcription factor.

7. CDK protein binds to cyclin and autophosphorylates the complex.

8. Cyclin binds a transcription factor so that the CDK protein can phosphorylate one of its amino acids.

9. Transcription factors are activated by phosphorylation of one or more amino acids.

10. Transcription factors are not inhibited. They are always inactive unless phosphorylated.

11. If a cell lost its receptor for a necessary signal to inhibit transcription, then the cell would always be "on." It would have uncontrolled proliferation.

12. If a a cell lost its receptor for a necessary signal to activate transcription, then the cell could never be activated to divide.

▶ CONCEPT MAP 24-3: MULTICELLULAR ORGANISMS

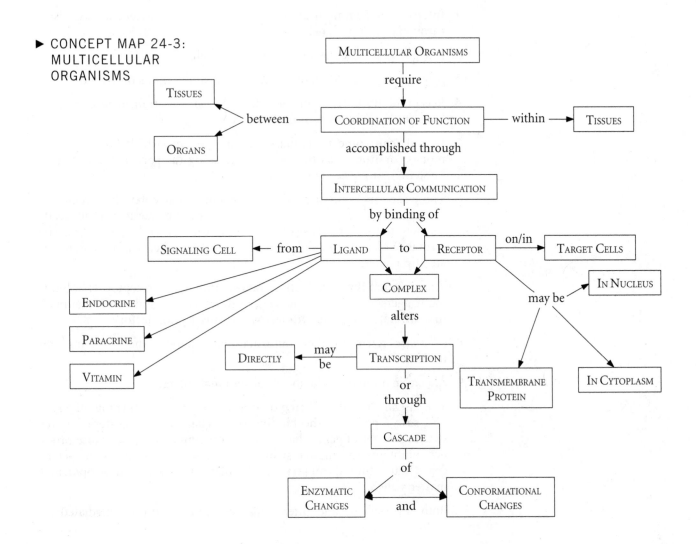

Questions for Concept Map 24-3

1. How is intercellular communication carried out in multicellular organisms?

2. What is a ligand?

3. What is a receptor?

4. Where are receptors located?

5. What are the sources of ligands?

6. Distinguish between *endocrine* and *paracrine.*

7. How is transcription altered?

8. What does *allosteric* mean?

9. What undergoes a conformational change?

10. What processes are part of the cascade?

11. How does intracellular communication differ from intercellular communication?

Answers to Concept Map 24-3 Questions

1. Intercellular communication is carried out in multicellular organisms through chemical signals.

2. A ligand is the chemical from a signaling cell.

3. A receptor is located on or in a target cell, and it binds the ligand.

4. Receptors are located on the cell membrane, as transmembrane proteins, in the cytoplasm, or in the nucleus.

5. Ligands can come from the endocrine system (which produces hormones) or from cells that are very close by (as part of the paracrine "system"), or they can be vitamins.

6. The endocrine system produces hormones that enter the circulatory system and affect quite remote cells. The paracrine "system" is not really a system in the same sense of the word as used for *endocrine system.* The term is used to denote the effect of one cell on other nearby cells, mediated by a ligand.

7. Transcription is altered by the ligand's binding to a receptor. That binding triggers a cascade of changes, both enzymatic and conformational, that ultimately activates or inactivates transcription factors.

8. *Allosteric* refers to the conformational changes that proteins can undergo.

9. Allosteric proteins undergo conformational change.

10. The cascade includes turning on one set of genes and turning off a second set of genes by the binding of a ligand to an allosteric master switch. The set of genes that is turned on contains a gene whose product maintains the master switch setting. Another gene in the set of genes that is turned can serve as a regulator for the next developmental master switch.

11. Both intracellular and intercellular communication are mediated by

allosteric proteins that, through conformational and enzymatic changes, regulate transcription within the cell. They differ only in the source of the initiating signal and the fact that in intercellular communication the signal must enter the cell in order to alter transcription.

▶ CONCEPT MAP 24-4:
CANCER AS A GENETIC
DISEASE

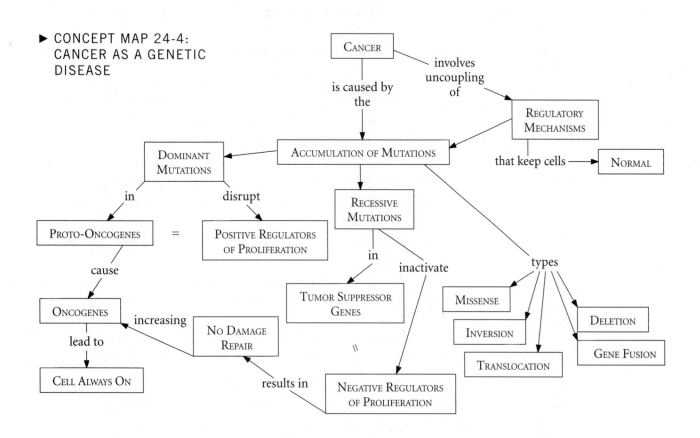

Questions for Concept Map 24-4

1. What is cancer?
2. What causes cancer?
3. What gene types are always involved in cancer?
4. In what genes do dominant mutations occur to produce cancer?
5. In what genes do recessive mutations occur to produce cancer?
6. What is a proto-oncogene?
7. What is an oncogene?
8. What types of genes are positive regulators of cell proliferation?
9. What is a positive regulator?
10. Give an example of how a positive regulator can be altered to produce cancer.
11. What types of genes accumulate recessive mutations in order to produce cancer?

12. What are the negative regulators of cell proliferation?

13. What is negative regulation?

14. How can a recessive germ-line mutation result in a dominant disease?

15. What types of damage to DNA can cause cancer?

16. What is one tumor-suppressing gene, and how does it function?

Answers to Concept Map 24-4 Questions

1. Cancer is an accumulation of mutations within a somatic cell that leads to an uncoupling of the regulatory mechanisms that keep cells normal.

2. Cancer is caused by mutations in specific genes that either regulate the cell cycle or repair damage.

3. The gene types that are always involved in cancer are cell cycle regulators or tumor suppressors.

4. Dominant mutations occur in genes that positively regulate cell cycle proliferation to produce cancer.

5. Recessive mutations occur in genes that are tumor suppressors to produce cancer.

6. A proto-oncogene is a gene sequence that acts as a positive regulator of the cell cycle. It is involved with cyclin, CDK proteins, and their interactions.

7. An oncogene is a mutation located in a proto-oncogene that results in a cell's always being turned on with regard to proliferation.

8. Proto-oncogenes are positive regulators of cell proliferation.

9. A positive regulator activates transcription through binding a protein.

10. If a positive regulator undergoes a mutation that permanently alters its configuration so that it no longer responds to p21, for example, the cell will always be turned on with regard to proliferation despite DNA damage that is unrepaired. This in turn will lead to further mutations.

11. Tumor-suppressing genes accumulate recessive mutations in order to produce cancer. These genes are normally negative regulators of cell proliferation.

12. The negative regulators of cell proliferation are tumor-suppressing genes.

13. Negative regulation is the turning off of transcription when binding a protein.

14. A recessive germ-line mutation can result in a dominant disease because any cell of a tissue that experiences a mutation in the same gene will result in a loss of function (of a negative regulator). Because the mutation rate is high and there are a large number of cells, almost always that second mutation occurs in one of the cells to produce the disease.

15. The types of damage to DNA that can cause cancer include missense mutations, nonsense mutations, deletions, duplications, and chromosome rearrangement.

16. One tumor-suppressing gene is p53. It detects mismatches and activates p21, which in turn inhibits the cyclin-CDK complex from phosphorylating an amino acid to activate a transcription factor.

Be sure that you have thoroughly read the entire chapter before you attempt any of the problems.

1. a. If the number of copies of a transcription factor is increased, transcription will occur at a higher rate. That could lead to an acceleration of cellular growth, which would be cancer.

   b. A nonsense mutation would lead to a decrease of the normal protein product. If that protein were part of the receptor for a growth factor, which stimulates cell proliferation, then cell division could not be triggered.

   c. If the mutant protein blocks the binding action of an inhibitory protein, then transcription would be increased. An increase in transcription rate could lead to unregulated growth, which is cancer.

   d. Cytoplasmic tyrosine-specific protein kinase phosphorylates proteins in response to signals received by the cell. The phosphorylations lead to activation of the transcription factors for the next step in the cell cycle. If the active site is disrupted, then phosphorylations will not occur and transcription factors for the next step will not be activated.

2. All cells of the tumor should express the same alleles of all the X-linked genes that are heterozygous within the organism and not express the alternative set of alleles.

3. a. There must be a diffusible substance produced by the anchor cell that affects development of the six cells. The 1° has the strongest response to the substance, and the 3° represents a lack of response due to a low concentration or absence of the diffusible substance.

   b. Remove the anchor cell and the six equivalent cells. Arrange the six cells in a circle around the anchor cell. All six cells will develop the same phenotype, which will depend on the distance from the anchor cell.

4. a. The results suggest that ABa and ABp are not determined at this point in their development. Also, future determination and differentiation of these cells are dependent on their position within the developing organism.

   b. Because an absence of EMS cells leads to a lack of determination and differentiation of AB cells, the EMS cells must be at least in part responsible for AB-cell development, either through direct contact or by the production of a diffusible substance.

   c. Most descendants of the AB cells do not become muscle cells when P2 is present; all descendants of the AB cells become muscle cells when P2 is absent. Therefore, P2 must prevent some AB descendants from becoming muscle cells.

5. Normal Ras is a G-protein that activates a protein kinase, which in turn phosphorylates a transcription factor. If it were simply deleted, no cancer could develop because cell division would not occur. If it were sim-

ply duplicated, an excess of the G-protein could not cause cancer because it must be activated before it can activate the protein kinase, and presumably the enzyme that activates normal Ras is closely regulated and would not make too many copies active. However, if it were to have a point mutation, it might lose its specificity for the protein kinase that it normally activates. If it were to activate other protein kinases out of sequence, it could lead to a malignancy as cell growth becomes more chaotic.

In contrast, normal *c-myc* is a transcription factor. If the gene were to be duplicated, too much transcription factor would lead to malignancy.

6. Because the receptor is defective, testosterone cannot enter the cell and initiate the cascade of developmental changes that will switch the embryo from the "default" female development to male development. Therefore, the phenotype will be female.

7. **a.** Mutations in a tumor suppressor gene are loss-of-function. The function can be restored by transgenes.

   b. If the mutations are dominant, they are gain-of-function mutations. The normal function cannot inhibit these mutants, and the transgene would be ineffective in restoring the normal phenotype.

8. **a.** Type A diabetes is most likely due to a defect in the pancreas. The pancreas normally makes insulin, and type A diabetes can be treated by supplying insulin. Type B diabetes is most likely due to a target cell defect because type B is unresponsive to exogenous insulin.

   b. Type B diabetes appears to be caused by a defect in the target cell. A number of genes are responsible for the receptor and the subsequent cascade of changes that occur in leading to a change in transcription. Any of these genes could have a mutant form.

9. In order to work this problem, first diagram the sequence of changes:

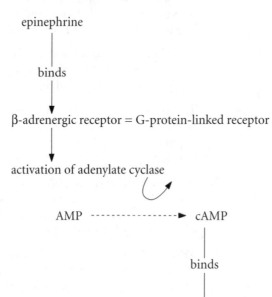

a. If epinephrine is constitutively produced, then the CPK activity is continuous. The system is always on.

b. If adenylate cyclase is not active, the system is always off.

c. If cAMP cannot bind to the regulatory subunit of CPK, the system is always off.

d. The double mutant constitutively produces epinephrine but no cAMP is made. The system is always off.

e. The double mutant has no active adenylate cyclase, and any that might be made cannot bind the regulatory subunit of CPK. The system is always off.

Developmental Genetics:
Cell Fate and
Pattern Formation

The **totipotent** egg gives rise to the mature adult through the sequence of **determination** and then **differentiation.** In humans and many other, but not all, species, the genetic information contained in each cell remains unchanged throughout this process. Some differentiated cells in some species also retain totipotency, which leads to **regeneration.**

Fate maps show the destinies of the descendants of specific cells during embryogenesis. Cells can also be marked genetically through the use of **mosaics.** In species with an invariant pattern of cell division, it is possible to derive a complete **lineage** of every cell from fertilized egg to the mature adult. **Cell lineage** studies in *C. elegans* have revealed that development is invariant for that organism.

The **pattern of determination** is apparently established in the cytoplasm of the egg in many species and is under the control of the maternal genotype.

Embryonic genes affecting determination for relatively large segments of the embryo do so within the restrictions imposed by the maternal genotype. Other embryonic genes affecting determination do so within the restrictions imposed for the large segment in which they function.

Determination studies in *Drosophila* have demonstrated that the egg is initially asymmetric because of maternal contributions. This chemical asymmetry is then followed by differential gene activity in the egg in response to the localized chemical asymmetry, which results in physical asymmetry as development proceeds. There is a **hierarchy** of gene expression that divides the embryo continually in ever more refined subdivisions on the anterior-posterior axis. The order is **maternal *bcd* gene, gap genes, pair-rule genes,** and **segment-polarity genes.** The **homeotic genes** are responsible for segment identity. This cascade of determination in part sets **cell fate.** Communication between cells is also required for this process because fixed cell lineage is not involved as it is in *C. elegans.*

The **homeobox** is a 180–base-pair DNA sequence that results in a 60–amino-acid polypeptide, the **homeodomain.** All homeotic genes share a similar homeobox sequence. The homeodomain binds DNA, altering transcription. The homeotic genes also maintain segmental identity in *Drosophila* after differentiation is completed.

Communication between cells can be **inductive**, leading to the acquisition of a new fate, or it can be **inhibitory**, suppressing the ability of a cell to adopt the same fate as that of its neighbor. In cases where the process has been defined at the molecular level, it has turned out to involve modulation of transcription factors.

The developmental principles discovered in *C. elegans* and *Drosophila* also can be found to apply in other organisms, including humans.

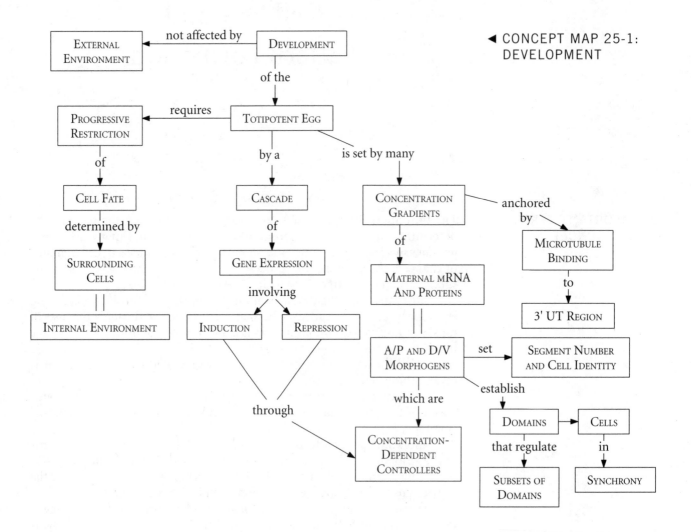

◄ CONCEPT MAP 25-1:
DEVELOPMENT

Questions for Concept Map 25-1

1. What does *totipotent* mean?

2. Is the egg uniform throughout the cytoplasm?

3. What is the location of maternally derived mRNAs?

4. For which proteins do these maternally derived mRNAs code?

5. What is the role of these proteins?

6. How do they remain localized in the oocyte?

7. What is the location of a protein from follicle cells?

8. What is one role of this protein?

9. What are morphogens, and what are some examples?

10. What is cell determination?

11. What is cell fate?

12. For most cells, how is cell fate established?

13. What is positional information?

14. What is a domain?

15. What is positive induction? Negative induction?

16. How many axes are established in the totipotent egg during development? What are they? What is their role in development?

17. How do regulatory genes within a domain function?

18. How are most developmental decisions remembered by a cell?

19. Is all cell determination within the developing embryo controlled by maternal-effect genes?

20. How long does it take to go from a fertilized egg to a hatched larva in *Drosophila?*

21. What is a nurse cell?

22. Which cells are responsible for the eggshell of a developing *Drosophila* egg?

23. What is a syncytium?

24. What is cellularization?

25. How does cellularization occur?

26. Rationalize the observation that cells within a domain are in synchrony with regard to cell division.

27. Where is the highest concentration of the NOS protein, and how is it maintained at that location?

28. What is one role of the NOS protein?

29. What genes are induced by the BCD, HB-M, and DL proteins?

30. What is the mechanism of this induction?

31. When these genes are induced, what do they do?

32. What function is controlled by pair-rule genes?

33. What function is controlled by segment-polarity genes?

34. Which two types of proteins are controlled by GAP proteins?

35. What do pair-rule proteins do?

36. What do homeotic proteins do?

37. What is homeosis?

38. What is a homeodomain?

39. What does the homeodomain do?

40. What is a zoo blot?

41. What does *paralogus* refer to?

42. The HOM-C gene cluster is in *Drosophila*. What is the equivalent in humans?

Answers to Concept Map 25-1 Questions

1. *Totipotent* refers to the ability of an egg to go through all stages of development and give rise to a normal adult.

2. The egg is not uniform throughout the cytoplasm, and the lack of uniformity allows for setting up the anterior-posterior axis and the dorsal-ventral axis.

3. The maternally derived mRNAs are located at the anterior pole of the egg, where they are called *morphogens*. At the posterior pole, maternally derived protein, rather than mRNA, is deposited.

4. These maternally derived mRNAs code for the BCD and HB-M proteins.

5. These proteins are transcription factors that set up a gradient of assorted products within the developing embryo.

6. The maternally derived mRNAs remain localized in the oocyte by being anchored at their 3′ untranslated region to the microtubule cytoskeleton.

7. The location of a protein from follicle cells is the posterior pole.

8. This protein is a transcription factor for the cardinal genes, and it activates proteins at the ventral midline through phosphorylation.

9. A morphogen is a concentration-dependent controller of development. Two examples are the BCD and HB-M mRNAs. Another example is the DL protein.

10. Cell determination is the series of changes that occur that commit a cell to develop along a certain pathway.

11. Cell fate is the capacity to differentiate to a specific cell.

12. Cell fate is established gradually for most cells, with a progressive narrowing of options during development as the internal environment of the developing embryo becomes more and more differentiated.

13. Positional information is information that has differential concentrations at different locations, specifically the A/P and D/V axes.

14. A domain is a region of development that is under the control of a specific gene.

15. Positive induction is the turning on of a gene by a transcription factor. Negative induction is the turning off of a gene by a transcription factor.

16. There are two axes that are established in the totipotent egg during development. They are the anterior-posterior axis, which is responsible for establishing head and tail ends of the developing embryo, and the dorsal-ventral axis, which is responsible for setting the front and back of the developing embryo.

17. Regulatory genes within a domain activate other regulatory genes in different subdivisions of the domain.

18. Most developmental decisions are remembered by a cell through the establishment of a feedback loop, so that transcription of a gene results in continued transcription of that gene.

19. Not all cell determination within the developing embryo is controlled by maternal-effect genes.

20. It takes approximately 24 hours to go from a fertilized egg to a hatched larva in *Drosophila*.

21. A nurse cell is one of 15 cells surrounding the egg. It has an identical genotype and common cytoplasm with the egg. Its role is to provide the morphogens and other products needed for rapid development.

22. The follicle cells are responsible for the eggshell of a developing *Drosophila* egg.

23. A syncytium is a single cell with many nuclei.

24. Cellularization is the process in *Drosophila* development whereby the invagination of the plasma membrane divides the original syncytium into diploid cells.

25. Cellularization occurs through the invagination of the plasma membrane.

26. In order to have the very precise regulation that occurs during development, the cells within a domain are very closely regulated by one gene, and it is not at all surprising that they are in synchrony with regard to cell division. If they were not, the developmental changes that occur in different cells would perhaps become uncoupled from one another.

27. The highest concentration of the NOS protein is at the posterior pole. The NOS mRNA is anchored by the positive end of the subunits of microtubules, which bind to the 3′ UT region.

28. The NOS protein blocks translation of the HB-M mRNA.

29. BCD, HB-M, and DL proteins induce the cardinal genes.

30. The mechanism of this induction is that the promoters of these genes are differentially sensitive to the concentration of one or more of these morphogens.

31. The cardinal genes induce the regulatory genes of each domain.

32. Pair-rule genes control the number of segments in the embryo.

33. The segment-polarity genes control the identity of a segment.

34. GAP proteins control the pair-rule and segment-polarity genes.

35. Pair-rule proteins set up the number of segments in the embryo.

36. The homeotic proteins are responsible for activating and repressing a battery of genes within a domain.

37. Homeosis is the conversion of one body part into another body part.

38. A homeodomain is the common 60 amino acids found in homeogenes.

39. The homeodomain activates and represses a battery of genes by binding to DNA sequences.

40. A zoo blot is a Southern blot that contains DNA from a number of species.

41. *Paralogus* refers to the sequence of genes in each cluster of genes that is very similar.

42. The equivalent genes to HOM-C genes in humans are the Hox genes.

► SOLUTIONS TO PROBLEMS

Be sure that you have thoroughly read the entire chapter before you attempt any of the problems.

1. Human development, like the development of all multicellular organisms, is a process that continues throughout life and ends only when death occurs. It is a result of the complex interaction of genetics, the environment, developmental noise (see Chapter 1), and chance. There is no "final phenotype" per se. The production of human clones in the laboratory would be the first step in an experiment that would have to last for the life span of all the clones, and it would be virtually impossible to tease out the various factors impinging upon each of the clones. Therefore, clonal reproduction of humans would likely be useless in determining the relative effects of heredity and environment even though the clones are genetically identical.

2. A number of experiments could be devised. A comparison of amino acid sequence between mammalian gene products and insect gene products would indicate which genes are most similar to each other. Using cloned cDNA sequences from mammalian genes for hybridization to insect DNA would also indicate which genes are most similar to each other.

3. **a.** The anterior 20% of the embryo is normally devoted to the head and thorax regions. The bicaudal phenotype results in the loss of these regions and in the loss of A1 through A3. The gap proteins are responsible for the induction of the pair-rule proteins, which ultimately set the number of segments, and the homeotic proteins, which set the identity of the segments. Obviously, the gap proteins are improperly regulated to produce the bicaudal phenotype.

   Normal regulation of the gap proteins is accomplished by differential sensitivity to the differing concentrations of the maternally derived morphogens. Because the anterior portion of the embryo has been removed, high concentrations of the morphogens in these regions have also been removed. This results in the abnormal segment number and identity that are observed.

   b. The *oskar* mutation results in the loss of the posterior localization of the *nos* mRNA and protein. Therefore, there is no repression of HB-M translation. The lack of repression of the HB-M transcription factor results in an excess of the HB-M protein. The normal shallow gradient, A to P, is therefore lost. Because the gap genes respond differentially to the BCD:HB-M ratio, no induction of the gap genes occurs, which leads to reduced segmentation. This results simply in a broader head and thorax, and no mirror image phenotype is possible.

4. If you diagram these results, you will see that deletion of a gene that functions posteriorly allows the next-most anterior segments to extend in a posterior direction. Deletion of an anterior gene does not allow

extension of the next-most posterior segment in an anterior direction. The gap genes activate *Ubx* in both thoracic and abdominal segments, whereas the *abd-A* and *Abd-B* genes are activated only in the middle and posterior abdominal segments. The functioning of the *abd-A* and *Abd-B* genes in those segments somehow prevents *Ubx* expression. However, if the *abd-A* and *Abd-B* genes are deleted, *Ubx* can be expressed in these regions.

5. Proper *ftz* expression requires *Kr* in the fourth and fifth segments and *kni* in the fifth and sixth segments.

6. It may be that the wild-type allele in the embryo produces a gene product that can inhibit the gene product of the rescuable maternal-effect lethal mutations, while the nonrescuable maternal-effect lethal mutations produce a product that cannot be inhibited.

   Alternatively, the nonrescuable maternal-effect lethal mutations may produce a product that is required very early in development, before the developing fly is producing any proteins, while the rescuable maternal-effect lethal mutations may act later in development when embryo protein production can compensate for the maternal mutation.

7. a. The determination of anterior-posterior portions of the embryo is governed by a concentration gradient of *bcd*. The concentration is highest in the anterior region and lowest in the posterior region. The furrow develops at a critical concentration of *bcd*. As *bcd*⁺ gene dosage (and, therefore, BCD concentration) decreases, the furrow shifts anteriorly; as the gene dosage increases, the furrow shifts posteriorly.

   b.

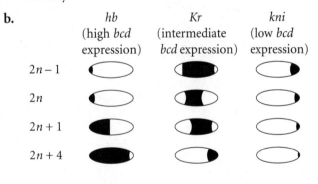

8. The anterior-posterior axis would be reversed.

Population Genetics

Among individuals in a population, there is **phenotypic variation**, or **polymorphism.** Offspring are phenotypically closer to their parents than they are to unrelated individuals. Some phenotypes **survive** to reproduce better than other phenotypes in a given environment. **Natural selection** of these more successful phenotypes in a given environment leads to a **reproductive advantage**, or an increased **fitness**, for them. This results in a change of allelic frequency. **Evolution** is a change in genotypic frequencies.

All variation ultimately comes from **mutation.**

The frequency of a given allele in a specific population is affected by recurrent mutation, selection, migration, and random sampling effects. In an idealized, randomly interbreeding population not subjected to any forces that alter genotypic frequencies, the genotypic frequencies do not change. They can be represented by the **Hardy-Weinberg equilibrium** equation, $p^2 + 2pq + q^2 = 1.0$.

Questions for Concept Map 26-1

1. Mendel's studies yielded the mechanisms of inheritance. What did Darwin's studies yield?

2. Why can Mendel's laws not be applied to populations, since all the individuals in the population obey his laws?

3. What are variants?

4. What is the source of variants?

5. What is a differential rate of reproduction?

6. What is a differential rate of survival?

7. What role does the environment play in differential rates of survival and reproduction?

8. There is a struggle for existence. Who struggles, and against what?

9. What is fitness?

10. An individual who dies takes out of the reproductive pool every allele that he or she has. Why do population geneticists speak in terms of the fitness of a specific allele?

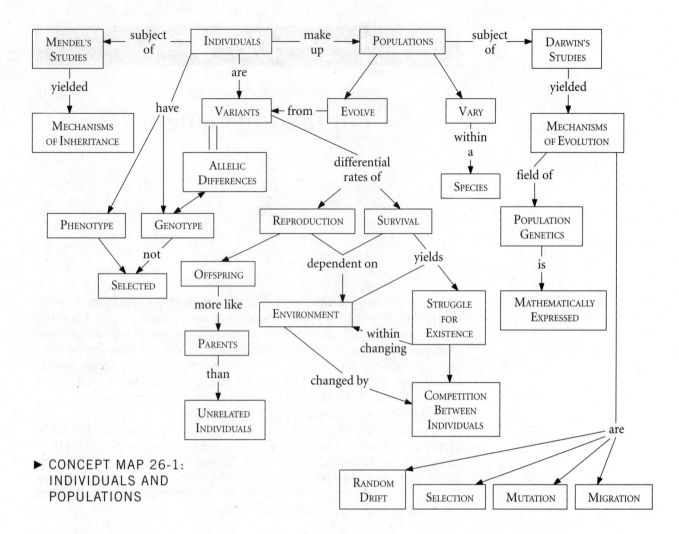

► CONCEPT MAP 26-1:
INDIVIDUALS AND
POPULATIONS

11. Which differ more: individuals within a population or different populations?

12. What are the basic population genetic mechanisms of evolution?

13. Do populations or individuals evolve?

14. What is evolution?

15. Do traits appear in response to an environmental change?

16. What is a haplotype?

17. What is the meaning of *polymorphism*?

18. What is natural selection?

19. What is random drift?

20. What is migration?

21. What role does migration have in evolution?

22. What role does mutation have in evolution?

23. Some claim that human evolution no longer exists because of advances in medical treatment. What is your response to this idea?

Answers to Concept Map 26-1 Questions

1. Darwin's studies yielded the mechanisms of evolution.

2. An individual either survives and reproduces or does not survive and reproduce. With the former, he follows the laws of Mendel; with the latter, the laws are irrelevant. Within a population, however, the survival and reproduction of one phenotype is a matter of probability, not yes or no. The probability is in part dependent on the other phenotypes present, the environment within which the phenotypes are existing, chance events, and the relative success one phenotype has in comparison with others in that specific environment. The laws of Mendel do not deal with these phenomena.

3. Variants are the phenotypic or genotypic alternatives that exist within a population.

4. The source of variants is ultimately mutation.

5. A differential rate of reproduction refers to the fact that some phenotypes are more successful at reproduction within a specific environment than other phenotypes.

6. A differential rate of survival refers to the fact that some phenotypes are more successful at survival within a specific environment than other phenotypes.

7. The environment is the major "selector" in determining differential rates of survival and reproduction.

8. The struggle for existence is by members of a population. They struggle both with the environment itself and against other members of the population in competition for limited resources.

9. Fitness is a measure of reproductive success within a specific environment.

10. Population geneticists speak in terms of a specific allele that is either expressed or not within the phenotype because it can be easily observed. All other alleles are considered to be background, and the assumption is that there is no difference between individuals for these alleles, which is not true. This is a simplification of the real situation, obviously, but it is a simplification that is necessary if one is to attempt to handle the fact of evolution mathematically. Any mathematical conclusion based on such simplification must be checked and evaluated against reality.

11. There is more difference (genetic variation) between individuals within a population than there is between different populations.

12. The mechanisms of evolution are random drift (basically, chance), selection, mutation, and migration.

13. Populations evolve.

14. Evolution is simply the change in allelic frequency over time within a population for a specific environment. Direction and "goodness" and "betterness" are not implied by the term. The more common perception of evolution is that one species leads to another species. *Speciation,* as this is called, requires both genetic and geographic barriers to reproduction.

15. Traits do not appear in response to an environmental change. Variation is a random occurrence. The traits are then selected for or against by

the environment. This was the entire point of the research involving replica plating.

16. A haplotype is the multilocus haploid genotype inherited from a parent.

17. *Polymorphism* simply means "many forms," and it is used to refer to variants for a trait.

18. Natural selection is the outcome of many factors that ultimately breaks down to the differential rate of survival and reproduction of phenotypes within a specific environment. It does not include differential survival and reproduction that are purely chance effects, however.

19. Random drift is the change in frequencies of phenotypes and genotypes that occurs because of random sampling of gametes for the next generation in a population of limited size.

20. Migration is the movement of individuals into and out of a population.

21. Migration has a major role in evolution because it is the fastest mechanism for producing changes in gene frequencies. It prevents the random divergence of isolated populations and retards the selective divergence of populations living in different environmetns.

22. Mutation is the ultimate source of variability within a population.

23. Human evolution is ongoing. Although the selection against some specific phenotypes may have been relaxed owing to medical treatment, other phenotypes are being subjected to selective forces. Consider the frequency of deaths of people who drink and then drive. Their death rate is high, death more frequently occurs prior to reproduction, and those individuals who combine a lower tolerance for alcohol along with an inability to use common sense are being selected against. Those with a quick reaction time and genetic ability to metabolize alcohol efficiently have positive selection. Consider AIDS. Here the strongest selection occurs for those who are genetically resistant to the virus and against individuals who combine a high rate of different sexual partners, the inability to learn from the experience of others, and the inability to think ahead some. To see human evolution, simply look at behaviors that lead to a higher death rate prior to reproduction. Just as urban young squirrels go through natural selection every spring, primarily in relationship to automobiles, so do urban young humans go through natural selection.

Questions for Concept Map 26-2

1. What are the two major types of variation?

2. What are some sources of variation?

3. What chromosomal polymorphisms exist?

4. What is an allelic polymorphism?

5. Is an allelic polymorphism a source or outcome of variation?

6. What is required for a population to be in Hardy-Weinberg equilibrium?

7. Are these conditions ever met in reality?

8. What does population genetics do with the fact that the conditions for

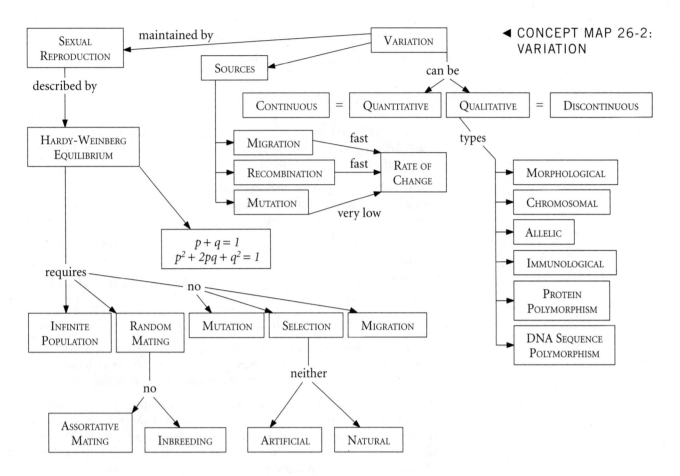

◄ CONCEPT MAP 26-2: VARIATION

Hardy-Weinberg equilibrium are never met in reality? How is this handled within population genetics?

9. What two types of assortative mating exist and what is their effect on a population?

10. What is the definition of *inbreeding*?

11. What is gene frequency?

12. What types of DNA sequence polymorphisms exist?

13. How are DNA sequence polymorphisms detected?

14. Is there more variation between individuals within a race than there is between races or is there less? What does this mean?

15. Are all variable traits heritable?

16. What are canalized characters?

17. What role does recombination have in evolution?

18. How do new functions develop?

19. What is adaptation?

20. What happens to the relative fitness of a phenotype with environmental change?

21. What is the meaning of balanced polymorphisms?

22. What is the meaning of multiple adaptive peaks?

23. What is neutral selection and how important is it?

24. What does it mean to become fixed?

Answers to Concept Map 26-2 Questions

1. The two major types of variation are quantitative and qualitative. Most traits are quantitative, but population genetics deals mostly with qualitative variation.

2. Some sources of genetic variation are migration, recombination, and mutation. Developmental variation comes from environmental variation, and developmental noise.

3. The chromosomal polymorphisms include translocation, inversion, deletion, and duplication.

4. An allelic polymorphism is two or more alleles at a locus that are present within a population.

5. Allelic polymorphism is an outcome of mutationally caused variation.

6. For a population to be in Hardy-Weinberg equilibrium, it must be of infinite size, random mating must occur within the population, and there must be no selection, mutation, or migration occurring.

7. No population is infinite in size. Nonrandom mating is always occurring. Mutation is ongoing. Selection is constant. Migration is always occurring. The conditions for a Hardy-Weinberg equilibrium are never met in reality, but they are usually weak enough so that a population is quite close to Hardy-Weinberg equilibrium.

8. Population genetics studies the effects on the Hardy-Weinberg equilibrium of these various "violations" of the requirements.

9. There are positive and negative assortative mating. The former, mating between "likes," reduces heterozygosity. The latter, mating between "unlikes," increases heterozygosity.

10. *Inbreeding* is the increased chance of being homozygous by descent as compared with the average in a population because of mating between individuals with a common ancestor.

11. Gene frequency is the relative frequency of a specific allele within an environment.

12. DNA sequence polymorphisms include restriction-site variation, tandem repeats, and silent mutations.

13. DNA sequence polymorphisms can be detected through DNA sequencing or RFLP analysis if there a duplication or deletion or if a restriction site has been created or destroyed.

14. There is more variation between individuals within a race than there is between races. This means that any two populations are more alike than they are different. Also, this means that human racial classification does not capture an important amount of human variation. As a result, anthropologists have abandoned the notion of races.

15. Not all variable traits are heritable.

16. Canalized characters are those traits of an organism that may have substantial genetic variation without an accompanying morphological variation except in extremely stressful environments, which then uncover the variation.

17. Recombination produces new combinations of traits that are then selected by the environment.

18. New functions develop by a two-step mutational process: gene duplication followed by divergence of the two sequences through mutation.

19. Adaptation refers to the balance that occurs between a phenotype and the environment in which it exists. Some phenotypes are better than others for specific environments, and they have a higher rate of survival and reproduction.

20. The relative fitness of a phenotype changes with environmental change because the same phenotype may be better or less adapted to the new environmental conditions.

21. Balanced polymorphisms are the stable maintenance within a population of two or more phenotypes. They may be due to the selection for or against the heterozygote, or to selection for rarer variants.

22. Multiple adaptive peaks are found for phenotypes within an environment. They refer to combinations of traits that can exist and reflect the history of the variants rather than solely the environment.

23. Neutral selection is the incorporation of new unselected mutations into a population by chance alone. It appears to be an important force in evolution.

24. An allele becomes fixed when it has a frequency of one or zero.

▶ SOLUTIONS TO PROBLEMS

Be sure that you have thoroughly read the entire chapter before you attempt any of the problems.

1. The frequency of an allele in a population can be altered by selection, mutation, migration, inbreeding, and random genetic drift.

2. There are a total of $(2)(384) + (2)(210) + (2)(260) = 1708$ alleles in the population. Of those, $(2)(384) + 210 = 978$ are $A1$ and $210 + (2)(260) = 730$ are $A2$. The frequency of $A1$ is $978/1708 = 0.57$, and the frequency of $A2$ is $730/1708 = 0.43$.

3. The given data are $q^2 = 0.04$ and $p^2 + 2pq = 0.96$. If $q^2 = 0.04$, q = 0.2 and $p = 0.8$. To check this, use these numbers in the second equation: $(0.8)^2 + (2)(0.8)(0.2) = 0.64 + 0.32 = 0.96$. The frequency of BB is 0.64, and the frequency of Bb is 0.32.

4. This problem assumes that there is no backward mutation. Use the following equation: $p_n = p_o e^{-n\mu}$. That is,

$$p_{50,000} = (0.8)e^{-(5 \times 10^4)(4 \times 10^{-6})} = (0.8)(0.81873) = 0.65$$

5. **a.** If the variants represent different alleles of gene X, a cross between any two variants should result in a 1:1 progeny ratio. All the vari-

ants should map to the same locus. Amino acid sequencing of the variants should reveal differences of one to just a few amino acids.

b. There could be another gene (gene Y), with five variants, that modifies the gene X product post-transcriptionally. If so, the easiest way to distinguish between the two explanations would be to find another mutation in X and do a dihybrid cross. For example, if there is independent assortment,

P $X^1 Y^1 \times X^2 Y^2$

F$_1$ $1\ X^1 Y^1 : 1\ X^1 Y^2 : 1\ X^2 Y^1 : 1\ X^2 Y^2$

If the mutation in X led to no enzyme activity, the ratio would be

2 no activity:1 variant one activity:1 variant two activity

The same mutant in a one-gene situation would yield 1 active:1 inactive.

6. **a.** If the population is in equilibrium, $p^2 + 2pq + q^2 = 1$. Use p from the data to predict the frequency of q, and then check the calculated values against the observed.

$p = [406 + 1/2(744)]/1482 = 0.5249$

$q = 1 - p = 0.4751 = $ predicted value

The phenotypes should be distributed as follows if the population is in equilibrium:

$L^M L^M = p^2(1482) = 408$

$L^M L^N = 2pq(1482) = 739$

$L^N L^N = q^2(1482) = 334$

The population is in equilibrium.

b. If mating is random with respect to blood type, then the following frequency of matings should occur.

$L^M L^M \times L^M L^M = (p^2)(p^2)(741) = 56.25$

$L^M L^M \times L^M L^N = (2p^2)(2pq)(741) = 203.6$

$L^M L^M \times L^N L^N = (2p^2)(q^2)(741) = 92$

$L^M L^N \times L^M L^N = (2pq)(2pq)(741) = 184.28$

$L^M L^N \times L^N L^N = 2(2pq)(q^2)(741) = 166.8$

$L^N L^N \times L^N L^N = (q^2)(q^2)(741) = 37.75$

The mating is random with respect to blood type.

7. **a.** When the allelic frequency differs between sexes for an X-linked gene, $p = 1/2(1/3p_m + 2/3p_f)$. At generation 0, $p = 1/2[1/3(0.8) + 2/3(0.2)] = 0.2$.

In generation 1, all males get their X from their mother, and therefore, the frequency of p in the males will be the same as it is in the mother, 0.2. All females get an X from each parent at the allelic frequency in each parent. Therefore, the daughters will have a frequency of $p_1 = 1/2(p_m + p_f) = 1/2(0.8 + 0.20) = 0.5$. The following table provides information for further generations.

GENERATION	P MALE	P FEMALE
0	0.8	0.2
1	0.2	0.5
2	0.5	0.35
3	0.35	0.425
.	.	.
.	.	.
.	.	.
n	$p_{f(n-1)}$	$p_{1/2[m(n-1) + f(n-1)]}$

where m = male and f = female.

b. Let p = frequency in males and p' = frequency in females. For any generation, $p_n = p'_{n-1}$ and $p'_n = 1/2(p_{n-1} + p'_{n-1})$. The difference between these two, d, is

$$d = 1/2(p_{n-1} + p'_{n-1}) - p'_{n-1}$$
$$= 1/2(p_{n-1} - p'_{n-1})$$

Given the initial values of p_0 and p'_0,

$$d_n = (1/2)^n (p_0 - p'_0)$$

8. **a.** and **b.**

POPULATION	p	q	EQUILIBRIUM?
1	1.0	0.0	yes
2	0.5	0.5	no
3	0.0	1.0	yes
4	0.625	0.375	no
5	0.375	0.625	no
6	0.5	0.5	yes
7	0.5	0.5	no
8	0.2	0.8	yes
9	0.8	0.2	yes
10	0.993	0.007	yes

c. $4.9 \times 10^{-6} = 5 \times 10^{-6}/s$; $s = 0.102$

d.

GENOTYPE	FREQUENCY	FITNESS	GAMETES	A	A
AA	0.25	1.0	0.25	0.25	0.0
AA	0.50	0.8	0.40	0.20	0.20
aa	0.25	0.6	0.15	0.0	0.15
				0.45	0.35

$$p = 0.45/(0.45 + 0.35) = 0.56$$
$$q = 0.35/(0.45 + 0.35) = 0.44$$

9. a. Assuming equilibrium, if $q = 0.1$, $q^2 = 0.01$.

   b. 10 times.

   c. Marriages in which half the children of both sexes would be colorblind are $X^B X^b \times X^b Y$. Such marriages occur with a frequency of $(2pq)(q) = 2pq^2 = 2(0.9)(0.1)^2 = 0.018$.

   d. All children would be normal if the female were homozygous normal. The frequency of such marriages is $p^2(p + q) = (0.9)^2(0.5 + 0.5) = 0.81$.

   e. Colorblind females result from two types of matings:

$$X^B X^b \times X^b Y = (2pq)(q') = 2(0.2)(0.8)(0.6) = 0.192$$
$$X^b X^b \times X^b Y = (q)^2(q') = (0.2)^2(0.6) = 0.024$$

   Half the females from the first mating and all the females from the second will be colorblind, so the frequency of colorblind female progeny is $1/2(0.192) + 0.024 = 0.12$.

   Colorblind males will result when the mother is either heterozygous (half the male offspring) or homozygous recessive (all the male offspring), regardless of the father's genotype. Therefore, the frequency of colorblind male progeny is

$$1/2(2pq) + q^2 = 1/2(2)(0.8)(0.2) + (0.2)^2 = 0.2$$

   f. The male frequency will be 0.2 (frequency of the female in the previous generation), and the female frequency will be $1/2(0.2 + 0.6) = 0.4$.

10. The frequency of a phenotype in a population is a function of the frequency of alleles that lead to that phenotype in the population. To determine dominance and recessiveness, do standard Mendelian crosses.

11. Assume that proper function results from the right gene products in the proper ratio to all other gene products. A mutation will change the gene product, eliminate the gene product, or change the ratio of it to all other gene products. All three outcomes upset a previously balanced system. While a new balance may be achieved, this is unlikely.

12. Dominance is usually wild-type because most detectable mutations in enzymes result in lowered or eliminated enzyme function. To be dominant, the heterozygote has approximately the same phenotype as the homozygous dominant. This will be true only when the wild-type allele produces a product and the mutant allele does not.

The chromosomal rearrangements are dominant mutations because so many genes are affected that it is highly unlikely that all their alleles will be dominant and "cover" for them.

13. Prior to migration, $q^A = 0.1$ and $q^B = 0.3$. Immediately after migration, $q^{A+B} = 1/2(q^A + q^B) = 1/2(0.1 + 0.3) = 0.2$. The frequency of affected males is 0.2, and the frequency of affected females is $(0.2)^2 = 0.04$.

14. The probability of homozygosity by descent (f) is $f = (1/2)^n$, where n = number of ancestors in a closed loop.

a. $f = (1/2)^2 = 1/4$

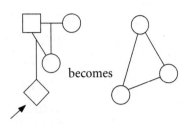

becomes

b. $f = (1/2)^5 + (1/2)^5 = 1/16$

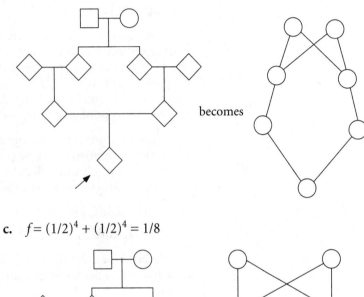

becomes

c. $f = (1/2)^4 + (1/2)^4 = 1/8$

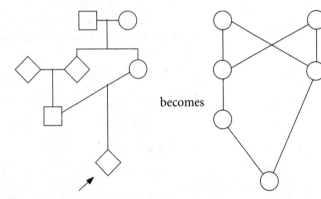

becomes

15. Albinos may have been considered lucky and encouraged to breed at very high levels in comparison with nonalbinos. They may also have been encouraged to mate with each other. Alternatively, in the tribes with a very low frequency, albinos may have been considered very unlucky and destroyed at birth or prevented from marriage.

16. The allele frequencies are

A: $0.2 + 1/2(0.60) = 50\%$

a: $1/2(0.60) + 0.2 = 50\%$

Positive assortative mating. The alleles will randomly unite within a phenotype. For $A–$, the mating population is $0.2\ AA + 0.6\ Aa$. The allele frequencies within this population are

A: $[0.2 + 1/2(0.6)]/0.8 = 0.625$

a: $1/2(0.6)/0.8 = 0.375$

The phenotypic frequencies that result are

A–: $p^2 + 2pq = (0.625)^2 + 2(0.625)(0.375) = 0.3906 + 0.4688 = 0.8594$

aa: $q^2 = (0.375)^2 = 0.1406$

However, because this subpopulation represents 0.8 of the total population, these figures must be adjusted to reflect that by multiplying by 0.8:

A–: $(0.8)(0.8594) = 0.6875$

aa: $(0.8)(0.1406) = 0.1125$

The *aa* contribution from the other subpopulation will remain unchanged because there is only one genotype, *aa.* The contribution to the total phenotypic frequency is 0.20. Therefore, the final phenotypic frequencies are *A–* = 0.6875 and *aa* = 0.20 + 0.1125 = 0.3125. These frequencies will remain unchanged over time, but the end result will be two separate populations, *AA* and *aa,* which will not interbreed.

Negative assortative mating: If assortative mating is between unlike phenotypes, the two types of progeny will be *Aa* and *aa. AA* will not exist. *Aa* will result from all *AA* × *aa* matings and half the *Aa* × *aa* matings. The matings will occur with the following frequencies:

AA × *aa* = $(0.2)(0.2) = 0.04$

Aa × *aa* = $(0.6)(0.2) = 0.12$

Because these are the only matings that will occur, they must be put on a 100 percent basis by dividing by the total frequency of matings that occur:

AA × *aa:* 0.04/0.16 = 0.25, all of which will be *Aa*

Aa × *aa:* 0.12/0.16 = 0.75, half *Aa* and half *aa*

The phenotypic frequencies in the next generation will be

Aa: 0.25 + 0.75/2 = 0.625

aa: 0.75/2 = 0.375

In the second generation, the same method will result in a final ratio of 0.5 *Aa*:0.5 *aa.* These values will remain unchanged after the second generation of negative assortative mating.

17. Many genes affect bristle number in *Drosophila.* The artificial selection resulted in lines with mostly high-bristle-number alleles. Some mutations may have occurred during the 20 generations of selective breeding, but most of the response was due to alleles present in the original population. Assortment and recombination generated lines with more high-bristle-number alleles.

Fixation of some alleles causing high bristle number would prevent complete reversal. Some high-bristle-number alleles would have no bad effects on fitness, so there would be no force pushing bristle number back down because of those loci.

The low fertility in the high-bristle-number line could have been due to pleiotropy or linkage. Some alleles that caused high bristle number may also have caused low fertility (pleiotropy). Chromosomes with high-bristle-number alleles may also carry alleles at different loci that caused low fertility (linkage). After artificial selection was relaxed, the low-fertility alleles would have been selected against through natural selection. A few generations of relaxed selection would have allowed low-fertility-linked alleles to recombine away, producing high-bristle-number chromosomes that did not contain low-fertility alleles. When selection was reapplied, the low-fertility alleles had been reduced in frequency or separated from the high-bristle loci, so this time there was much less of a fertility problem.

18. **a.** The needed equation is $p' = pW_a/\overline{W}$, where $\overline{W}_a = p\overline{W}_{AA} + qW_{Aa}$ and $W = p^2W_{AA} + 2pqW_{Aa} + q^2W_{aa}$.

$$p' = \frac{(0.5)[(0.5)(0.9) + (0.5)(1.0)]}{(0.5)^2(0.9) + 2(0.5)(0.5)(1.0) + (0.5)^2(0.7)} = 0.528$$

b. $\hat{p} = (W_{aa} - W_{Aa})/[(W_{aa} - W_{Aa}) + (W_{AA} - W_{Aa})]$

$$\mu\hat{p} = \frac{0.7 - 1.0}{0.7 - 1.0) + (0.9 - 1.0)} = 0.75$$

19. The equation needed is $\hat{q} = \sqrt{\mu/s}$

or $s = \mu/\hat{q}^2 = \mu/\text{recessive frequency} = 10^{-5}/10^{-3} = 0.01$

20. Affected individuals $= Bb = 2pq = 4 \times 10^{-6}$. Because q is almost equal to 1.0, $2p = 4 \times 10^{-6}$. Therefore, $p = 2 \times 10^{-6}$.

$\mu = hsp = (1.0)(0.7)(2 \times 10^{-6}) = 1.4 \times 10^{-6}$

21. The probability of not getting a recessive lethal genotype for one gene is $1 - 1/8 = 7/8$. If there are n lethal genes, the probability of not being homozygous for any of them is $(7/8)^n = 13/31$. From log tables, $n = 6.5$, or an average of 6.5 recessive lethals in the human genome.

22. **a.** $\hat{q} = \sqrt{\mu/s} = \sqrt{10^{-5}/0.5} = 4.47 \times 10^{-3}$

$sq^2 = 0.5(4.47 \times 10^{-3})^2 = 10^{-5}$

b. $\hat{q} = \sqrt{\mu/s} = \sqrt{2 \times 10^{-5}/0.5} = 6.32 \times 10^{-3}$

$sq^2 = 0.5(6.32 \times 10^{-3})^2 = 2 \times 10^{-5}$

c $\hat{q} = \sqrt{\mu/s} = \sqrt{10^{-3}/0.3} = 5.77 \times 10^{-3}$

$sq^2 = 0.3(5.77 \times 10^{-3})^{-2} = 10^{-5}$

Quantitative Genetics

The **genotype** can be identified and studied only through its **phenotypic** effects. The study of **genetics** is the study of allelic substitutions that cause **qualitative** differences in the phenotype. However, the actual variation among organisms is usually **quantitative** rather than qualitative. Quantitative variation gives rise to **continuous** variation among members of a species, rather than discrete differences.

Continuous variation is the result of a **norm of reaction** for each genotype and the fact that most traits are controlled by more than one locus. Two individuals with the same genotype can have different phenotypes. Two individuals with different genotypes can have the same phenotype.

Quantitative traits have a **statistical distribution**. This can be presented as a **frequency histogram** or a **distribution function**. A distribution of phenotypes can be described by its **mode**, which is the most frequent class. Some distributions are **bimodal**. The distribution can also be described by the **mean**, the arithmetic average. The **variance** is the spread around the central class. The **standard deviation** is the square root of the variance. Two variables may be described by their **correlation**.

A collection of observations constitutes a **sample** from the **universe** of all observations. This sample may be **biased** or **unbiased**.

The **heritability** of a trait is the proportion of phenotypic variation that can be attributed to genetic variation. The estimates of genetic and environmental variance are specific to the population and environment in which the estimates were made.

Questions for Concept Map 27-1

1. What is variation?

2. Is all variation due to genetic differences?

3. Distinguish between quantitative and qualitative variation.

4. Is there a genetic basis to quantitative variation?

5. Who first successfully studied the genetic basis of qualitative variation?

6. Can the techniques devised by Mendel be used to study quantitative variation? Why?

7. What techniques are used to study quantitative variation?

8. What are the factors that result in quantitative variation?

► CONCEPT MAP 27-1:
QUANTITATIVE
VARIATION

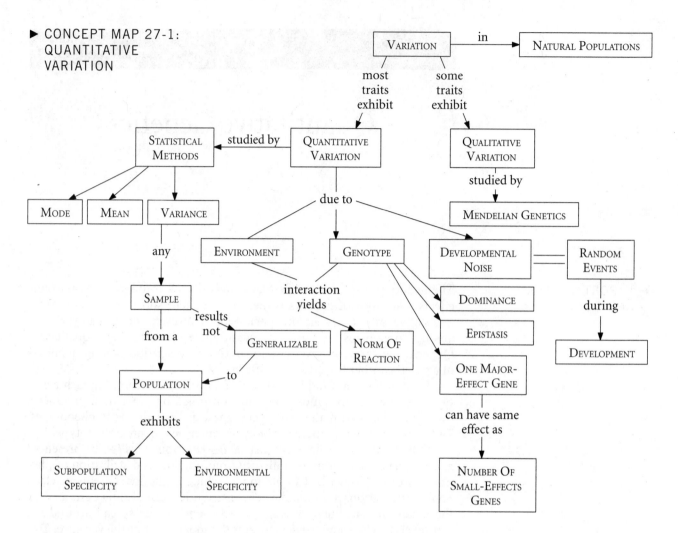

9. What is developmental noise?

10. What is the norm of reaction?

11. What is epistasis?

12. Distinguish between a major-effect gene and minor-effects genes.

13. Why can the heritability of a sample not be generalized to an entire population?

14. What is a mean?

15. What is a mode?

16. What is variance?

17. What is standard deviation?

Answers to Concept Map 27-1 Questions

1. Variation is the observable phenotypic differences that can be detected for a specific trait.

2. Variation is due to genetic, environmental, and chance differences that occur during the entire life of an organism.

3. Quantitative variation is the variation observed along a continuum such as height and weight. Qualitative variation is the variation seen that is discrete; that is, a trait may be present or absent; a shape may be round or oval or square.

4. There usually is a genetic basis to quantitative variation, but it can be due entirely to environmental differences and developmental noise.

5. Mendel first studied the genetic basis of qualitative variation.

6. The techniques devised by Mendel usually cannot be used to study quantitative variation. This is because such mating techniques often cannot distinguish the multiple genotypes that give rise to the same phenotype.

7. Statistical methods are used to study quantitative variation.

8. Quantitative variation arises from genetic differences, interactions among different genes, environmental differences, and developmental noise.

9. Developmental noise is the variation that arises during development owing to random events.

10. The norm of reaction is the series of different phenotypes that can arise from the same genotype as the environment is varied.

11. Epistasis is the effect on expression an allele of one gene has on an allele of a different gene.

12. A major-effect gene strongly affects a specific trait, and minor-effects genes each contribute a small effect to a trait.

13. Heritability is a function of the specific genotypes, and their proportions, that compose a sample in a specific environment. The entire population from which the sample was taken will have other genotypes present, the proportions of the various genotypes will undoubtedly be at least somewhat different from the sample proportions, and the entire population will be in more environments than the sample from the population.

14. A mean is the arithmetic average.

15. A mode is the most frequent class.

16. Variance is a measure of the spread around the mean.

17. Standard deviation is the square root of variance.

► SOLUTIONS TO PROBLEMS

Be sure that you have thoroughly read the entire chapter before you attempt any of the problems.

1. Continuous variation can be represented by a bell-shaped curve. Examples are height and weight. Discontinuous variation results in easily classifiable, discrete entities. Examples are red versus white and taster versus nontaster.

2. a. Broad heritability is $H^2 = s_g^2/(s_g^2 + s_e^2)$. Narrow heritability is
$$h^2 = s_a^2/(s_a^2 + s_d^2 + s_e^2), \text{ where } s_g^2 = s_a^2 + s_d^2$$

Shank length:

$H^2 = (46.5 + 15.6)/(46.5 + 15.6 + 248.1) = 0.200$

$h^2 = 46.5/(46.5 + 15.6 + 248.1) = 0.150$

Neck length:

$H^2 = (73.0 + 365.2)/(73.0 + 365.2 + 292.2) = 0.600$

$h^2 = 73.0/(73.0 + 365.2 + 292.2) = 0.010$

Fat content:

$H^2 = (42.4 + 10.6)/(42.4 + 10.6 + 53.0) = 0.500$

$h^2 = 42.4/(42.4 + 10.6 + 53.0) = 0.400$

b. The larger the h^2 value, the more that characteristic will respond to selection. Therefore, fat content would respond best to selection.

c. The formula needed is

$h^2 \times$ selection differential = selection response

Therefore, selection response $= (0.400)(4.0) = 1.6\%$ decrease in fat content, or 8.9% fat content.

3. a. Homozygotes at one locus can be homozygotic at A or at B or at C. The probability of being homozygotic is 1/2 (for A: AA or aa), and the probability of being heterozygotic is 1/2. Putting this all together,

p(homozygotic at 1 locus) $= 3(1/2)^3 = 3/8$

p(homozygotic at 2 loci) $= 3(1/2)^3 = 3/8$

p(homozygotic at 3 loci) $= (1/2)^3 = 1/8$

b. p(0 capital letters)

    $= p$(all homozygous recessive)

    $= (1/4)^3 = 1/64$

p(1 capital letter)

    $= p$(1 heterozygote and 2 homozygous recessive)

    $= 3(1/2)(1/4)(1/4) = 3/32$

p(2 capital letters)

    $= p$(1 homozygous dominant and 2 homozygous recessive)

    or

p(2 heterozygotes and 1 homozygous recessive)

    $= 3(1/4)^3 + 3(1/4)(1/2)^2 = 15/64$

p(3 capital letters)

    $= p$(all heterozygous)

    or

p(1 homozygous dominant, 1 heterozygous, and 1 homozygous recessive)

$$= (1/2)^3 + 6(1/4)(1/2)(1/4) = 10/32$$

$p(4$ capital letters$)$

  $= p(2$ homozygous dominant and 1 homozygous recessive$)$

  or

  $p(1$ homozygous dominant and 2 heterozygous$)$

  $= 3(1/4)^3 + 3(1/4)(1/2)^2 = 15/64$

$p(5$ capital letters$)$

  $= p(2$ homozygous dominant and 1 heterozygote$)$

  $= 3(1/4)^2(1/2) = 3/32$

$p(6$ capital letters$)$

  $= p($all homozygous dominant$) = (1/4)^3 = 1/64$

4. For three genes there are a total of 27 genotypes that will occur in predictable proportions. For example, there are three genotypes that have two heterozygotes and a homozygote recessive (*Aa Bb cc, Aa bb Cc, aa Bb Cc*). The frequency of this combination is $3(1/2)(1/2)(1/4) = 3/16$, and the phenotypic score is $3 + 3 + 1 = 7$. The distribution of scores is as follows:

SCORE	PROPORTION
3	1/64
5	6/64
6	3/64
7	12/64
8	12/64
9	11/64
10	12/64
11	6/64
12	1/64

5. The population described would be distributed as follows:

3 bristles   19/64

2 bristles   44/64

1 bristle   1/63

Note that the 3-bristle class contains 7 different genotypes, the 2-bristle class contains 19 different genotypes, and the 1-bristle class contains only 1 genotype. It would be very difficult to determine the underlying genetic situation by doing controlled crosses and determining progeny frequencies.

6. **a.** First, solve the formula for values of x over the range for each genotype. Plot those values. Next, add the three values, one for each

genotype, for each phenotypic value and plot these summed values. This will give you the overall population distribution.

| PHENOTYPIC VALUE | GENOTYPE | | | TOTAL |
	1	2	3	
0.03	0.9037			0.9037
0.04	0.9147			0.9147
0.05	0.9250			0.9250
0.06	0.9347			0.9347
0.07	0.9437			0.9437
0.08	0.9520			0.9520
0.09	0.9597			0.9597
0.10	0.9667		0.9020	1.8687
0.11	0.9730		0.9155	1.8885
0.12	0.9787	0.0000	0.9280	1.9067
0.13	0.9837	0.1900	0.9395	2.1132
0.14	0.9880	0.3600	0.9500	2.2980
0.15	0.9917	0.5100	0.9595	2.4612
0.16	0.9947	0.6400	0.9680	2.6027
0.17	0.9970	0.7500	0.9755	2.7225
0.18	0.9987	0.8400	0.9820	2.8207
0.19	0.9997	0.9199	0.9875	2.8972
0.20	1.0000	0.9600	0.9920	2.9520
0.21	0.9997	0.9900	0.9955	2.9852
0.22	0.9987	1.0000	0.9980	2.9967
0.23	0.9970	0.9900	0.9995	2.9865
0.24	0.9947	0.9600	1.0000	2.9547
0.25	0.9917		0.9995	1.9912
0.26	0.9880		0.9980	1.9860
0.27	0.9837		0.9955	1.9792
0.28	0.9787		0.9920	1.9707
0.29	0.9730		0.9875	1.9605
0.30	0.9667		0.9820	1.9487
0.31	0.9597		0.9755	1.9352
0.32	0.9520		0.9680	1.9200
0.33	0.9437		0.9595	1.9032

	GENOTYPE			
PHENOTYPIC VALUE	1	2	3	TOTAL
0.34	0.9347		0.9500	1.8847
0.35	0.9250		0.9395	1.8645
0.36	0.9147		0.9280	1.8427
0.37	0.9037		0.9155	1.8192
0.38			0.9020	0.9020

b. The overall population distribution will not result in three distinct modes. With sufficient variation within genotypes, there is a continuous distribution of phenotypes.

7. mean = sum of all measurements/number of measurements

$$= \frac{1 + 2(4) + 3(7) + 4(31) + 5(56) + 6(17) + 7(4)}{1 + 4 + 7 + 31 + 56 + 17 + 4}$$

= 564/120 = 4.7

variance = average squared deviation from the mean

$$= 1/120 \; \Sigma \left[\begin{array}{l} (1 - 4.7)^2 + (2 - 4.7)^2 + (3 - 4.7)^2 \\ + \; (4 - 4.7)^2 + (5 - 4.7)^2 + (6 - 4.7)^2 \\ + \; (7 - 4.7)^2 \end{array} \right]$$

= 31.43/120 = 0.2619

standard deviation = square root of variance = 0.5117

8.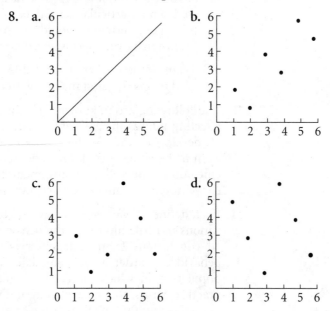

Use the following formula to calculate the correlation between x and y:

$$\text{correlation} = r_{xy} = \frac{\text{cov } xy}{s_x s_y} \, ,$$

where

$$\text{cov } xy = 1/N \, \Sigma \, x_i \, y_i - \overline{xy}$$

a. $\text{cov } xy = 1/6[(1)(1) + (2)(2) + (3)(3) + (4)(4) + (5)(5) + (6)(6)] - (21/6)(21/6)$

$= 2.92$

$$s_x = \sqrt{1/N \Sigma x_i^2 - \overline{x}^2} = \sqrt{1/6(1^2 + 2^2 + 3^2 + 4^2 + 5^2 + 6^2) - (21/6)^2}$$

$$= \sqrt{2.92} = 1.71$$

$$s_y = \sqrt{1/N \Sigma y_i^2 - \overline{y}^2} = \sqrt{1/6(1^2 + 2^2 + 3^2 + 4^2 + 5^2 + 6^2) - (21/6)^2}$$

$$= \sqrt{2.92} = 1.71$$

Therefore, $r_{xy} = 2.92/(1.71)(1.71) = 1.0$. The other correlation coefficients are calculated in a like manner.

b.   0.88

c.   0.70

d. −0.21

9. **a.** H^2 has meaning only with respect to the population that was studied in the environment in which it was studied. Otherwise, it has no meaning.

  b. Neither H^2 and h^2 are reliable measures that can be used to generalize from a particular sample to a "universe" of the human population. They certainly should not be used in social decision making (as implied by the terms *eugenics* and *dysgenics*).

  c. Again, H^2 and h^2 are not reliable measures, and they should not be used in any decision making with regard to social problems.

10. The following are unknown: (1) norms of reaction for the genotypes affecting IQ; (2) the environmental distribution in which the individuals developed; and (3) the genotypic distributions in the populations. Even if the above were known, because heritability is specific to a specific population and its environment, the difference between two different populations cannot be given a value of heritability.

11. First, define *alcoholism* in behavioral terms. Next, realize that all observations must be limited to the behavior you used in the definition and all conclusions from your observations are applicable only to that behavior. In order to do your data gathering, you must work with a population in which familiarity is distinguished from heritability. In practical terms, this means using individuals who are genetically close but who are found in all environments possible.

12. Before beginning, it is necessary to understand the data. The first entry, *h/h h/h*, refers to the II and III chromosomes, respectively. Thus, there are four *h* sets of alleles in two or more genes on separate chromosomes. The next entry is *h/l h/h*. Chromosome II is heterozygous and chromosome III is homozygous.

The effect of substituting a low for a high chromosome II can be seen within each row. In the first row, the differences are 25.1 − 22.2 = 2.9 and 22.2 − 19.0 = 3.2. In the second row the differences are 3.1 and 5.2. They are 2.7 and 6.8 in the third row. The average difference is 23.9/6 = 3.98, which actually tells you very little.

The effect of substituting one *l* chromosome for an *h* chromosome in chromosome II, and therefore going from homozygous *hh* to heterozygous *hl*, can be seen in the differences along the rows in the first two columns. The average change is (2.9 + 3.1 + 2.7)/3 = 2.9. When chromosome II goes from heterozygous *hl* to homozygous *ll*, the average change is (3.2 + 5.2 + 6.8)/3 = 5.1.

The effect of substituting one *l* chromosome for an *h* chromosome in chromosome III, and therefore going from homozygous *hh* to heterozygous *hl*, can be seen in the differences between rows: 25.1 − 23.0 = 2.1; 22.2 − 19.9 = 2.3; 19.0 − 14.7 = 4.3; 23.0 − 11.8 = 11.2; 19.9 − 9.1 = 10.8; 14.7 − 2.3 = 12.4. When chromosome III goes from homozygous *hh* to heterozygous *hl*, the average change is (2.1 + 2.3 + 4.3)/3 = 2.9. When it goes from heterozygous *hl* to homozygous *ll*, the average change is (11.2 + 10.8 + 12.4)/3 = 11.5.

Here is a summary of these results:

	CHROMOSOME II	CHROMOSOME III	TOTAL
hh to *hl*	2.9	2.9	5.8
hl to *ll*	5.1	11.5	16.6

Now it should be clear that each set of alleles for both chromosomes is expressed in the phenotype, but that expression varies with the chromosome. Chromosome III appears to have a stronger effect on the phenotype than does chromosome II (compare total amount of change). There is some dominance of *h* over *l* for both chromosomes because the change from *hh* to *hl* is less than the change from *hl* to *ll*. Finally, there is definitely some epistasis occurring. Compare *h/h h/h* with both *l/l h/h* and *h/h l/l*. The difference in the first case is 6.1 and, in the second case, 13.3. The expected amount of change in going from *h/h h/h* to *l/l l/l* is therefore 6.1 + 13.3 = 19.4. The *l/l l/l* phenotype should be 25.1 − 19.4 = 5.7, but the observed value is 2.3.

13. a. b.

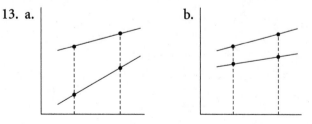

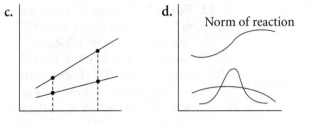

14. a. If you assume that individuals at the extreme of any spectrum are homozygous, then their offspring are more likely to be heterozygous than are the original individuals. That is, they will be less extreme.

   b. For Galton's data, regression is an estimate of heritability (h^2), assuming that there were few environmental differences between father and son.